MATHEMATICAL PHYSICS

Editor

Elmo Kistler

scitus
academics

Mathematical Physics

Edited by **Elmo Kistler**

This Book is Printed on Acid-Free Paper. ♾

ISBN: 9781681179636

Library of Congress Control Number: 2017949105

SCITUS Academics LLC,

616, Corporate Way, Suite 2, 4766,

Valley Cottage, NY 10989

www.scitusacademics.com

Printed in United States of America.

Preface

Mathematical physics refers to development of mathematical methods for application to problems in physics. Physics, as in the language of a physicist, is the study of matter and energy and to a layman it is simply the study of Nature. Physics deals into the deeper insights and truths about the universe. As it is known globally, 'Mathematics is the language for physicists', and without proper mathematical knowledge it is nearly impossible to learn or understand higher physics. Mathematics is required not to understand the physical phenomenon but to convey and exchange the knowledge and information about it among the scientific community and the common masses in a proper structure. The effort to put physical theories on a mathematically rigorous footing has inspired many mathematical developments, such as, the development of quantum mechanics and some aspects of functional analysis parallel each other in many ways. The mathematical study of quantum mechanics, quantum field theory and quantum statistical mechanics has motivated results in operator algebras. The attempt to construct a rigorous quantum field theory has also brought about progress in fields such as representation theory. Use of geometry and topology plays an important role in string theory.

Mathematical Physics provides a broad coverage of the field of mathematical physics, from dominantly mathematical subjects to particle physics, condensed matter, and application of mathematical physics methods in various areas such as astrophysics, together with functional analysis, linear and partial differential equations, algebras, modern differential and algebraic geometry and topology, representations of Lie groups, calculus of variations, asymptotic methods, random process theory, dynamical systems, and control theory. This book aims to expose the reader to the indispensable role that mathematics plays in modern physics.

Table of Contents

CHAPTER 1

Mathematical Physics in Diffusion Problems

Takahisa Okino

Department of Applied Mathematics, Faculty of Engineering, Oita University, Oita City, Japan

ABSTRACT

Using the divergence theorem and the coordinate transformation theory for the general Fickian second law, fundamental diffusion problems are investigated. As a result, the new findings are obtained as follows. The unified diffusion theory is reasonably established, including a self-diffusion theory and an N ($N \geq 2$) elements system interdiffusion one. The Fickian first law is incomplete without a constant diffusion flux corresponding to the Brown motion in the localized space. The cause of Kirkendall effect and the nonexistence of intrinsic diffusion concept are theoretically revealed. In the parabolic space, an elegant analytical method of the diffusion equation is mathematically established, including a nonlinear diffusion equation. From the Schrödinger equation and the diffusion equation, the universal expression of diffusivity proportional to the Planck constant is reasonably obtained. The material wave equation proposed by de Broglie is also derived in relation to the Brown motion. The fundamental diffusion theories discussed here will be highly useful as a standard theory for the basic study of actual interdiffusion problems such as an alloy, a compound semiconductor, a multilayer thin film, and a microstructure material.

Keywords: Brown Particle, Boltzmann Factor, Markov Process, Parabolic Law, Error Function

1. INTRODUCTION

First of all, we state that the basic diffusion equation of the general nonlinear Fickian second law is discussed in accordance with the fundamental mathematical physics in the present work. The extended diffusion equations in detail are not thus discussed. Nevertheless, the new findings, which are extremely dominant in the diffusion study, are reasonably obtained. In the diffusion history, the problems relevant to the coordinate transformation of diffusion equation had not been discussed in accordance with the Gauss divergence theorem until recently. That is just a reason why the new diffusion theories are discussed in the present study. It will be gradually clarified in the text that the coordinate transformation theory is essentially indispensable for the diffusion study. It is obvious that analyzing the extended diffusion equation must be based on the fundamental diffusion theory. The new fundamental findings different from the existing diffusion theories obtained here will thus exert a great influence on the actual diffusion problems in detail, just because of fundamental ones.

A great many phenomena in various science fields are expressed by using the well-known evolution equations. The diffusion equation is one of them and mathematically corresponds to the Markov process in relation to the normal distribution rule [1]. In other words, the motion of diffusion particles corresponds to the well-known Brown movement satisfying the parabolic law [2] [3]. It is widely accepted that the Brown problem is a general term of investigating subjects in various science fields relevant to the Markov process, such as material science, information science, life science, and social science [4] - [9].

In physics, we can also understand the diffusion equation in accordance with the Gauss divergence theorem [10]. If we apply the divergence theorem to the diffusion problem for a material under the condition of no sink and source of the material, it is found that the material conservation law is valid for the diffusion particles, regardless of a thermodynamic state of material. The diffusion equation is also called "the continuous equation" and is extremely fundamental one in physics. In history, the heat conduction equation, which is mathematically equivalent to the diffusion equation, was proposed by Fourier, regardless of the Markov process and the divergence theorem [11].

In accordance with the industrial requirement, the solid materials, such as alloys, semiconductors, and multilayer materials, have been widely fabricated. The heat treatment is indispensable for their fabrication processes then. The migration of particles in a material is caused by the heat treatment. In relation to the migration of their particles, the diffusion problems of various solid materials have been thus widely investigated [12]. Therefore, the diffusion problem is a fundamental study subject in the materials science including the cases of liquid and gas states.

In the present work, the fundamental problems of the general Fickian second law where a driving force affects the diffusion system are discussed in accordance with the mathematical theory. The present analytical method is applicable to interdiffusion problems of an N elements system of every material in an arbitrary thermodynamic state. Although the physical validity of the present method is investigated by using the diffusion data concerning the solid metals, the mathematical generality discussed here is still kept.

The heat conduction equation proposed by Fourier in 1822 has been applied to investigating the temperature distribution in materials [11]. In 1827, the so-called Brown motion was found, where the self-diffusion of water was visualized by pollen micro particle motions [2] [3]. In 1855, Fick applied the heat conduction equation to diffusion phenomena as it had been [13]. Nevertheless, the Brown motion had not been recognized as a diffusion problem until the Einstein theory of Brown motion in 1905, although it was a typical diffusion problem [3].

Although the concentration of diffusion particles is a real quantity in physics, the temperature is a thermodynamic state quantity. As far as the shape of heat conduction material is unchangeable during a thermal treatment, the coordinate system of heat conduction equation set in a material is a fixed one, since the coordinate system is not influenced by variations of the material internal structure. On the other hand, strictly speaking, the coordinate system of diffusion equation set in the diffusion field (solvent) is a moving one, since it is generally influenced by such variations.

When the Fickian first and second laws (F1 law and F2 law) were proposed, the Gauss divergence theorem had been already reported in 1840 [10]. Nevertheless, the problem of coordinate system of diffusion equation was not mathematically investigated in accordance with the divergence theorem.

In general, it is indispensable for understanding the diffusion problems to discuss their coordinate systems, since it is, strictly speaking, considered that the diffusion particles, solvent particles and also the diffusion region space simultaneously move against the experimentation system in the diffusion region outside.

Although the Fickian laws were still widely applied to various diffusion phenomena as essential equations in physics, the problem between coordinate systems was not discussed. Recently, the diffusion equation was thus mathematically investigated in accordance with the divergence theorem and the coordinate transformation theory [14] - [16]. It is revealed in the text that the diffusion flux should be determined by taking account of the concerned coordinate system of diffusion equation.

Using the corresponding diffusion flux to the coordinate system of diffusion equation for interdiffusion, one way diffusion, impurity diffusion and self-diffusion, they are uniformly discussed in the text. As a result, we found that the foundation of diffusion problems is included in interdiffusion problems. The interdiffusion theory of an N ($N \geq 2$) elements system applicable to every material was thus reasonably established [16]. In the analysis of interdiffusion problems, the only difference between a binary system and an N elements system is whether the solvent material is one element or (N - 1) elements.

The coordinate transformation theory reveals that the corresponding F1 law to the F2 law is incomplete without a constant diffusion flux relevant to the Brown motion in the localized space. Further, it was also found that the Kirkendall effect (K effect) is caused by a shift between the coordinate systems of the diffusion equation like the Doppler effect relevant to a wave equation is caused by a shift between the fixed coordinate system and the moving one for the wave equation [14] - [17]. At the same time, this means that the concept of intrinsic diffusion is unnecessary for understanding the K effect.

As far as the sink and source of a material is nonexistent in the diffusion system, the Gauss divergence theory shows that the general F2 law satisfies the material conservation law even if a driving force affects the diffusion system. The driving force affects entropy and jump frequencies of diffusion particles in the material. The influence of a driving force is thus all incorporated into the diffusivity of diffusion equation then. In the existing diffusion theory, the diffusion fluxes influenced by a driving force are discussed incorporating not influence of both entropy and jump frequencies but only that of jump frequencies on the diffusivity into them, for example such as a drift velocity. However, such concept as a drift velocity is not necessary for analyzing the diffusion equation, since we can solve a nonlinear diffusion equation in accordance with the fundamental theory in the mathematical physics, regardless of the discussion about diffusion fluxes.

The diffusivity is obtained from the Taylor expansion of a probability density function, satisfying the parabolic law [3]. This indicates that the analysis of diffusion equation is essentially possible in the parabolic space. The diffusivity is defined by an interaction between a diffusion particle and the diffusion field near the diffusion particle itself. This indicates that the diffusivity should be essentially investigated in the quantum mechanics, since the behavior of a micro particle should be investigated by analyzing the Schrödinger Equation (S equation) [18].

Based on the suggestions mentioned above, the elegant analytical method in the parabolic space is reasonably established in the text [19] [20]. As a result, the solutions of the Boltzmann transformation equation for nonlinear interdiffusion phenomena were, for the first time, obtained as analytical expressions in the parabolic space [21].

From applying the diffusion equation to a problem of diffusion elementary process, we derived the S equation [22]. It is revealed in the text that the diffusivity corresponds to the angular momentum operator in the quantum mechanics. As a result, the universal expression of diffusivity, which is applicable to every material in an arbitrary thermodynamic state, was obtained in proportional to the Planck constant $\hbar$.

It was also found that the well-known material wave relation proposed by de Broglie in 1923, which is the most fundamental one in materials science, is obtained from a relation between diffusivity expressions [23]. This gives evidence for the theory discussed in the present study.

In the present work, we review the fundamental diffusion theories relevant to the general F2 law, where they are systematically reframed in points of view different from the previous works, adding some new discussions to them. The new findings obtained here will be widely applicable to fundamental problems as a standard theory in various actual diffusion phenomena.

2. FUNDAMENTAL THEORY OF DIFFUSION EQUATION

As far as a material is conserved in the given region, the divergence theory shows that the diffusion equation is applicable to diffusion phenomena of every material in an arbitrary thermodynamic state. The diffusion information of a material, such as crystal material or amorphous material, and/or solid, liquid and gas states, is all incorporated into the diffusivity in the given diffusion equation. The diffusion equation with such an arbitrary diffusivity is thus investigated in the following.

In the present study, an abbreviate differential notation for an arbitrary independent variable ξ and the well- known Dirac's bracket notation for an arbitrary vector are used as follows [24] ;

$$\partial_\xi = \frac{\partial}{\partial \xi}, \quad \langle \tilde{\nabla} | = \left(\partial_x, \partial_y, \partial_z \right).$$

If an operator Q is Hermite one, $\langle Q| = \{|Q\rangle\}^{\dagger}$ is valid in the Hermite conjugate †. Here, the notation $\langle\tilde{\nabla}|$ is thus defined as $\langle\tilde{\nabla}| = -\{|\nabla\rangle\}^{\dagger}$ because of $\partial_{\xi} = -\{\partial_{\xi}\}^{\dagger}$.

2.1. Diffusion Equation

The function $C(t_j,|r_j\rangle)$ is defined as a normalized concentration where a diffusion particle in the initial state $(t_0,|r_0\rangle)$ exists in the time and space state $(t_j,|r_j\rangle)$ after j times jumps. *A* diffusion particle moves at random. It is, therefore, considered that the jump frequency $1/\Delta t$ and jump displacement $\Delta r = \left\| |r_j\rangle - |r_{j-1}\rangle \right\| = \| \Delta r \rangle \|$ are equivalent in probability to their mean values of all diffusion particles in the collective system.

Since it is also considered that the probability of diffusion-jump from the state of $(t_{j-1},|r_{j-1}\rangle)$ to $(t_j,|r_{j-2}\rangle)$ is equivalent to one from the same state to $(t_j,|r_j\rangle)$ in the isotropic space, the relation of

$$C(t+\Delta t,|r\rangle) = \{C(t,|r\rangle - |\Delta r\rangle) + C(t,|r\rangle + |\Delta r\rangle)\}/2 \tag{2-1}$$

is thus valid. The Taylor expansion of both sides of Equation (2-1) yields

$$C(t+\Delta t,|r\rangle) = C(t,|r\rangle) + \Delta t \partial_t C(t,|r\rangle) + \cdots, \tag{2-2}$$

$$C(t,|r\rangle \pm |\Delta r\rangle) = C(t,|r\rangle) \pm \langle \Delta r | \nabla C(t,|r\rangle)\rangle + \frac{(\Delta r)^2}{2} \langle \tilde{\nabla} | \nabla C(t,|r\rangle)\rangle \pm \cdots. \tag{2-3}$$

The substitution of Equations (2-2) and (2-3) into Equation (2-1) gives

$$\partial_t C = \frac{(\Delta r)^2}{2\Delta t} \langle \tilde{\nabla} | \nabla C \rangle. \tag{2-4}$$

Since the averaged Δt and Δr are physically finite, it is considered that $(\Delta r)^2/2\Delta t$ becomes a constant value $\boldsymbol{D_0}$ given by

$$D_0 = \frac{(\Delta r)^2}{2\Delta t}, \tag{2-5}$$

satisfying the well-known parabolic law [3]. Substituting Equations (2-5) into Equations (2-4), the diffusion equation is obtained as

$$\partial_t C = D_0 \langle \tilde{\nabla} | \nabla C \rangle = \langle \tilde{\nabla} | D_0 \nabla C \rangle = \langle \tilde{\nabla} | D \nabla C \rangle, \tag{2-6}$$

where $\boldsymbol{D_0}$ is rewritten as D when the diffusion-jump direction is not isotropic resulting from the existence of a driving force in the diffusion system.

The existence of a driving force affects diffusivity $\boldsymbol{D_0}$ and the diffusivity D becomes a function of the independent variables t and $|r\rangle$. In that case, Equation (2-6) gives the diffusion flux yielding

$$|J\rangle = -|D\nabla C\rangle = -|\{D_0 + D_{\mathrm{df}}(t,|r\rangle)\}\nabla C\rangle, \tag{2-7}$$

where $D_{\mathrm{df}}(t,|r\rangle)$ is a diffusivity caused by a driving force.

On the other hand, using a drift velocity $|v_{\mathrm{F}}\rangle$ caused by a driving force F, the approximate diffusion flux is expressed as

$$|J\rangle = -|D^* \nabla C\rangle + |v_{\mathrm{F}} C\rangle \tag{2-8}$$

in the usual textbooks (see Appendix A) [25]. However, the diffusion flux $|J_{\mathrm{F}}\rangle = |v_{\mathrm{F}} C\rangle$ discussed here is different from one given by

$$|J_{\mathrm{df}}\rangle = -|D_{\mathrm{df}} \nabla C\rangle. \tag{2-9}$$

In the present study, the discussion about diffusion fluxes is not necessary for analyzing the diffusion equation, since we can solve the nonlinear diffusion equation in accordance with the fundamental theory in the mathematical physics, regardless of diffusion fluxes.

In the following, therefore, Equation (2-6) is directly solved in accordance with the mathematical physics. It is in the nature of things that its analytical method is applicable to the diffusion problems relevant to a driving force then.

For a collective system composed of N $(j = 1, 2, \cdots, N)$ elements in the region $V(x, y, z)$ within the closed surface $S(x, y, z)$, the divergence theorem yields

$$\iint_S \left\{ \int \partial_t C^j \mathrm{d}x \right\} \mathrm{d}y\mathrm{d}z = \iint_S \left\{ \int \langle \tilde{\nabla} || D^j \nabla C^j \rangle \mathrm{d}x \right\} \mathrm{d}y\mathrm{d}z, \tag{2-10}$$

where $C^j(t, x, y, z)$ and D^j are a concentration and a diffusivity for an element j under the condition of no sink and source of the element. The integral calculation of {} in Equation (2-10) means an inflow flux from the surface outside through the surface element dydz, and *y* and *z* are then accepted as constant values because of the multiple integral calculation. Therefore, Equation (2-10) yields the diffusion equation of

$$\partial_t C^j(t,x) = \partial_x \left\{ D^j \partial_x C^j(t,x) \right\} \tag{2-11}$$

in the diffusion field. Defining the outflow flux as a plus value, the diffusion flux is

$$J_x^j(t,x) = -\int \partial_t C^j(t,x)\,\mathrm{d}x = -\int \partial_x \left\{ D^j \partial_x C^j(t,x) \right\} \mathrm{d}x. \tag{2-12}$$

In mathematics, Equation (2-12) yields

$$J_x^j(t,x) = J_{\mathrm{F}}^j(t,x) + J_x^j(t) + J_{\mathrm{eq}}^j \text{ for } J_{\mathrm{F}}^j(t,x) = -D^j \partial_x C^j(t,x) \tag{2-13}$$

where $J_{\mathrm{F}}^j(t,x)$ is the F1 law and $J_x^j(t) + J_{\mathrm{eq}}^j$ is an integral constant against x. In physics, $J_x^j(t)$ means a movement of a diffusion region space caused by a movement of the diffusion field. In the coordinate system of diffusion region inside, it should be thus physically accepted as $J_x^j(t) = 0$. However, $J_x^j(t) \neq 0$ must be taken into account in the coordinate system of diffusion region outside. The intrinsic diffusion flux J_{eq}^j independent of the time and space is relevant to the Brown motion in the localized space. It plays an important role for understanding a self-diffusion mechanism.

2.2. Coordinate System of Diffusion Equation

As discussed later, the unified theory of diffusion problems shows that the foundation of diffusion is in the interdiffusion problems. In the interdiffusion problems, solvent particles as well as diffusion particles move in the diffusion region space. When a micro particle in a material jumps from a site *M* to an interstice space *N*, if we call such an interstice space "micro hole" including a vacancy in case of a crystal material, the micro hole of site N is annihilated and a new micro hole is generated at the site M after jumping.

The diffusion region space interacts with the space of diffusion system outside in accordance with the annihilation and/or generation of micro holes like the diffusion system becomes a thermal equilibrium state resulting from an antinomy between the principle of increase of entropy and that of minimum of free energy.

The coordinate system of the basic diffusion equation must be set in the diffusion field, since the diffusivity depends on an interaction between a diffusion particle and the diffusion field near the diffusion particle itself. In general, therefore, the diffusion problem should be investigated among three coordinate systems, where the origins of two coordinate systems are a point P of the coordinate system (*x*,*y*,*z*) set in the diffusion field (solvent) and a point Q of the coordinate system $(\tilde{x},\tilde{y},\tilde{z})$ set in the diffusion region space composed of micro holes and the third origin is a point R of the coordinate system (ξ,η,ζ) set in the diffusion region outside.

In the present work, we investigate an interdiffusion problem applicable to every material in an arbitrary thermodynamic state. It is thus investigated using the original diffusion equation (2-11) of the coordinate system (*t*, *x*) and the diffusion equations transformed into the coordinate systems of $(\tilde{t},\tilde{x})$ and (τ,ξ) under the condition of $t=\tilde{t}=\tau$.

The origins P (*t*, *x*) = (0, 0) and Q $(\tilde{t},\tilde{x})$ = (0, 0) are set at a point of solvent material and at a point of micro holes on the initial interface of a diffusion couple. The origin R (τ,ξ) = (0, 0) is set at a point of the diffusion system outside. The x axis is perpendicular to the interface then. The $\tilde{x}$ axis and also ξ axis are parallel to the x axis under the condition of $x=\tilde{x}=\xi=0$ at $t=\tilde{t}=\tau=0$.

It is generally considered that P (*t*, *x*) = (0, 0) and Q $(\tilde{t},\tilde{x})$ = (0, 0) move with a velocity $v_{\mathrm{RQ}}(\tau)$ and with a velocity $v_{\mathrm{RQ}}(\tau)$ against R (τ,ξ) = (0, 0), respectively and also that P (*t*, *x*) = (0, 0) moves with a velocity $v_{\mathrm{QP}}(\tilde{t})$ against Q $(\tilde{t},\tilde{x})$ = (0, 0), including a possibility of $v_{\mathrm{RP}}(\tau)v_{\mathrm{RQ}}(\tau)v_{\mathrm{QP}}(\tilde{t})=0$. The relation among these velocities of

$$v_{\mathrm{RP}}=v_{\mathrm{RQ}}+v_{\mathrm{QP}} \tag{2-14}$$

must be physically valid.

Hereafter, we use the suffix P, Q and R for a physical quantity relevant to (*t*, *x*), $(\tilde{t},\tilde{x})$and (τ,ξ). Equation (2-13) is thus rewritten as

$$J_{\mathrm{P}}^{j}(t,x)=J_{\mathrm{F}}^{j}(t,x)+J_{\mathrm{eq}}^{j}\quad\text{for}\quad J_{\mathrm{F}}^{j}(t,x)=-D^{j}\partial_{x}C^{j}(t,x). \tag{2-13}$$

When the coordinate origin of (*t*, *x*) moves from that of $(\tilde{t},\tilde{x})$ with a velocity v_{QP}, the relations of

$$t=\tilde{t}\,,\quad x=\tilde{x}-\Delta x_{\rm sft}\quad \text{for}\quad \Delta x_{\rm sft}=\int_0^t v_{\rm QP}\,{\rm d}\tilde{t} \tag{2-15}$$

are valid under the condition of $\tilde{x}=x=0$ at $\tilde{t}=t=0$. The relations of

$$\frac{\partial}{\partial t}=\frac{\partial \tilde{t}}{\partial t}\frac{\partial}{\partial \tilde{t}}+\frac{\partial \tilde{x}}{\partial t}\frac{\partial}{\partial \tilde{x}}=\frac{\partial}{\partial \tilde{t}}+v_{\rm QP}\frac{\partial}{\partial \tilde{x}}\,,\quad \frac{\partial}{\partial x}=\frac{\partial \tilde{t}}{\partial x}\frac{\partial}{\partial \tilde{t}}+\frac{\partial \tilde{x}}{\partial x}\frac{\partial}{\partial \tilde{x}}=\frac{\partial}{\partial \tilde{x}} \tag{2-16}$$

between differential operators are obtained then.

Using Equation (2-16) for Equation (2-11), the diffusion equations of $(\tilde{t},\tilde{x})$ is obtained as

$$\partial_{\tilde{t}}C^j=\partial_{\tilde{x}}\left(D^j\partial_{\tilde{x}}C^j-v_{\rm QP}C^j\right), \tag{2-17}$$

where $C^j=C^j(\tilde{t},\tilde{x})$ and $D^j=D^j(\tilde{t},\tilde{x})$. Equation (2-17) leads to the diffusion flux of

$$J_{\rm Q}^j(\tilde{t},\tilde{x})=J_{\rm F}^j(\tilde{t},\tilde{x})+v_{\rm QP}C^j+J_{\rm eq}^j\,,\quad J_{\rm F}^j(\tilde{t},\tilde{x})=-D^j\partial_{\tilde{x}}C^j\,, \tag{2-18}$$

since $(\tilde{t},\tilde{x})$ is in the diffusion region inside.

In the same manner of the above, the diffusion equation of (τ,ξ) is obtained as

$$\partial_\tau C^j=\partial_\xi\left(D^j\partial_\xi C^j-v_{\rm RP}C^j\right). \tag{2-19}$$

Since (τ,ξ) is in the diffusion region outside, the diffusion flux becomes

$$J_{\rm R}^j(\tau,\xi)=J_{\rm F}^j(\tau,\xi)+v_{\rm RP}C^j+J_{\rm R}^j(\tau)+J_{\rm eq}^j\,,\quad J_{\rm F}^j(\tau,\xi)=-D^j\partial_\xi C^j\,, \tag{2-20}$$

where $J_{\rm R}^j(\tau)$ is the diffusion flux of micro holes caused by the movement of the diffusion field against (τ,ξ) = (0, 0) .

3. INTERDIFFUSION PROBLEMS

The interdiffusion problems of an *N* elements system in a diffusion couple composed of arbitrary materials A and B are discussed as follows.

In an amorphous material, we then estimate a specimen cross section perpendicular to the *x* axis where each cross section interval *l* corresponds to the averaged jump distance in the solid state or to the averaged collision distance in the gas or liquid state. In a crystal material, we estimate the jump distance l between the nearest neighbor crystal cross sections perpendicular to the *x* axis.

When materials A and B are in the solid state during a thermal diffusion, we conceive that the cross section *S* of the material A is uniform and equal to that of the material B. If materials A and B are in the fluid state, we conceive that they are in the receptacle corresponding to the case of solid materials. The interface of diffusion couple between materials A and B is smoothly joined. In that case, the origin $x = 0$ of coordinate system (t, x) is set at a point on the initial interface in the diffusion field and the orientation of x axis is defined as A→B.

The material A is composed of *N* elements and we define the normalized concentration C_{A}^{j} for $j = 1, 2, \ldots, N$ in the initial state, also including a possibility of $C_{\mathrm{A}}^{m} = 0$ for $j = m$. The normalized concentration in the material B is similarly defined as C_{B}^{j}.

The concentration and diffusivity of an element *j* are C^j and D^j at a diffusion time t in the diffusion region $x_{\mathrm{A}} \leq x \leq x_{\mathrm{B}}$. The boundary values of concentration are $C^j(t, x_{\mathrm{A}}) = C_{\mathrm{A}}^{j}$ and $C^j(t, x_{\mathrm{B}}) = C_{\mathrm{B}}^{j}$.

The remained (*N*-1) elements are considered to be a solvent, also including a possibility where C^j becomes a solvent itself in accordance with a given diffusion system.

The cross section of diffusion system is uniform and the shape variation of diffusion couple is usually negligible during the thermal diffusion. In that case, the relation of normalized concentrations yielding

$$\sum_{j=1}^{N} C^j = 1 \tag{3-1}$$

is usually accepted on each cross section between $x_{\mathrm{A}} \leq x \leq x_{\mathrm{B}}$. Here, Equation (3-1) is physically valid regardless of a coordinate system.

3.1. Correlation of Diffusion Fluxes with Coordinate Systems

Equations (2-17) and (3-1) yield

$$\partial_{\tilde{x}}\sum_{j=1}^{N}\left(D^j\partial_{\tilde{x}}C^j - v_{\mathrm{QP}}C^j\right) = \partial_{\tilde{x}}\left[\sum_{j=1,\,j\neq k}^{N}\left\{\left(D^j - D^k\right)\partial_{\tilde{x}}C^j\right\} - v_{\mathrm{QP}}\right] = 0 \tag{3-2}$$

because of $\sum_{j=1}^{N}\partial_{\tilde{t}}C^j = \partial_{\tilde{t}}\sum_{j=1}^{N}C^j = 0$. In mathematics, the partial differential equation (3-2) of $\tilde{t}$ and $\tilde{x}$ is rewritten as

$$\Gamma(\tilde{t}) = \sum_{j=1,\,j\neq k}^{N}\left\{\left(D^j - D^k\right)\partial_{\tilde{x}}C^j\right\} - v_{\mathrm{QP}},$$

using an arbitrary function $\Gamma(\tilde{t})$ of $\tilde{t}$ because of $\partial_{\tilde{x}}\Gamma(\tilde{t}) = 0$. Here, $\left(D^j - D^k\right)\partial_{\tilde{x}}C^j$ means a diffusion flux dependent on $\tilde{t}$ and $\tilde{x}$. On the other hand, $\Gamma(\tilde{t}) + v_{\mathrm{QP}}$ means a diffusion flux only dependent only on $\tilde{t}$ Therefore, Equation (3-2) means D^j-D^k = 0 $D^j - D^k = 0$ for an arbitrary *j*. The interdiffusion coefficient $\tilde{D}$ is thus defined as

$$\tilde{D} = D^1 = D^2 = \cdots = D^{\mathrm{N}} \tag{3-3}$$

in the diffusion region $\tilde{x}_{\mathrm{A}} \leq \tilde{x} \leq \tilde{x}_{\mathrm{B}}$. We can easily confirm that Equation (3-3) does not depend on a coordinate system. Their boundary values are thus obtained as

$$\tilde{D}(\tau,\xi_{\mathrm{A}}) = \tilde{D}_{\mathrm{A}} \text{and } \tilde{D}(\tau,\xi_{\mathrm{B}}) = \tilde{D}_{\mathrm{B}} \tag{3-4}$$

Using Equation (3-3) for the F1 law, the following relation of

$$\sum_{j=1}^{N}J_{\mathrm{F}}^{j}(\tau,\xi) = -\sum_{j=1}^{N}D^j\partial_{\xi}C^j = -\tilde{D}\partial_{\xi}\sum_{j=1}^{N}C^j = 0 \tag{3-5}$$

is valid regardless of a coordinate system. The relation of

$$\sum_{j=1}^{N}J_{\mathrm{eq}}^{j} = 0 \tag{3-6}$$

is also valid as discussed later in the self-diffusion theory.

Using Equations (3-5) and (3-6) for Equation (2-20), the relation of

$$J_{\mathrm{R}} = \sum_{j=1}^{N} J_{\mathrm{R}}^{j}(\tau,\xi) = v_{\mathrm{RP}} + J_{\mathrm{R}}(\tau) \quad \text{for} \quad J_{R}(\tau) = \sum_{j=1}^{N} J_{\mathrm{R}}^{j}(\tau) \tag{3-7}$$

is obtained. Here, $J_{\mathrm{R}}(\tau)$ means the total diffusion flux of micro holes against the origin of (τ,ξ). In that situation, $J_{\mathrm{R}}(\tau) = -v_{\mathrm{RQ}}$ is valid in the present coordinate system because of $\langle J_{\mathrm{R}}(\tau) | v_{\mathrm{RQ}} \rangle < 0$, using a velocity v_{RQ} of the origin of $(\tilde{t},\tilde{x})$ against the origin of (τ,ξ). Equation (3-7) is thus rewritten as

$$J_{\mathrm{R}} = v_{\mathrm{RP}} - v_{\mathrm{RQ}}. \tag{3-7}$$

In the same manner, the following relations

$$J_{\mathrm{P}} = \sum_{j=1}^{N} J_{\mathrm{P}}^{j}(t,x) = 0 \tag{3-8}$$

$$J_{\mathrm{Q}} = \sum_{j=1}^{N} J_{\mathrm{Q}}^{j}(\tilde{t},\tilde{x}) = v_{\mathrm{QP}} \tag{3-9}$$

are also obtained from Equations (2-13) and (2-18). Equations (3-7)-(3-9) show that the diffusion flux independent of the space coordinate corresponds to the velocity between the coordinate systems. The relation of

$$J_{\mathrm{R}} = J_{\mathrm{O}} + J_{\mathrm{P}} \tag{3-10}$$

must be physically valid because of $t = \tilde{t} = \tau$. Equation (3-10) then yields the previous equation of

$$v_{\mathrm{RP}}(\tau) = v_{\mathrm{RQ}}(\tau) + v_{\mathrm{QP}}(\tau) \tag{2-14}$$

because of Equations (3-7)-(3-10).

In the following, $J_{\mathrm{Q}}(= v_{\mathrm{QP}})$ is investigated by using the concentration difference $\Delta C^{j} = C_{\mathrm{B}}^{j} - C_{\mathrm{A}}^{j}$ and the diffusion junction depth $\Delta\tilde{x}_{\mathrm{jun}} = \mu\sqrt{\tilde{D}_{\gamma}\tilde{t}}$

between $\tilde{x} = \tilde{x}_A$ and $\tilde{x} = \tilde{x}_B$. Taking account of their concentration gradients, the flux of diffusion field for an element k ($j \neq k$) is obtained as

$$J_Q^k(\tilde{t}) = -\sum_{j=1,\, j\neq k}^{N} \tilde{D}_\gamma \frac{\Delta C^j}{\Delta x_{\mathrm{jun}}} = \frac{1}{2\sqrt{\tilde{t}}} \sum_{j=1,\, j\neq k}^{N} \sqrt{\tilde{D}_\gamma}\left\{C_A^j - C_B^j\right\}, \tag{3-11}$$

where the suffix γ means $\gamma \to A$ if $C_A^j \geq C_B^j$ or $\gamma \to B$ if $C_A^j \geq C_B^j$ and $\mu = 2$ is tentatively adopted. Here, $J_Q^k(\tilde{t})$ is adopted instead of $J_Q^k(\tilde{t}, \tilde{x})$ as an approximate equation because of $J_Q = v_{QP}(\tilde{t})$.

Equation (3-11) yields

$$J_Q = v_{QP}(\tilde{t}) = \sum_{k=1}^{N} J_Q^k(\tilde{t}) = \frac{N-1}{2\sqrt{\tilde{t}}} \sum_{j=1}^{N} \sqrt{\tilde{D}_\gamma}\left\{C_A^j - C_B^j\right\}. \tag{3-12}$$

Equation (3-12) shows the movement of (t, x) against $(\tilde{t}, \tilde{x}) = (0,0)$. Substituting Equation (3-12) into Equation (2-14) yields

$$v_{RP} = v_{RQ} + \frac{N-1}{2\sqrt{\tau}} \sum_{j=1}^{N} \sqrt{\tilde{D}_\gamma}\left\{C_A^j - C_B^j\right\} \tag{3-13}$$

because of $t = \tilde{t} = \tau$.

3.2. Kirkendall Effect in Interdiffusion Problems

In the metallurgy field, the interface of (t, x) = (0,0) is the so-called Matano interface (M interface) and that of $(\tilde{t}, \tilde{x}) = (0,0)$ is the Kirkendall interface (K interface) [17] [26]. In 1947, Kirkendall found in the binary system interdiffusion experimentation that the inert marker set on the K interface in the initial state moves from the M interface during a thermal diffusion and also that the M interface does not move during thermal diffusion, i.e., $v_{RP} = 0$. Since then, the displacement between the M interface and the K interface has been called the Kirkendall effect (K effect). Using an experimental value m, the K effect Δx_{eff} is obtained as

$$\Delta x_{\mathrm{eff}} = m\sqrt{t}, \tag{3-14}$$

regardless of the space coordinate and it satisfies the parabolic law.

C^{II}As far as the material shape of a diffusion couple is unchangeable, the K effect also reveals that the total quantity of element I diffusing across the K interface is different from that of element II. In other words, for the micro hole concentration C^{H} and its thermal equilibrium value C_0^{H} between $x_A \le x \le x_B$, the relation of $C_0^{\text{H}} - C^{\text{H}} < 0$ is valid between $x_A \le x \le x_B$ if $C_0^{\text{H}} - C^{\text{H}} < 0$ is valid between $0 \le x \le x_B$ during thermal diffusion, and vice versa. The value of $C_0^{\text{H}} - C^{\text{H}}$ becomes gradually zero in accordance with the thermodynamic principle.

In the binary system interdiffusion problems, the F1 law shows that the relation of

$$J^{\text{I}} + J^{\text{II}} = \left(D^{\text{I}} - D^{\text{II}}\right)\partial_x C^{\text{II}} = 0$$

is valid because of $C^{\text{I}} + C^{\text{II}} = 1$, and $\tilde{D} = D^{\text{I}} = D$ is valid then for an interdiffusion coefficient $\tilde{D}$. On the other hand, it was also considered that the K effect is caused by a difference between the diffusion flux of element I and that of element II across the K interface. In other words, the so-called intrinsic diffusion coefficient satisfying the relation of $D_{\text{int}}^{\text{I}} \neq D_{\text{int}}^{\text{II}}$ was newly conceived to understand the K effect then.

However, the definition of diffusivity shows that the diffusivity is proportional to a statistical jump frequency of a diffusion particle at a point (t, x) occupied by the diffusion particle itself. It is, therefore, physically difficult to accept what a diffusion particle of the element I or II has two diffusivities D^{I} and D^{I} or D^{II} and $D_{\text{int}}^{\text{II}}$ at the same point (t, x) in the concerned diffusion field under the condition of $\tilde{D} = D^{\text{I}} = D^{\text{II}}$, since the statistical jump frequency value of a diffusion particle should be physically only one in the concerned diffusion field.

In the present diffusion theory, the behavior of micro holes is visualized by the movement of an inert marker in the diffusion region because of the inert characteristic. Therefore, the K effect means a shift between the coordinate systems (t, x) and $(\tilde{t}, \tilde{x})$ after the thermal diffusion. In other words, Equation (3-12) yields the displacement between the K interface and the M interface because of $v_{\text{RP}} = 0$ in accordance with the experimental results. The K effect is thus obtained as

$$\Delta x_{\text{eff}} = \int_0^t v_{\text{QP}}(\tilde{t})\,\mathrm{d}\tilde{t} = (N-1)\sum_{j=1}^{N}\sqrt{\tilde{D}_\gamma}\left\{C_{\text{A}}^{j} - C_{\text{B}}^{j}\right\}\sqrt{t}, \tag{3-15}$$

satisfying the parabolic law.

The concept of intrinsic diffusion is thus unnecessary for understanding the K effect. Further, the K effect occurs in the interdiffusion not only of metal crystal but also of every material, since it is caused by a shift between the coordinate systems.

Equation (3-15) gives evidence that the K effect is caused by a shift between the coordinate systems, since it does not depend on the space coordinate. Substituting N = 2 into Equation (3-15) and comparing it with Equation (3-14) give

$$m=\left(\tilde{D}_{\mathrm{A}}-\tilde{D}_{\mathrm{B}}\right)\left(C_{\mathrm{A}}^{\mathrm{I}}-C_{\mathrm{B}}^{\mathrm{I}}\right). \tag{3-16}$$

Here, we must notice that μ = 2 is tentatively adopted in the diffusion junction depth $\Delta\xi_{\mathrm{jun}}=\mu\sqrt{\tilde{D}_{\gamma}\tau}$ in the present case. Therefore, the ? value in the relation of

$$m=\frac{2}{\mu}\left(\tilde{D}_{\mathrm{A}}-\tilde{D}_{\mathrm{B}}\right)\left(C_{\mathrm{A}}^{\mathrm{I}}-C_{\mathrm{B}}^{\mathrm{I}}\right)$$

must be concretely investigated so as to be suitable for the concerned interdiffusion problems, since the m value is experimentally obtained.

3.3. Unified Theory of Diffusion Problems

The experimental results show that the relation of v_{RP} = 0 is valid in the metallurgy field. This means that the shape of diffusion system is unchangeable before and after diffusion treatment. In that case, Equation (2-19) is thus rewritten as

$$\partial_{\tau}C^{j}=\partial_{\xi}\left(D^{j}\partial_{\xi}C^{j}\right) \tag{3-17}$$

Equation (3-17) of (τ,ξ) is thus equivalent to Equation (2-11) of (t,x). However, the diffusion flux of Equation (2-20) is rewritten as

$$J_{\mathrm{R}}^{j}\left(\tau,\xi\right)=J_{\mathrm{F}}^{j}\left(\tau,\xi\right)+J_{\mathrm{R}}^{j}\left(\tau\right)+J_{\mathrm{eq}}^{j},\quad J_{\mathrm{F}}^{j}\left(\tau,\xi\right)=-D^{j}\partial_{\xi}C^{j},$$

In the following, various diffusion problems are systematically investigated by applying the diffusion flux of Equation (2-18) to them for j = I, II, under the condition of

$$v_{QP} = \left(\sqrt{\tilde{D}_A} - \sqrt{\tilde{D}_B}\right)\left(C_A^I - C_B^I\right)/2\sqrt{\tilde{t}}$$.

Equation (2-18) is then expressed as

$$\begin{cases} (i)\ J_Q^I(\tilde{t},\tilde{x}) = J_F^I(\tilde{t},\tilde{x}) + v_{QP}C^I(\tilde{t},\tilde{x}) + J_{eq}^I, & J_F^I(\tilde{t},\tilde{x}) = -D^I\partial_{\tilde{x}}C^I(\tilde{t},\tilde{x}), \\ (ii)\ J_Q^{II}(\tilde{t},\tilde{x}) = J_F^{II}(\tilde{t},\tilde{x}) + v_{QP}C^{II}(\tilde{t},\tilde{x}) + J_{eq}^{II}, & J_F^{II}(\tilde{t},\tilde{x}) = -D^{II}\partial_{\tilde{x}}C^{II}(\tilde{t},\tilde{x}). \end{cases} \quad (3\text{-}18)$$

(a) Interdiffusion:

Under the condition of $\tilde{D} = D^I = D^{II}$, Equation (3-18) becomes

(i) $J_Q^I(\tilde{t},\tilde{x}) = -\tilde{D}\partial_{\tilde{x}}C^I(\tilde{t},\tilde{x}) + v_{QP}C^I(\tilde{t},\tilde{x}) + J_{eq}^I$,

(ii) $J_Q^{II}(\tilde{t},\tilde{x}) = -\tilde{D}\partial_{\tilde{x}}C^{II}(\tilde{t},\tilde{x}) + v_{QP}C^{II}(\tilde{t},\tilde{x}) + J_{eq}^{II}$.

(b) One-way diffusion:

Under the condition of $D^{II} = 0$, Equation (3-18) becomes

(i) $J_Q^I(\tilde{t},\tilde{x}) = -D^I\partial_{\tilde{x}}C^I(\tilde{t},\tilde{x}) + v_{QP}C^I(\tilde{t},\tilde{x}) + J_{eq}^I$,

(ii) $J_Q^{II}(\tilde{t},\tilde{x}) = v_{QP}^{II}C^{II}(\tilde{t},\tilde{x}) + J_{eq}^{II}$.

(c) *Impurity diffusion:*

Under the condition of $D^{II} = 0$ and $C_A^{II} = C_B^{II}$, Equation (3-18) becomes

(i) $J_Q^I(\tilde{t},\tilde{x}) = -D^I\partial_{\tilde{x}}C^I(\tilde{t},\tilde{x}) + J_{eq}^I$, (ii) $J_Q^{II}(\tilde{t},\tilde{x}) = J_{eq}^{II}$.

(d) *Self-diffusion:*

Under the condition of $C^I = C_A^I = C_B^I$ and $C^{II} = C_A^{II} = C_B^{II}$, Equation (3-18) becomes

(i) $J_Q^I(\tilde{t},\tilde{x}) = J_{eq}^I$, (ii) $J_Q^{II}(\tilde{t},\tilde{x}) = J_{eq}^{II}$.

If we rewrite $J_Q^I(\tilde{t},\tilde{x})$ and $J_Q^{II}(\tilde{t},\tilde{x})$ as

$J_Q^I(\tilde{t},\tilde{x}) \to J_Q^k(\tilde{t},\tilde{x})$ and $J_Q^{II}(\tilde{t},\tilde{x}) \to \sum_{j=1,\, j\neq k}^{N} J_Q^j(\tilde{t},\tilde{x})$ for $j = 1,2,\ldots,k,\ldots,N$

the above theory is still valid in the N elements system. In the diffusion problems, it was thus found that the behavior of diffusion region space plays an important role for understanding diffusion mechanisms.

3.4. Self-Diffusion Theory

The diffusivity expression of Equation (2-5) is valid regardless of pure materials [3]. The Einstein theory of Brown motion shows that the self-diffusion of water is visualized by diffusion phenomena of pollen micro particles. It is thus obvious that the self-diffusion occurs in a pure material. However, even if we apply the F2 law to the self-diffusion problems in a pure material, we cannot understand whether the self-diffusion occurs or not, since the concentration is unchangeable in the self-diffusion system.

Langevin theoretically investigated not the diffusion equation of collective motion but the motion equation of a diffusion particle with a viscosity resistance in a pure liquid material [27]. It was thus revealed that the diffusion particle moves in accordance with the normal distribution rule.

The theories of Langevin as well as Einstein are relevant to diffusion particles with a driving force in a pure material. In other words, the diffusion phenomena of a pure material itself are not directly investigated. However, Equation (2-5) shows that the self-diffusion occurs in a pure material even if there is no driving force in the diffusion system.

The self-diffusion problems have been also experimentally investigated as an impurity diffusion problem by using small quantities of the isotope material. Strictly speaking, however, the self-diffusion problems are obviously different from the impurity ones, as can be seen from the difference of (i) between (c) and (d) in Section 3.3.

As discussed in Section 2.1, the corresponding F 1 law to the F2 law is incomplete without the intrinsic diffusion flux. In the following, therefore, the self-diffusion problems of a pure material are directly investigated in accordance with the intrinsic diffusion flux, regardless of a driving force.

(1) *Application of intrinsic diffusion flux to self-diffusion problem*

We cannot experimentally confirm J_{eq}^{j} in Equation (2-13). The theory of Brown motion reveals that the diffusion phenomena occur in a material even if the macro concentration gradient is zero. The concept of intrinsic diffusion flux should be thus valid in a localized region space. The coordinate notation (τ, ξ) is, for convenience, used instead of $(\tilde{t},\tilde{x})$ and/or (t,x) in the following.

Using the coordinate notation (τ, ξ) for the equation of $J_{\mathrm{Q}}^{j}(\tilde{t},\tilde{x})=J_{\mathrm{eq}}^{j}$ relevant to the self-diffusion, the relation of

$$J_{\mathrm{eq}}^{j}\Delta\tau=-\int_{\xi_0}^{\xi_0+\Delta\xi}\left\{C^{j}\left(\tau_0+\Delta\tau,\xi\right)-C^{j}\left(\tau_0,\xi\right)\right\}\mathrm{d}\xi \tag{3-19}$$

is obtained. If the concentration difference between (τ_0, ξ_0) and $(\tau_0 + \Delta\tau, \xi_0 + \Delta\xi)$ is defined as, ΔC^j Equation (3-19) is rewritten as

$$\Delta C^j = -J^j_{\mathrm{eq}} \, \Delta\tau / \Delta\xi \,. \tag{3-20}$$

Taking account of $(\tau_0 + \Delta\tau, \xi_0 + \Delta\xi)$ in Equation (3-19) and using Equation (2-5) for Equation (3-20), we have the relation of

$$\frac{\Delta C^j}{\Delta\xi} = -J^j_{\mathrm{eq}} / D^j_0 \,. \tag{3-21}$$

Equation (3-21) shows an instantaneous concentration gradient in the localized time $\Delta\tau$ and space $\Delta\xi$ in the cause of the thermal fluctuation, even if the macro concentration gradient is zero. In other words, although the concentration gradient of a pure material is zero in macro behavior, it is nonzero in the minute region.

The behavior of concentration distribution in the local space is thus obtained as

$$C^j(\xi) = -\frac{J^j_{\mathrm{eq}}}{D^j_0}\xi + C^j(0). \tag{3-22}$$

In a pure material or a material in the thermal equilibrium state, Brownian particles move in accordance with Equation (3-22). If, $J^j_{\mathrm{eq}} = 0$ the Brownian particles get a standstill state inconsistent with the Einstein theory of Brown motion. The intrinsic diffusion flux J^j_{eq} is thus an important concept in the fundamental diffusion theory. At the same time, the F1 law is incomplete without J_{eq}.

If we apply Equation (3-22) to the interdiffusion problems, the relation of

$$\sum_{j=1}^{N} C^j(\xi) = -\frac{\xi}{\tilde{D}} \sum_{j=1}^{N} J^j_{\mathrm{eq}} + \sum_{j=1}^{N} C^j(0) \tag{3-23}$$

is valid in accordance with Equation (3-3). Equation (3-23) reveals that Equation (3-6) is valid because of

$$\sum_{j=1}^{N} C^j(\xi) = \sum_{j=1}^{N} C^j(0) = 1$$

Using the notation $\alpha^j = J^j_{\rm eq}/\tilde{D}$ for Equation (3-22), the concentration behavior of an element j in the thermal equilibrium state is expressed as

$$C^j(\xi) = -\alpha^j \xi + C^j(0) \tag{3-24}$$

under the condition of $\sum_{j=1}^{N} \alpha^j = 0$ in the localized space.

The diffusion region of self-diffusion problem of a pure material is, for convenience, divided into 2 regions as a binary system interdiffusion problem composed of a pure material I of $\xi > 0$ and the pure material II of $\xi > 0$ at $\tau = 0$. The diffusion region is between $\xi_A \le \xi \le \xi_B$ at a time τ then. The boundary values are $C^{\rm I}_{\rm A} = 2C_0$ for $\xi \le \xi_A$ and $C^{\rm II}_{\rm B} = 2C_0$ for $\xi \le \xi_B$ under the condition of $C_0 = 0.5$.

Using Equation (3-24) for j = I,II, the relations of

$$C^{\rm I}(\xi) = \alpha\xi + C_0 \text{ and } C^{\rm II}(\xi) = -\alpha\xi + C_0 \tag{3-25}$$

are obtained, where $\alpha = J_{\rm eq}/\tilde{D}$. Equation (3-25) shows that $C^{\rm I}(\xi)$ and $C^{\rm II}(\xi)$ depend on ξ and that $C^{\rm I}(\xi) \ne C^{\rm II}(\xi)$ is valid if $\xi \ne 0$. If we pay attention only to one side of $C^{\rm I}$ or $C^{\rm II}$, it is obvious that the particle migration occurs via the random movement in the pure material. In other words, the diffusion occurs in a pure material, even if the macro concentration gradient is zero.

On the other hand, if we pay attention to both sides of $C^{\rm I}$ and $C^{\rm II}$ in Equation (3-25), $C^{\rm I}(\tau,\xi) = C^{\rm II}(\tau,\xi) = C_0$ seems to be valid, since we cannot actually distinguish their diffusion particles between $\xi_A \le \xi \le \xi_B$ whether they were ones of $C^{\rm I}$ or $C^{\rm II}$ in the initial state at $\tau = 0$.

(2) Physical meaning of intrinsic diffusion flux

Here, we estimate the $J_{\rm eq}$ value in the following. When the averaged jump distance or the averaged collision distance is written as l in a pure material, the substitution of $\xi_A = -l$ and $\xi_B = l$ into Equation (3-25) yields

$$J_{\rm eq} = -D_{\rm self} C_0 / l, \tag{3-26}$$

where $\tilde{D}$ is rewritten as D_{self} and C_0/l corresponds to the concentration gradient. It is considered that the order of l is several angstroms and the diffusion junction depth in a usual experiment is several micrometers. Therefore, the absolute value of $J_F = -\tilde{D}\partial_\xi C$ is considerably small, compared with that of $J_{eq} = -D_{self}C_0/l$.

We cannot actually observe J_{eq} in the macro state because of Equation (3-6). However, we can approximately estimate it by using the relation of

$$J_{eq} = -D^*_{self}C_0/l$$

where D^*_{self} is a self-diffusion coefficient obtained from the tracer diffusion experiment for the concerned material, C_0 = 0.5and l corresponds to the averaged jump distance or the averaged collision distance of the diffusion particle.

Equation (3-25) independent of τ means the time-averaged concentration profile caused by the Brown motion in the localized region space. The diffusion occurs even in a pure material, resulting from the chain reaction of instantaneous jumping of diffusion particle into micro holes caused by a thermal fluctuation in a material minute region. Equation (3-26) thus gives evidence that the Brown motion occurs in a pure material, even if an external force does not exist in the diffusion system.

From the historical point of view, if the F1 law were investigated at an early stage in accordance with the Gauss divergence theory, we might understand the Brown motion behavior before the Einstein theory [2] [3] [14].

3.5. Diffusion Equation of Micro Holes

The space coordinate dependence of micro holes has been neglected in the diffusion theory mentioned above. In the metallurgy field, however, the experimental results of multiple markers show that their movements depend on the space coordinate in accordance with a marker position set in the initial diffusion couple [28] - [30]. This indicates that we must investigate the space coordinate dependence of micro holes in the diffusion study in detail, since the behavior of micro holes is visualized by the characteristic of inert markers.

In that case, Equation (3-1) is rewritten as

$$C^H + \sum_{j=1}^{N} C^j = 1, \tag{3-27}$$

where C^{H} is the normalized concentration of micro holes between $x_A \le x \le x_B$. In general, there is a difference $\Delta C^{H} = -\left(C_A^{H} - C_B^{H}\right)$ between the thermal equilibrium concentration of micro holes $C_A^{H}\left(x \le x_A\right)$ in the material A and that of $C_B^{H}\left(x \ge x_B\right)$ in the material B. When the difference ΔC^{H} is not negligible, the diffusion of micro holes occurs between $x_A \le x \le x_B$.

For the thermal equilibrium concentration of micro holes C_0^{H} in the diffusion region $x_A \le x \le x_B$, the generation/annihilation term of the diffusion equation is expressed as

$$k_H\left(C_0^{H} - C^{H}\right) \tag{3-28}$$

where k_H is a chemical reaction constant. Taking account of Equation (3-28), the diffusion equation of micro holes is expressed as

$$\partial_t C^{H} = \partial_x\left(D^{H}\partial_x C^{H}\right) + k_H\left(C_0^{H} - C^{H}\right), \tag{3-29}$$

where D^{H} is a diffusivity of micro holes between $x_A \le x \le x_B$.

The K-effect shows that the concentration of micro holes yields a supersaturated section and an unsaturated one in the diffusion region. Therefore, the averaged value of diffusion flux yielding

$$\left(x_B - x_A\right)^{-1}\int_{x_A}^{x_B} k_H\left(C_0^{H} - C^{H}\right)\mathrm{d}x \tag{3-30}$$

is considered to be negligible. We thus assume that the averaged contribution of the generation/annihilation term to the total diffusion flux in the diffusion system is negligible.

Here, if we integrate Equation (3-29) with respect to x, the diffusion flux relation of

$$J_x^{H} = -\int \partial_t C^{H}\mathrm{d}x = -D^{H}\partial_x C^{H} - \int k_H\left(C_0^{H} - C^{H}\right)\mathrm{d}x \cong -D^{H}\partial_x C^{H} \tag{3-31}$$

is obtained by neglecting the contribution of Equation (3-30). In that case, Equations (3-27), (2-11) and (3-31) yield

$$\partial_x\left(\sum_{j=1}^{N+1} D^{j}\partial_x C^{j}\right) = \partial_x\left(\sum_{j=1}^{N}\left(D^{j} - D^{H}\right)\partial_x C^{j}\right) = 0,$$

under the condition of N+1 = H. Equation (3-3) is thus rewritten as

$$\tilde{D} = D^{\mathrm{H}} = D^{1} = D^{2} = \cdots = D^{\mathrm{N}} \tag{3-32}$$

Equation (3-32) shows that we can accept a micro hole as a virtual diffusion particle.

The diffusion equation of $j \neq$ H in the N elements system interdiffusion becomes

$$\partial_t C^j = \partial_x \left(\tilde{D} \partial_x C^j \right), \tag{3-33}$$

using a common interdiffusion coefficient $\tilde{D}$. It seems that Equation (3-33) is an independent equation of C^j. However, it is not independent because of Equation (3-27). When we must investigate the influence of ΔC^{H} on the interdiffusion problems in detail, the diffusion equation of micro holes yielding

$$\partial_t C^{\mathrm{H}} = \partial_x \left(\tilde{D} \partial_x C^{\mathrm{H}} \right) + k_{\mathrm{H}} \left(C_0^{\mathrm{H}} - C^{\mathrm{H}} \right) \tag{3-34}$$

should be investigated. Equation (3-34) shows that the behavior of multiple markers depends on the time and space, since the obtained C^{H} dependent on the time and space is visualized by the inert markers.

3.6. Application of Present Theory to Actual Diffusion Problems

Hereinbefore, we discussed some fundamental problems in the bulk diffusion of an ideal diffusion couple. The interdiffusion problems for a many elements system, such as an alloy, a compound semiconductor, a multilayer thin film, a microstructure material and so on, have been widely investigated [4] [31] - [34]. It is then fundamentally indispensable for the material science investigation to understand the N ($N \geq 2$) elements system interdiffusion theory, since the migration of each element in a material occurs during the thermal treatment in the material fabrication process.

In the diffusion study history, the concept of intrinsic diffusion has been accepted in order to understand the K effect. In the existing theory of interdiffusion problems, the Darken equation derived from the intrinsic diffusion concept has been widely used for analyzing them [35]. The application of the Darken theory to a many elements system interdiffusion was extremely complicated even if it was a ternary elements system [36].

On the other hand, the simple relation among interdiffusion coefficients obtained here will be extremely useful for analyzing various interdiffusion problems.

As can be seen from the Appendix [B], we revealed that the Darken equation is not mathematically valid. It is also reported that the analytical results obtained by using the Darken equation are not consistent with the experimental results [37]. This fact is consistent with the present theory where there is no such concept of intrinsic diffusion in the diffusion phenomena.

Under the condition of Equation (3-27), therefore, it is necessary for analyzing a many elements system interdiffusion problem to solve the diffusion equations (3-33) and (3-34) of each element in accordance with their initial and boundary conditions. Here, it seems that these diffusion equations are independent of each other since they have a common interdiffusion coefficient. However, it is not independent because of Equation (3-27).

In the parabolic space discussed later, the analytical method of a nonlinear diffusion equation is established. The analytical solutions of Equations (4-48) and (4-49) are thus obtained then. Using them for analyzing each of the above equations, we can fundamentally solve an N ($N \geq 2$) elements system interdiffusion problems. The theory discussed here is thus widely applicable to analyzing the fundamental problems of N elements system interdiffusion.

Since the present analytical theory is fundamental, it will be necessary for actual diffusion problems to modify it partly in accordance with the concerned problems like the Boyle Charles law for an ideal gas state is modified into the van der Waals equation in the actual case by incorporating an interaction between gas particles into it.

4. ANALYSIS OF DIFFUSION EQUATION IN PARABOLIC SPACE

Einstein theoretically revealed that the Brownian particles randomly move in accordance with the parabolic law [3]. After that, it was experimentally confirmed by Perrin [38]. The parabolic law is universally shown in phenomena relevant to the normal distribution in a probability problem. The Brown problem is a study subject not only in the material science but also in the complex-system sciences relevant to the Markov process. Therefore, the analytical method of the Brown problem is mathematically common in various science fields.

The diffusion junction depth is directly relevant to the parabolic law. The diffusion problem is justifiably one of the Brown problems. This indicates that the solutions of diffusion problems are possible in the parabolic space. In other words, the analytical method discussed here is widely applicable to various Brown problems.

The F2 law is a continuous equation in the conservation system under the condition of no sink and source of diffusion particles, and it is one of the basic equations in physics. When the existence of a driving force or the sink and/or source of diffusion particles is negligible in a given diffusion system, Equation (2-6) is a parabolic type linear partial differential equation in the evolution equations. A lot of diffusion problems have been investigated by solving the diffusion equation of the time and space (*t,x,y,z*) since 1855.

When the diffusivity depends on the concentration, the diffusivity becomes a function of the independent variables (*t,x,y,z*) via the concentration. Even if the diffusion equation depends only on(*t,x*), the mathematical solutions of the nonlinear partial differential equation are almost impossible. A new analytical method in the parabolic space, which is extremely superior in analyzing to the existing ones, is discussed in the following.

4.1. Definition of Parabolic Space

For convenience, hereafter we use the coordinate notation $\langle\theta(\xi_u)| = (\xi_1, \xi_2, \xi_3)$ transformed from the time and space (*t,x,y,z*) into the parabolic space and they are correlated to $\xi_i = x_i/\sqrt{t}$ and $\tau = t$ under the condition of $(t_1, x_1, y_2, z_3) \equiv (t,x,y,z)$. The relations of differential operators are obtained as

$$\frac{\partial}{\partial x_1} = \frac{\partial \tau}{\partial x_1}\frac{\partial}{\partial \tau} + \frac{\partial \xi_1}{\partial x_1}\frac{\partial}{\partial \xi_1} + \frac{\partial \xi_2}{\partial x_1}\frac{\partial}{\partial \xi_2} + \frac{\partial \xi_3}{\partial x_1}\frac{\partial}{\partial \xi_3} = \frac{1}{\sqrt{\tau}}\frac{\partial}{\partial \xi_1}$$

and

$$\frac{\partial}{\partial t} = \frac{\partial \tau}{\partial t}\frac{\partial}{\partial \tau} + \frac{\partial \xi_1}{\partial t}\frac{\partial}{\partial \xi_1} + \frac{\partial \xi_2}{\partial t}\frac{\partial}{\partial \xi_2} + \frac{\partial \xi_3}{\partial t}\frac{\partial}{\partial \xi_3} = \frac{\partial}{\partial \tau} - \frac{1}{2\tau}\left(\xi_1\frac{\partial}{\partial \xi_1} + \xi_2\frac{\partial}{\partial \xi_2} + \xi_3\frac{\partial}{\partial \xi_3}\right).$$

From the similar calculation, the differential operators in the parabolic space are obtained as

$$\langle\tilde{\nabla}| = \frac{1}{\sqrt{\tau}}(\partial_1, \partial_2, \partial_3) = \frac{1}{\sqrt{\tau}}\langle\tilde{\nabla}_\sigma|, \quad \frac{\partial}{\partial t} = \frac{\partial}{\partial \tau} - \frac{1}{2\tau}\langle\theta|\nabla_\sigma\rangle, \qquad (4\text{-}1)$$

where $\partial_i = \partial_{\xi_i} = \partial/\partial_{\xi_i}$ and $\langle \tilde{\nabla}_\sigma | = -\{|\nabla_\sigma\rangle\}^\dagger$ because of the Hermite characteristic.

4.2. Diffusion Equation in Parabolic Space

If we directly apply Equation (4-1) to homogeneous partial differential Equation (2-6), the ellipse type differential equation is obtained as

$$-\frac{1}{2}\langle\theta|\nabla_\sigma C(\xi_1,\xi_2,\xi_3)\rangle = \langle\tilde{\nabla}_\sigma|D\nabla_\sigma C(\xi_1,\xi_2,\xi_3)\rangle, \tag{4-2}$$

where $\partial_t C(\xi_1,\xi_2,\xi_3)=0$ is physically valid in accordance with initial conditions [21].

If $n = 1$ for $\langle\theta(\xi_n)|$, Equation (4-2) becomes

$$\frac{\mathrm{d}}{\mathrm{d}\xi_1}\left(D\frac{\mathrm{d}C(\xi_1)}{\mathrm{d}\xi_1}\right) = -\frac{\xi_1}{2}\frac{\mathrm{d}C(\xi_1)}{\mathrm{d}\xi_1} \tag{4-3}$$

which is the well-known Boltzmann transformation equation [21]. Further, it is rewritten as

$$-\frac{\xi_1}{2D} = \frac{1}{D}\frac{\mathrm{d}D}{\mathrm{d}\xi_1} + \left(\frac{\mathrm{d}C(\xi_1)}{\mathrm{d}\xi_1}\right)^{-1}\frac{\mathrm{d}^2C(\xi_1)}{\mathrm{d}\xi_1^{\,2}} \tag{4-4}$$

The integral calculation of Equation (4-4) yields the integro-differential equation given by

$$D(\xi_1)\frac{\mathrm{d}C(\xi_1)}{\mathrm{d}\xi_1} = J_0^1 \exp\left[-\int_0^{\xi_1}\frac{\eta}{2D(\eta)}\mathrm{d}\eta\right], \tag{4-5}$$

Where $J_0^1 = D(\xi_1)\partial_1 C(\xi_1)\big|_{\xi_1=0}$.

Here, if we define

$$J(\xi_1) = -J_0^1 \exp\left[-\int_0^{\xi_1}\frac{\eta}{2D(\eta)}\mathrm{d}\eta\right], \tag{4-6}$$

the expression of

$$J(\xi_1) = -D(\xi_1)\frac{\mathrm{d}C(\xi_1)}{\mathrm{d}\xi_1} \qquad (4\text{-}7)$$

is obtained from using Equation (4-6) for Equation (4-5). Boltzmann transformation equation (4-3) is essentially equivalent to Equation (4-7).

We can accept Equation (4-7) as a diffusion flux in the parabolic space. The F1 law in the time and space cannot be used directly for analyzing a diffusion problem. However, the diffusion flux of Equation (4-7) is possible for analyzing a diffusion problem under the initial condition of Equation (4-6).

Further, as described later, the solutions of three dimensions problems in the parabolic space (ξ_1, ξ_2, ξ_3) are obtained as a linear combination among the solutions of each one dimension diffusion problem. Therefore, Equation (4-7) is the most important essential equation relevant to the diffusion problems in the parabolic space.

In the parabolic space (ξ_1, ξ_2, ξ_3), the diffusion flux is defined as

$$|J(\xi_1, \xi_2, \xi_3)\rangle = -D|\nabla_\sigma C(\xi_1, \xi_2, \xi_3)\rangle \qquad (4\text{-}8)$$

where the component of $\langle J(\xi_1, \xi_2, \xi_3)| = (J_{\xi_1}, J_{\xi_2}, J_{\xi_3})$ is expressed as

$$J_{\xi_j} = -J_0^j \exp\left[-\int_0^{\xi_j} \frac{\eta}{2D}\mathrm{d}\eta\right] \text{ for } \eta = \xi_j \qquad (4\text{-}9)$$

using $J_0^j = D\partial_j C(\xi_n)\big|_{\xi_j=0}$ for n = 1,2,3. Here, note that the initial and/or boundary condition is taken into account in the diffusion flux of Equation (4-6) or (4-8), although it is not taken into account in the F1 law.

Equation (4-8) in the parabolic space (ξ_1, ξ_2, ξ_3) is equivalent to Equation (2-6) in the time and space (*t,x,y,z*). In mathematics, the following relation of

$$\frac{\mathrm{d}C}{\mathrm{d}\xi_j} = \partial_j(\xi_n) + \frac{\partial C}{\partial D}\partial_j D \quad \text{for } j = 1,2,3 \qquad (4\text{-}10)$$

is generally valid, if the diffusivity D depends on independent variables. Equation (4-10) is extremely useful for analyzing a nonlinear diffusion equation.

4.3. Analytical Solutions of Linear Diffusion Equation

The utility of the present method is confirmed through solving some typical diffusion problems by concrete calculations. The mathematical method used here is also widely applicable to the Brown problems in various science fields.

(1) One dimension parabolic space

We first solve the linear diffusion equation given by

$$\partial_t C(t,x) = D_0 \partial_x^{\ 2} C(t,x) \tag{4-11}$$

under the initial and boundary conditions of

$$\begin{cases} t=0, \quad x<0: C(0,x)=C_{\mathrm{A}} \quad x>0: C(0,x)=C_{\mathrm{B}}, \\ t>0, \quad C(t,-\infty)=C_{\mathrm{A}}, \quad C(t,\infty)=C_{\mathrm{B}}. \end{cases} \tag{4-12}$$

The initial condition of Equation (4-12) is rewritten as

$$C(-\infty)=C_{\mathrm{A}}, \quad C(\infty)=C_{\mathrm{B}}, \tag{4-13}$$

in the parabolic space ξ_1.

Substituting $D = D_0$ and $\xi_1 = 0$ into Equation (4-5), the initial value of

$$J_0^1 = D_0 C_0^{(1)} \text{ for } C_0^{(1)} = \left. \frac{\mathrm{d}C(\xi_1)}{\mathrm{d}\xi_1} \right|_{\xi_1=0} \tag{4-14}$$

is obtained. Using Equation (4-14) for Equation (4-5), the ordinary differential equation of

$$\frac{\mathrm{d}C(\xi_1)}{\mathrm{d}\xi_1} = C_0^{(1)} \exp\left[-\frac{\xi_1^{\ 2}}{4D_0}\right] \tag{4-15}$$

is obtained.

From the definition of the well-known error function, the general solution of Equation (4-15) is obtained as

$$C(\xi_1) = \sqrt{\pi D_0} C_0^{(1)} \operatorname{erf}\left(\frac{\xi_1}{2\sqrt{D_0}}\right) + C_r, \tag{4-16}$$

where C_r is an integral constant. Using the initial condition of Equation (4-13) for the general solution, the well-known solution is obtained as

$$C(\xi_1) = \frac{C_A + C_B}{2} - \frac{C_A - C_B}{2} \operatorname{erf}\left(\frac{\xi_1}{2\sqrt{D_0}}\right). \tag{4-17}$$

In the analytical method in the parabolic space, the solution is thus easily obtained only by the elementary integral. It is, therefore, obvious that the present analytical method is extremely superior in analyzing diffusion equations to the existing methods.

(2) Parabolic space of n dimensions

In case of $D = D_0$, Equation (4-9) is rewritten as

$$J_{\xi_j} = -D_0 C_{j0}^{(1)} \exp\left[-\frac{\xi_j^2}{4D_0}\right] \quad \text{for} \quad C_{j0}^{(1)} = \partial_j C(\xi_n)\Big|_{\xi_j=0}.$$

Using this equation for $n = 3$, the relation of

$$\left|\nabla_\sigma(\partial_n) C(\xi_n)\right\rangle = \begin{pmatrix} C_{10}^{(1)} \exp\left[-\xi_1^2/4D_0\right] \\ C_{20}^{(1)} \exp\left[-\xi_2^2/4D_0\right] \\ C_{30}^{(1)} \exp\left[-\xi_3^2/4D_0\right] \end{pmatrix} \tag{4-18}$$

is obtained from Equation (4-8).

The operation of $\langle\tilde{\nabla}_\sigma|$ on both sides of Equation (4-18) yields the Poisson equation given by

$$\sum_{j=1}^{n} \partial_j^2 C(\xi_n) = -\frac{1}{2D_0} \sum_{j=1}^{n} C_{j0}^{(1)} \xi_j \exp\left[-\xi_j^2/4D_0\right]. \tag{4-19}$$

We solve the ellipse type differential equation (4-19) corresponding to Equation (2-6) under the initial and boundary conditions in the following.

The initial and boundary values of Equation (2-6) given by

$$
\begin{aligned}
&C(t,x,y,z)=C_1^0 \quad \text{for } t\geq 0,\ y>0,\ z>0 \text{ and } x=0 \\
&C(t,x,y,z)=C_2^0 \quad \text{for } t\geq 0,\ x>0,\ z>0 \text{ and } y=0 \\
&C(t,x,y,z)=C_3^0 \quad \text{for } t\geq 0,\ x>0,\ y>0 \text{ and } z=0 \\
&C(t,x,y,z)=0 \quad \text{for } t=0,\ x>0,\ y>0 \text{ and } z>0
\end{aligned} \tag{4-20}
$$

correspond to

$$
C(0,\infty,\infty)=C_1^0,\ C(\infty,0,\infty)=C_2^0,\ C(\infty,\infty,0)=C_3^0,\ C(\infty,\infty,\infty)=0 \tag{4-21}
$$

in the parabolic space (ξ_1,ξ_2,ξ_3).

In the analysis of Equation (4-19), it is easily found that the equation of

$$
C_S(\xi_n)=\sqrt{\pi D_0}\sum_{j=1}^{n} C_{j0}^{(1)}\operatorname{erf}\left[\xi_j/2\sqrt{D_0}\right] \tag{4-22}
$$

satisfies Equation (4-19). In other words, Equation (4-22) is the particular solution of Equation (4-19). When the general solution of Laplace equation of

$$
\sum_{i=1}^{n}\partial_i^2 C(\xi_n)=0 \tag{4-23}
$$

is obtained as $C(\xi_n)=C_L(\xi_n)$, the general solution of Equation (4-19) is expressed as

$$
C(\xi_n)=C_L(\xi_n)+C_S(\xi_n) \tag{4-24}
$$

in accordance with the mathematical theory.

Assuming the equation of

$$C_L(\xi_n)=\prod_{j=1}^{n}F_j(\xi_j) \tag{4-25}$$

and using the variable separation method, the solution of Equation (4-23) is possible. Substituting Equation (4-25) into Equation (4-23), the relation of

$$\sum_{j=1}^{n}\frac{1}{F_j(\xi_j)}\frac{\mathrm{d}^2F_j(\xi_j)}{\mathrm{d}\xi_j^2}=0$$

is valid. In order to satisfy this equation for an arbitrary ξ_j, the equation of

$$\left(\frac{\mathrm{d}^2}{\mathrm{d}\xi_j^2}-\lambda_j^2\right)F_j(\xi_j)=0 \tag{4-26}$$

must be valid under the condition of

$$\sum_{j=1}^{n}\lambda_j^2=0\,. \tag{4-27}$$

Equation (4-27) yields $\lambda_1 = 0$ for $n = 1$ and solution of Equation (4-26) is expressed as

$$C_L(\xi_1)=A_{1+}\xi_1+A_{1-}\,,$$

where A_{1+} and A_{1-} are arbitrary constants. $A_{1+}=0$ and $A_{1-}=C_1^0\left(=-\sqrt{\pi D_0}C_{10}^{(1)}\right)$ are determined by the initial condition. The well-known solution of Equation (4-19) is thus obtained as

$$C(\xi_1)=C_1^0\left(1-\mathrm{erf}\left[\xi_1/2\sqrt{D_0}\right]\right) \tag{4-28}$$

In case of n = 2,3 and $\prod_{j=1}^{n}\lambda_j\neq 0$, substituting the solution of Equation (4-26) given by

$$C_j(\xi_j)=A_{j+}\mathrm{e}^{\lambda_j\xi_j}+A_{j-}\mathrm{e}^{-\lambda_j\xi_j}$$

into Equation (4-25), the general solution of Equation (4-23) is obtained as

$$C_L(\xi_n) = \sum_{\lambda_j} \left\{ \prod_{j=1}^{n} \left(A_{j+} e^{\lambda_j \xi_j} + A_{j-} e^{-\lambda_j \xi_j} \right) \right\}, \tag{4-29}$$

where A_{j+} and A_{j-} are arbitrary constants. However, there is no solution satisfying the initial and boundary conditions because of Equation (4-27). As a result, the general solution of Equation (4-23) is obtained as

$$C_L(\xi_n) = \sum_{\lambda_j} \left\{ \prod_{j=1}^{n} \left(A_{j+} \xi_j + A_{j-} \right) \right\} \tag{4-30}$$

under the condition of $\lambda_1 = \lambda_2 = \lambda_3 = 0$. Since the initial and boundary conditions yield $A_{j+} = 0$, the present solution of Equation (4-23) is thus obtained as

$$C_L(\xi_n) = \sum_{j=1}^{n} C_j^0,$$

where $C_j^0 = -\sqrt{\pi D_0} C_{j0}^{(1)}$ in accordance with the initial and boundary conditions.

The solution of Equation (4-19) is thus obtained as

$$C(\xi_n) = \sum_{j=1}^{n} C_j^0 \left(1 - \text{erf}\left[\xi_j / 2\sqrt{D_0} \right] \right). \tag{4-31}$$

If we use $(\eta) = 1 - \text{erf}(\eta)$ for $n = 3$, the solution is also presented as

$$C(t,x,y,z) = C_1^0 \,\text{erfc}\left[\frac{x}{2\sqrt{D_0 t}} \right] + C_2^0 \,\text{erfc}\left[\frac{y}{2\sqrt{D_0 t}} \right] + C_3^0 \,\text{erfc}\left[\frac{z}{2\sqrt{D_0 t}} \right]. \tag{4-32}$$

The present study reveals that the particular solution is easily possible and the complementary function of Equation (4-19) is a constant value to satisfy the given initial and boundary conditions because of the singular characteristic of initial conditions. In other words, the diffusion behavior is incorporated into the inhomogeneous term of the Poisson equation. As a result, the solution is obtained as a linear combination of the error functions.

Although the diffusion problems have not ever been investigated in the parabolic space, it is obvious that the present solution is thus exceedingly simple and elegant, compared with the usual solution of

$$C(t,x,y,z)=\sum_{\mu}\left\{e^{-\mu t}\prod_{j=1}^{3}\left(A_{j+}e^{k_j x_j}+A_{j-}e^{-k_j x_j}\right)\right\} \tag{4-33}$$

For $\mu=\sum_{j=1}^{3}k_j^2$, where $x_1 = x$, $x_2 = y$, $x_3 = z$.

(3) Analysis of inhomogeneous diffusion equation

The research subjects of linear diffusion phenomena in detail are often expressed by an inhomogeneous partial differential equation. Therefore, the application of the present method to those problems is investigated. Generally, such linear diffusion equations are expressed as

$$Ly=g(t,x,y,z) \tag{4-34}$$

where L is the linear operator of

$$L=\partial_t - D_0\left\langle\tilde{\nabla}\middle|\nabla\right\rangle \tag{4-35}$$

From the mathematical theory, the solution of inhomogeneous Equation (4-34) is obtained as

$$y=f_C(t,x,y,z)+f_S(t,x,y,z), \tag{4-36}$$

where f_c (t,x,y,z) is the complementary function, i.e., the general solution of homogeneous equation given by

$$Ly=0 \tag{4-37}$$

and f_s (t,x,y,z) is a particular solution of

$$f_S(t,x,y,z)=L^{-1}g(t,x,y,z). \tag{4-38}$$

Equation (4-1) shows that the differential operators in the parabolic space are not able to apply to Equation (4-34). However, the linear operator in the parabolic space given by

$$L_{\mathrm{P}} = \frac{1}{2}\langle \theta | \nabla_{\sigma} \rangle + D_0 \langle \tilde{\nabla}_{\sigma} | \nabla_{\sigma} \rangle \tag{4-39}$$

is possible for analyzing the homogeneous Equation (4-37). In other words, we can obtain the complementary function by the present method. In order to obtain a solution of an inhomogeneous diffusion equation, therefore, we must obtain a particular solution by the existing methods.

4.4. Analytical Solutions of Nonlinear Diffusion Equation

Using the experimental profile $C(\xi_1)$ for Equation (4-3), Matano obtained the $\tilde{D}(\xi_1)$ profile in the interdiffusion problems between solid metals in 1933 [26]. The empirical Boltzmann Matano method has been widely applied to the analysis of the interdiffusion experiments between solid metals. However, the mathematical solutions of Equation (4-3) had not yet been obtained since 1894 for such a long time until the recent work [19]. If the analytical solutions of Equation (4-2) or (4-8) obtained, we can understand the effect of a driving force on the diffusivity. In other words, D_{df} of Equation (2-7) will be analytically obtained.

(1) Analytical method in parabolic space

When the diffusivity is affected by a driving force, the mathematical solutions of the nonlinear equations (2-11) and (4-3) or (4-7) are impossible if no other relation between diffusivity and concentration is given. In that case, the diffusivity depends on the independent variables and Equation (4-10) is generally valid in mathematics then. Equations (4-7) and (4-10) are thus simultaneously solved in the following.

The integro-differential equation (4-5) is superior in the approximate calculation to the second order differential equation (4-3). For example, using an effective diffusivity $\tilde{D}_{\mathrm{eff}}$ for the exponential part in Equation (4-5), the relation of

$$\int_0^{\xi_1} \frac{\eta}{2\tilde{D}(\eta)} \mathrm{d}\eta = \frac{\xi_1^2}{4\tilde{D}_{\mathrm{eff}}}$$

is valid because of the characteristic of the integral calculation, where D is rewritten as an interdiffusion coefficient $\tilde{D}$.

In order to clarify the validity of the present method, we applied the present method to the typical problems of a binary system interdiffusion where the complete solid solution is formed in the interdiffusion region between solid metals. The reason for that is as follows. The Boltzmann Matano method has been widely used for the analysis of various interdiffusion problems. Uncountable papers have been reported and useful findings have been thus accumulated. Even if the present results are thus evaluated by comparing them with the Boltzmann Matano results, the generality of present method is still kept from the viewpoints of mathematical physics.

In the following analysis, the relations of

$$C(-\infty)=C_{\mathrm{A}},\ C(\infty)=C_{\mathrm{B}},\ \tilde{D}(-\infty)=\tilde{D}_{\mathrm{A}},\ \tilde{D}(\infty)=\tilde{D}_{\mathrm{B}} \tag{4-40}$$

are adopted as an initial condition in the parabolic space. The coordinate origin $x = 0$ is set at a point on the initial interface of a diffusion couple between materials A and B similar to the one under the condition of $N = 2$ in Section 3. The interdiffusion region of $x_{\mathrm{A}} \le x \le x_{\mathrm{B}}$ corresponds to $\xi_{1\mathrm{A}} \le \xi_1 \le \xi_{1\mathrm{B}}$ in the parabolic space because of $\xi_1 = x/\sqrt{t}$. Here, the suffix I or II of C and $\tilde{D}$ is deleted since the diffusion equation of I is equal to that of II because of $C^{\mathrm{I}} + C^{\mathrm{II}} = 1$.

The countless experimental results always reveal that the $C(\xi_1)$ profile becomes an S-letter curve or a reverse one similar to the profile of Equation (4-17) [39]. In the typical interdiffusion problems, the Boltzmann Matano method shows that the $\tilde{D}(\xi_1)$ profile is also an S-letter curve or a reverse one. These indicate that $C(\xi_1)$ and $\tilde{D}(\xi_1)$ are expressed as a superposition of the error functions with various inflection points.

The relation of $\tilde{D}_{\mathrm{A}} < \tilde{D}_{\mathrm{B}}$ is conveniently adopted in this section. In such a case, the exponential part of Equation (4-5) satisfies

$$\exp\left[-\frac{\xi_1^2}{4\tilde{D}_{\mathrm{A}}}\right] < \exp\left[-\int_0^{\xi_1}\frac{\eta}{2\tilde{D}(\eta)}\mathrm{d}\eta\right] < \exp\left[-\frac{\xi_1^2}{4\tilde{D}_{\mathrm{B}}}\right]$$

Using a constant value $\tilde{D}_{\mathrm{int}}$ between $\tilde{D}_{\mathrm{A}} < \tilde{D}_{\mathrm{int}} < \tilde{D}_{\mathrm{B}}$, it is thus defined as

$$\exp\left[-\int_0^{\xi_1}\frac{\eta}{2\tilde{D}(\eta)}\mathrm{d}\eta\right] = \exp\left[-\frac{\xi_1^2}{4\tilde{D}_{\mathrm{int}}} - \alpha(\xi_1)\right], \tag{4-41}$$

where $\alpha(\xi_1)$ is a function to correct the error caused by using $\tilde{D}_{\text{int}}$ instead of $\tilde{D}(\xi_1)$.

Substituting Equations (4-5) and (4-41) into Equation (4-10) yields the diffusion flux relation of

$$\tilde{D}(\xi_1)\frac{\partial C}{\partial \xi_1}+\tilde{D}(\xi_1)\frac{\partial C}{\partial \tilde{D}}\frac{\mathrm{d}\tilde{D}}{\mathrm{d}\xi_1}=J_0^1\mathrm{e}^{-\alpha(\xi_1)}\exp\left[-\frac{\xi_1^2}{4\tilde{D}_{\text{int}}}\right]. \tag{4-42}$$

For the diffusion flux, the physical speculation produces the relation yielding

$$\tilde{D}(\xi_1)\frac{\partial C}{\partial \xi_1}=J_0^1\beta(\xi_1), \tag{4-43}$$

where $\beta(\xi_1)$ is a function of ξ_1 satisfying $\lim_{\xi_1\to\pm\infty}\beta(\xi_1)=0$. Equations (4-42) and (4-43) yield

$$\tilde{D}(\xi_1)\frac{\partial C}{\partial \tilde{D}}\frac{\mathrm{d}\tilde{D}}{\mathrm{d}\xi_1}=J_0^1\mathrm{e}^{-\alpha(\xi_1)}\left\{\exp\left[-\frac{\xi_1^2}{4\tilde{D}_{\text{int}}}\right]-\mathrm{e}^{\alpha(\xi_1)}\beta(\xi_1)\right\}. \tag{4-44}$$

In relation to a behavior of the error function, Equation (4-44) is divided into following two equations. One is

$$\frac{\mathrm{d}\tilde{D}}{\mathrm{d}\xi_1}=\gamma_1\left\{\exp\left[-\frac{(\xi_1-\varepsilon)^2}{4\tilde{D}_{\text{int}}}\right]-S(\xi_1)\right\} \tag{4-45}$$

and the other is

$$\frac{\partial C}{\partial \tilde{D}}=\frac{\gamma_2}{\tilde{D}(\xi_1)}\exp\left[-\alpha(\xi_1)-\frac{2\varepsilon\xi_1-\varepsilon^2}{4\tilde{D}_{\text{int}}}\right], \tag{4-46}$$

where ε is a shift parameter caused by the dependence of $\tilde{D}(\xi_1)$ on $C(\xi_1)$ and

$$S(\xi_1) = \exp\left[\alpha(\xi_1) + \frac{2\varepsilon\xi_1 - \varepsilon^2}{4\tilde{D}_{\mathrm{int}}}\right]\beta(\xi_1) \text{ and } \gamma_1\gamma_2 = J_0^1.$$

There is evidence for the validity of the above division as shown in the following. When the relation of

$$\left|\alpha(\xi_1) + \frac{2\varepsilon\xi_1 - \varepsilon^2}{4\tilde{D}_{\mathrm{int}}}\right| \ll 1$$

is valid, Equation (4-46) is approximately rewritten as $\tilde{D}\mathrm{d}C = \gamma_2 \mathrm{d}\tilde{D}$ and its integral calculation yields

$$C(\xi_1) = \frac{C_{\mathrm{A}} + C_{\mathrm{B}}}{2} + \frac{C_{\mathrm{A}} - C_{\mathrm{B}}}{\ln\tilde{D}_{\mathrm{A}} - \ln\tilde{D}_{\mathrm{B}}} \ln\left(\frac{\tilde{D}(\xi_1)}{\sqrt{\tilde{D}_{\mathrm{A}}\tilde{D}_{\mathrm{B}}}}\right) \tag{4-47}$$

under the condition of Equation (4-40). In the typical interdiffusion problems between solid metals, Equation (4-47) has been widely accepted [40] - [43]. The derivation of Equation (4-47) gives evidence for the validity of division of Equation (4-44) into Equations (4-45) and (4-46).

Since it is thus confirmed that the division of Equation (4-44) into Equations (4-45) and (4-46) is reasonable, Equation (4-45) is analyzed in accordance with mathematical theory. As can be seen from the analyzing process in Ref. [19] , the calculations are considerably complex. As a result, the general solution of

$$\tilde{D}(\xi_1) = \tilde{D}_m - \tilde{D}_\Delta \operatorname{erf}\left(\frac{\xi_1}{2\sqrt{\tilde{D}_{\mathrm{int}}}} - \frac{\xi_{\mathrm{IF}}}{2\sqrt{\tilde{D}_{\mathrm{int+}}}} + \operatorname{erf}^{-1}\left(\frac{\tilde{D}_{\mathrm{m}} - \tilde{D}_{\mathrm{IF}}}{\tilde{D}_\Delta}\right)\right) \tag{4-48}$$

was obtained as an analytical expression using initial values, where

$$\tilde{D}_{\mathrm{int}} = \tilde{D}_{\mathrm{int+}} \text{ for } \xi \geq 0 \text{ and } \tilde{D}_{\mathrm{int}} = \tilde{D}_{\mathrm{int-}} \text{ for } \xi < 0.$$

Here, the relation of $D_m = D_0$ for $\tilde{D} = D^{\mathrm{I}} = D^{\mathrm{II}} = D_0 + D_{\mathrm{df}}$ is valid in the present diffusion system. Therefore, the effect of a driving force and/ or a concentration dependence on the diffusivity, which is discussed in Section 2.1, is obtained as

$$D_{\mathrm{df}} = -\tilde{D}_{\Delta}\,\mathrm{erf}\left(\frac{\xi_1}{2\sqrt{\tilde{D}_{\mathrm{int}}}} - \frac{\xi_{\mathrm{IF}}}{2\sqrt{\tilde{D}_{\mathrm{int+}}}} + \mathrm{erf}^{-1}\left(\frac{\tilde{D}_{\mathrm{m}} - \tilde{D}_{\mathrm{IF}}}{\tilde{D}_{\Delta}}\right)\right). \tag{4-49}$$

Subsequently, using the obtained solution of Equation (4-48) for Equation (4-46), Equation (4-46) was analyzed also through considerably complex calculations. The general solution of

$$C(\xi_1) = C_{\mathrm{m}} - C_{\Delta}\,\mathrm{erf}\left(\frac{\xi_1}{2\sqrt{\tilde{D}_{\mathrm{int}}}} - \frac{\xi_{\mathrm{IN}}}{2\sqrt{\tilde{D}_{\mathrm{int-}}}} + \mathrm{erf}^{-1}\left(\frac{C_{\mathrm{m}} - C_{\mathrm{IN}}}{C_{\Delta}}\right)\right) \tag{4-50}$$

was thus obtained also as an analytical expression.

Here, the notations used for Equations (4-48) and (4-40) are as follows:

$$\begin{cases} \tilde{D}_{m} = (\tilde{D}_{\mathrm{A}} + \tilde{D}_{\mathrm{B}})/2, \quad \tilde{D}_{\Delta} = (\tilde{D}_{\mathrm{A}} - \tilde{D}_{\mathrm{B}})/2 \\ C_{m} = (C_{\mathrm{A}} + C_{\mathrm{B}})/2, \quad C_{\Delta} = (C_{\mathrm{A}} - C_{\mathrm{B}})/2 \\ \tilde{D}_{\mathrm{IF}} = (\tilde{D}_{\mathrm{A}} - \tilde{D}_{\mathrm{B}})/(\ln\tilde{D}_{\mathrm{A}} - \ln\tilde{D}_{\mathrm{B}}) \\ \xi_{\mathrm{IF}} = 0, \quad \xi_{\mathrm{IN}} = 2\sqrt{\tilde{D}_{\mathrm{A}}\tilde{D}_{\mathrm{B}}}\left(\sqrt{\tilde{D}_{\mathrm{A}}} - \sqrt{\tilde{D}_{\mathrm{B}}}\right)/\left(\sqrt{\tilde{D}_{\mathrm{A}}} + \sqrt{\tilde{D}_{\mathrm{B}}}\right) \\ C_{\mathrm{IN}} = C_{\mathrm{m}} - C_{\Delta}(\tilde{D}_{\mathrm{m}} - \tilde{D}_{\mathrm{IF}})/\tilde{D}_{\Delta} \\ \tilde{D}_{\mathrm{int+}} = (\tilde{D}_{\mathrm{A}} + \tilde{D}_{\mathrm{B}})/2 \text{ and } \tilde{D}_{\mathrm{int-}} = \sqrt{\tilde{D}_{\mathrm{A}}\tilde{D}_{\mathrm{B}}} \end{cases}$$

The validity of the present analytical method is numerically investigated by comparing Equations (4-48) and (4-50) with the results of the Boltzmann Matano method. As a result, we found that the present solutions agree well with those of the Boltzmann Matano method [19].

When the diffusivity does not depend on independent variables, the relation of $\tilde{D}_{\mathrm{A}} = \tilde{D}_{\mathrm{B}} = D_0$ is valid. In that case, the above relations become

$$\tilde{D}_{\mathrm{int}} = \tilde{D}_{\mathrm{int+}} = \tilde{D}_{\mathrm{int-}} = \tilde{D}_{\mathrm{m}} = D_0, \quad \tilde{D}_{\Delta} = \tilde{D}_{\mathrm{IF}} = 0, \quad \xi_{\mathrm{IF}} = \xi_{\mathrm{IN}} = 0 \text{ and } C_{\mathrm{IN}} = C_{\mathrm{m}}.$$

Equations (4-48) and (4-50) agree with $\tilde{D}(\xi_1) = D_0$ and Equation (4-17) then, respectively. The obtained solution of Equations (4-48) and (4-50) are thus accepted as a generalized ones.

The K effect $\Delta\sigma_{\text{eff}}$ in the parabolic space is easily obtained as

$$\Delta\sigma_{\text{eff}} = \Delta x_{\text{eff}} / \sqrt{t} = \left(\sqrt{\tilde{D}_{\text{A}}} - \sqrt{\tilde{D}_{\text{B}}}\right)\left(C_{\text{A}} - C_{\text{B}}\right)$$

by transforming Equation (3-15) into the parabolic space under the condition of N = 2. The K effect in the parabolic space is thus determined only by the initial values.

(2) Meaning of general solutions obtained as analytical expressions

Since the general solutions are obtained as Equations (4-48) and (4-50), the interdiffusion problems are fundamentally solved as follows:

We denote an impurity diffusivity $D_{\text{imp}}^{\text{I}}$ of a diffusion element I in a pure material B composed of an element II and also denote an impurity diffusivity $D_{\text{imp}}^{\text{II}}$ of the diffusion element II in the pure material A com-

posed of the element I. We first investigate an interdiffusion problem of the diffusion couple between pure materials A and B. It is then well-known that the relations of

$$\tilde{D}_{\text{A}} = \tilde{D}(\xi_{\text{IA}}) = D_{\text{imp}}^{\text{II}} \quad \text{and} \quad \tilde{D}_{\text{B}} = \tilde{D}(\xi_{\text{IB}}) = D_{\text{imp}}^{\text{I}} \tag{4-51}$$

are physically valid.

The solution $C(\xi_{\text{I}})$ is determined by using $C_{\text{A}}^{\text{I}} = 1$, $C_{\text{B}}^{\text{I}} = 0$ and the values of Equation (4-51) for Equation (4-50). It gives two independent variable values satisfying $\xi_{\text{I}} = \xi_{\alpha}$ at $C(\xi_{\text{I}}) = \alpha$ and $\xi_{\text{I}} = \xi_{\beta}$ at $C(\xi_{\text{I}}) = \beta$ as follows:

$$\xi_{\alpha} = 2\sqrt{\tilde{D}_{\text{int}}}\left\{\text{erf}^{-1}(1-2C_{\text{IN}}) - \text{erf}^{-1}(1-2\alpha)\right\} + \xi_{\text{IN}}\sqrt{\frac{\tilde{D}_{\text{int}}}{\tilde{D}_{\text{int}-}}} \tag{4-52}$$

$$\xi_{\beta} = 2\sqrt{\tilde{D}_{\text{int}}}\left\{\text{erf}^{-1}(1-2C_{\text{IN}}) - \text{erf}^{-1}(1-2\beta)\right\} + \xi_{\text{IN}}\sqrt{\frac{\tilde{D}_{\text{int}}}{\tilde{D}_{\text{int}-}}} \tag{4-53}$$

Substituting these $\xi_{\text{I}} = \xi_{\alpha}$ and $\xi_{\text{I}} = \xi_{\beta}$ and the values of Equation (4-51) into Equation (4-48), we can determine $\tilde{D}_{\text{A}}(\xi_{\alpha})$ and $\tilde{D}_{\text{B}}(\xi_{\beta})$ for $C(\xi_{\text{I}}) = \alpha$ and $C(\xi_{\text{I}}) = \beta$.

Subsequently, using $\tilde{D}_{\mathrm{A}}(\xi_\alpha)$ and $\tilde{D}_{\mathrm{B}}(\xi_\beta)$ for Equations (4-48) and (4-50) as boundary values of interdiffusion coefficients in the binary system, we can obtain the solutions of the interdiffusion problems for the diffusion couple between the material A composed of $C_{\mathrm{A}}^{\mathrm{I}}=\alpha$, $C_{\mathrm{A}}^{\mathrm{II}}=1-\alpha$ and the material B composed of $C_{\mathrm{B}}^{\mathrm{I}}=\beta$, $C_{\mathrm{B}}^{\mathrm{II}}=1-\beta$.

The solutions of an arbitrary binary system interdiffusion problem are thus possible by using impurity diffusivities in the concerned diffusion system. In the N elements system interdiffusion problems, using an impurity diffusivity of an element in the concerned diffusion couple, their solutions are also possible.

In the diffusion history, the nonlinear F2 equation had not been analytically solved until recently, although the numerical solutions were possible. In order to understand the diffusion problems, the corresponding diffusion fluxes to the F2 equation had been thus widely discussed in the existing theories, since it was considered that the mathematical solutions of nonlinear diffusion equation are impossible. For example, such a diffusion flux in the Appendix A has been discussed.

However, the discussion about such a diffusion flux is now unnecessary for understanding the diffusion problems, since the analytical solutions of Equations (4-48) and (4-50) relevant to the nonlinear F2 equation are obtained, regardless of the diffusion fluxes. In other words, as can be seem from Equation (4-49), we can understand the influence of a driving force on the diffusion flux from the obtained solutions.

Based on the above discussions, it is extremely meaningful that the solutions of the nonlinear F2 equation are obtained as analytical expressions.

(3) Nonlinear problems of three dimensions in parabolic space

The analytical solutions of linear diffusion equations were obtained in the parabolic space. Further, we could also obtain the physically meaningful solutions of nonlinear diffusion equation as analytical expressions in the one dimension parabolic space.

In mathematics, it is almost impossible to solve two or three dimensional nonlinear diffusion equation. However, we investigate two or three dimensional problems in the parabolic space in the following.

We rewrite Equation (4-2) into

$$\sum_{j=1}^{n}\partial_j^2 C(\xi_n)=W(\xi_n) \tag{4-54}$$

where

$$W(\xi_n) = -\sum_{j=1}^{n} \frac{1}{\tilde{D}} \left\{ \partial_j \tilde{D} \partial_j C(\xi_n) + \frac{1}{2} \xi_j \partial_j C(\xi_n) \right\}. \tag{4-55}$$

It is impossible to analyze Equation (4-54) exactly since $\tilde{D}(\xi_n)$ exists in $W(\xi_n)$. By reference to the one dimensional case, if we use the effective diffusivity $\tilde{D}^j_{\text{int}}$ independent of ξ_n instead of $\tilde{D}(\xi_n)$ for Equation (4-55) as an approximation, Equation (4-54) becomes

$$\sum_{j=1}^{n} \left\{ \partial_j^2 C(\xi_n) + \frac{1}{2\tilde{D}^j_{\text{int}}} \xi_j \partial_j C(\xi_n) \right\} = 0 \tag{4-56}$$

because of $\partial_j \tilde{D} = \partial_j \tilde{D}^j_{\text{int}} = 0$.

By reference to Equation (4-19), the solution of Equation (4-56) is obtained as

$$C(\xi_n) = \sum_{j=1}^{n} \left\{ A_j + B_j \operatorname{erf}\left(\frac{\xi_j}{2\sqrt{\tilde{D}^j_{\text{int}}}} - \frac{\xi^j_{IN}}{2\sqrt{\tilde{D}^j_{\text{int}-}}} + \operatorname{erf}^{-1}\left(\frac{C^j_m - C^j_{IN}}{C^j_\Delta} \right) \right) \right\}, \tag{4-57}$$

where A_j and B_j are integral constants.

5. UNIVERSAL EXPRESSION OF DIFFUSIVITY

From applying the diffusion equation to a collision problem between two micro particles in the diffusion elementary process, the S equation is reasonably derived. It is revealed that the diffusivity corresponds to the angular momentum operator in the quantum theory.

The universal diffusivity expression, which is applicable to every material in an arbitrary thermodynamic state, is obtained using an essential diffusion constant then. Further, the correlation between the well-known material wave relation and the diffusivity expression is discussed in a minute time and space region.

5.1. Derivation of Schrödinger Equation

If the movement of a material particle is not the Markov process but the continuous process, Equation (2-1) is rewritten as

$$C(t \pm \Delta t, |r\rangle) = \{C(t, |r\rangle - |\Delta r\rangle) + C(t, |r\rangle + |\Delta r\rangle)\}/2, \tag{5-1}$$

Where $\langle r| = (r_1, r_2, r_3) \equiv (x, y, z)$.

From the Taylor expansion of both sides of Equation (5-1), the well-known wave equation is obtained as

$$\partial_t^2 C = v^2 \langle \tilde{\nabla} | \nabla C \rangle, \tag{5-2}$$

where $v = \Delta r / \Delta t$ is a phase velocity under the condition of $\Delta t \to 0$. The general solution of Equation (5-2) is given by

$$C(t, r) = A \exp\left[i\left(\langle k | r \rangle - \omega t + \alpha \right) \right], \tag{5-3}$$

where A and α are a wave amplitude and an initial phase, and $k_j = 2\pi/\lambda_j$ and $\omega = 2\pi/T$ mean a component of the wave number vector $\langle k| = (k_1, k_2, k_3)$ relevant to a wavelength component ⍰$_j$ and an angular velocity relevant to a period *T*.

The collective behavior of micro particles relevant to the Markov process is expressed by Equation (2-6) derived from their random movements. However, it is revealed in the quantum mechanics that a micro particle has a wave characteristic. The relation of material wave yielding

$$|p\rangle = \hbar |k\rangle \tag{5-4}$$

was thus proposed by de Broglie in 1923, where $|p\rangle$ and $\hbar$ are the momentum of a material particle and the Planck constant $\hbar = h/2\pi$ [23]. The motion of a micro particle relevant to Equation (5-4) should be thus essentially investigated by using the S equation of

$$i\hbar \partial_t \psi = \mathrm{H} \psi, \tag{5-5}$$

where ψ is the wave function and H the Hamiltonian meaning the total energy in the given physical system [18]. In case of a free particle, the Hamiltonian H is given by

$$\mathrm{H} = \frac{\langle p|p\rangle}{2m}, \tag{5-6}$$

where m is the particle mass and $|p\rangle$ the momentum or its operator.

When Equation (5-4) is applied to a micro particle behavior, the left-hand side of Equation (5-4) means the particle motion in accordance with the Markov process. On the other hand, the right-hand side means the continuous process resulting from the wave characteristic of the micro particle during a free motion because of the wave factor $k_j = 2\pi/\lambda_j$. This suggests that the S equation correlates with the diffusion elementary process resulting from a random movement caused by a collision between a micro particle and a force field near the micro particle itself.

An activation energy caused by a local thermal fluctuation generates micro holes in a material and at the same time micro particles are possible to jump instantaneously to the nearest neighbor micro holes through each energy barrier. The macro behavior of their collective motion is investigated by using the diffusion equation (2-6).

On the other hand, behavior of a micro particle of mass m in the diffusion field is investigated by the well-known Langevin equation of

$$m\frac{\mathrm{d}|v\rangle}{\mathrm{d}t} = -k_{\mathrm{L}}|v\rangle + |F(t)\rangle, \tag{5-7}$$

where $|v\rangle = \mathrm{d}|r\rangle/\mathrm{d}t$ and $|f\rangle = -k_{\mathrm{L}}|v\rangle$ are the velocity and the viscosity resistance using a resistance factor k_{L} [27]. Here, $|F(t)\rangle$ is a force caused by a collision and the time-averaged value is considered to be $\langle|F(t)\rangle\rangle = 0$.

The Langevin equation shows that the behavior of a Brown particle in the minute time and space region is incorporated into the force $|F(t)\rangle$. Therefore, we cannot discuss the Brown motion in the minute region by using the Langevin equation because of $\langle|F(t)\rangle\rangle = 0$. Hereafter, we do not discuss $|F(t)\rangle$ but the acceleration of

$$|a\rangle = \frac{\mathrm{d}^2|r\rangle}{\mathrm{d}t^2} = \lim_{\Delta t\to 0}\frac{\Delta^2|r\rangle}{(\Delta t)^2} \tag{5-8}$$

caused by a collision between two micro particles in the minute time and space region.

Since the physical essence is still kept even if we consider the simplest collision problem of one dimensional case, we thus investigate a perfect elastic collision problem between micro particles A and B of the same kind. We consider that the particle A moves with a velocity $|v\rangle$ for a $|r\rangle$ direction and collides at time $t = 0$ with the particle B in the standstill state.

If we can then clarify the distinction between A and B after the collision, the particle A decelerates from the velocity $|v\rangle$ to the velocity zero and the particle B accelerates from the velocity zero to the velocity $|v\rangle$ between $0 \le t \le \Delta t$.

On the other hand, if we cannot clarify the distinction between the particles A and B after the collision, it seems that the particle A decelerates from the velocity $|v\rangle$ to the velocity zero between $0 \le t \le \varepsilon_t$ and subsequently accelerates again from the velocity zero to the velocity $|v\rangle$ between $\varepsilon_t \le t \le \Delta t$. In other words, it seems as if there is no collision process in the particle motion.

If we accept the above latter case, the relation of $-|a\rangle \to \gamma|a\rangle \; (\gamma > 0)$ must be valid in Equation (5-8) between $0 \le t \le \varepsilon_t$ and $\varepsilon_t \le t \le \Delta t$ in the collision process. This indicates that the impossibility of discrimination between the particles A and B yields $t \to \pm it$ between $\varepsilon_t \le t \le \Delta t$, as can be seen from the expression of Equation (5-8).

It is considered that the above former case corresponds to the Markov process caused by a collision and the latter case corresponds to a continuous process caused by the impossibility of identification between the micro particles. It is thus mathematically necessary for the transformation from the Markov process into the continuous process to introduce the imaginary time into a collision problem between micro particles.

In physics, it is difficult to accept the imaginary time in accordance with the continuity of time. In mathematics, however, the product of a physical quantity Q and an imaginary time it is equal to iQt and also that of a physical operator iQ and a real time t is equal to the same result. It is thus considered that the latter case is physically reasonable. However, it is possible that we mathematically analyze the diffusion equation using the imaginary time in accordance with the mathematical equivalence relation. The diffusion equation is thus applied to an elementary diffusion problem using the imaginary time $\tau = it$ in the following.

Rewriting the concentration C (t,r_1,r_2,r_3) of diffusion particles into a quantity of state expressed by a complex function φ (τ,r_1,r_2,r_3), Equation (2-6) yields

$$\partial_\tau \varphi = D_0 \langle \tilde{\nabla} | \nabla \varphi \rangle. \tag{5-9}$$

Assuming φ $(\tau,r_1,r_2,r_3) = T(\tau)\, S(r_1,r_2,r_3)$, Equation (5-9) can be solved by the separation method of variables. Using complex numbers k_j^n, A_{j+} and A_{j-} determined by the initial and boundary conditions, the general solution of Equation (5-9) is obtained as

$$\varphi = \sum_{n=1}^{\infty} \exp[\mu_n \tau] \prod_{j=1}^{3} \left(A_{j+} \exp\left[k_j^n r_j \right] + A_{j-} \exp\left[-k_j^n r_j \right] \right), \tag{5-10}$$

Where $\mu_n = \sum_{j=1}^{3} \left(k_j^n\right)^2$. Substituting $\tau = it$ into Equation (5-10) yields

$$\varphi = \sum_{n=1}^{\infty} \prod_{j=1}^{3} \left(A_{j+} \exp\left[k_j^n r_j + i\mu_n t \right] + A_{j-} \exp\left[-k_j^n r_j + i\mu_n t \right] \right).$$

Here, using the real functions ψ_1 (t,r_1,r_2,r_3) and ψ_2 (t,r_1,r_2,r_3), we rewrite the complex function φ into the complex-value function yielding

$$\varphi \equiv \psi = \psi_1 + i\psi_2. \tag{5-11}$$

Further, substituting Equation (5-11) into Equation (5-9) and multiplying both sides by $\hbar$, we have

$$i\hbar\partial_t \psi = -\hbar D_0 \langle \tilde{\nabla} | \nabla \psi \rangle \tag{5-12}$$

because of $\Delta\tau = i\Delta t$.

From the partial differential equation of probability, the diffusivity of

$$D_0 = \frac{(\Delta r)^2}{2\Delta t} \tag{2-5}$$

is obtained. When a micro particle randomly jumps from a position to another one, the jump orientation becomes the spherical symmetry in probability. Using the relation of $\Delta L = \Delta\left\||r \times p\rangle\right| = \Delta r\left\||\delta \times p\rangle\right|$ relevant to the angular momentum $|L\rangle = |r \times p\rangle$ defined by a position vector $|r\rangle$ and a momentum $|p\rangle = m|v\rangle$, Equation (2-5) is rewritten as

$$D_0 = \frac{(\Delta r)^2}{2\Delta t} = \frac{\Delta r}{2m}\left|m\frac{\mathrm{d}|r\rangle}{\mathrm{d}t}\right| = \frac{\Delta r}{2m}\left\||\delta \times p\rangle\right| = \frac{\Delta L}{2m}.$$

Where $\Delta L = \Delta L_{r_1} = \Delta L_{r_2} = \Delta L_{r_3}$, $r = \sqrt{\langle r|r\rangle}$, $\langle \delta|r\rangle = r$, $\langle \delta|p\rangle = 0$ is valid in the spherical symmetry space.

The relation of $\Delta L\psi = \hbar\psi$ leads to an operator relation of

$$D_0 \to \frac{|\Delta L\rangle}{2m} \to \frac{\hbar}{2m}. \tag{5-13}$$

Substituting Equation (5-13) into Equation (5-12) gives

$$i\hbar\partial_t\psi = -\frac{\hbar^2}{2m}\langle\tilde{\nabla}|\nabla\rangle\psi. \tag{5-14}$$

Here, if we define the imaginary differential operator given by

$$|p\rangle = -i\hbar|\nabla\rangle,\ \langle p| = \{|p\rangle\}^{\dagger} = -i\hbar\langle\tilde{\nabla}|, \tag{5-15}$$

Equation (5-14) is rewritten as

$$i\hbar\partial_t\psi = \frac{\langle p|p\rangle}{2m}\psi. \tag{5-16}$$

Further, the substitution of Equation (5-6) into Equation (5-16) yields the S Equation (5-5). The defined Equation (5-15) is one of the essential operators in the quantum mechanics.

It is considered that the relation of $\lambda = 2\pi\Delta r$ and $T = 2\pi\Delta t$ is valid for a wave packet existing in the minute time and space. In that case, the de Broglie equation

$$|p\rangle = \hbar|k\rangle \tag{5-4}$$

is thus easily obtained because of

$$\frac{(\Delta r)^2}{2\Delta t} = \frac{\hbar}{2m}\left(= D_0\right), \quad p = m\,\Delta r/\Delta t \quad \text{and} \quad \lambda = 2\pi\Delta r.$$

At the same time, this gives evidence for the validity of Equation (5-13) derived from the correlation between the diffusion equation and the S equation. It is, therefore, considered that the diffusivity is fundamentally relevant to the Planck constant $\hbar$ and it should be determined from the quantum mechanics.

5.2. New Universal Diffusivity

The definition of diffusivity shows that the diffusivity should be essentially determined by analyzing the *S* equation, since the diffusion particle such as an atom and/or a molecule moves as a material wave packet in the minute space. However, the existing diffusivity has been defined by the classical mechanics. In the following, therefore, a new diffusivity is investigated.

The existence probability *P* of a micro particle in the collective system of a heat quantity *Q* and an absolute temperature *T* is given by the well-known Boltzmann factor of

$$P = \exp\left[-Q/k_B T\right], \tag{5-17}$$

where k_B is the Boltzmann constant [44]. It is necessary for a jumping of diffusion particle to obtain an activation energy Q caused by a thermal fluctuation. In a collective system composed of micro particles, the diffusivity is thus directly proportional to the probability factor of Equation (5-17).

The jump frequency of a diffusion particle is relevant to a factor ρ derived from an atomic configuration in the diffusion field near the diffusion particle itself, and also it depends on entropy *S* relevant to a state change of diffusion field. If we define these effect on a diffusivity as $\Omega(\rho, S)$, the diffusivity expression D_0 of Equation (2-5) or (5-13) is rewritten as

$$D = D_N \frac{\Omega(\rho, S)}{n} \exp\left[-\frac{Q}{k_B T}\right], \tag{5-18}$$

where the molecular or the atomic weight *n* is expressed as $n = 10^3\, m N_A$ using the particle mass m and the Avogadro constant N_A. Here, D_N is an essential constant in the diffusion phenomena and it is

$$D_{\mathrm{N}} = 5\times10^{2} N_{\mathrm{A}} h = 3.18\times10^{-8}\left[\mathrm{m}^{2}\cdot\mathrm{s}^{-1}\right]. \tag{5-19}$$

Equation (3-32) shows that diffusion particles move randomly with a common diffusivity in the concerned diffusion field. On the other hand, Equation (5-18) shows that the diffusivity depends on the mass of diffusion particle. For the diffusion experimentation in a super gravity field, therefore, it is considered that an individuality of a micro particle resulting from the mass effect will be shown in the diffusivity, compared with one in a usual laboratory.

Hereafter, the investigation of $\Omega(\rho, S)$ will become a dominant subject in the diffusion study as a problem of a few many bodies system in the quantum mechanics, since the essential diffusion constant of Equation (5-19) is thus reasonably obtained.

5.3. Influence of a Driving Force on Diffusivity

The new diffusivity of Equation (5-18) is derived regardless of the bulk diffusion problems. However, it is considerably difficult to determine $\Omega(\rho, S)$ of Equation (5-18) from the quantum theory in a physical system with an interaction between micro particles [45] [46]. Further, if we consider Equation (5-18) in accordance with the existing diffusivity theory, it seems that there is no room to incorporate a driving force effect into the diffusivity expression of Equation (5-18), since the essential constant D_{N} has already the diffusivity dimension. On the other hand, the diffusivity expression of Equation (4-48) is derived from analyzing the nonlinear interdiffusion equation with a driving force.

If we incorporate the potential energy Q_{df} caused by a driving force into the activation energy Q in the expression of Equation (5-18), the relation is obtained as

$$\Omega(\rho, S) = n\,\tilde{D}/D_{\mathrm{N}} \exp\left[\frac{Q_{1} - Q_{\mathrm{df}}}{k_{\mathrm{B}} T_{1}}\right], \tag{5-20}$$

for a material composed of micro particles in the state of an absolute temperature $T = T_1$ and a heat quantity $Q = Q_1$, since Equation (5-18) corresponds to Equation (4-48) then. In the actual interdiffusion problems, $\Omega(\rho, S)$ obtained here is useful for fundamental knowledge relevant to a microstructure of material.

Equation (5-18) is thus valid when a driving force is nonexistent in the diffusion system. In other words, the relation of

$$D_0 = D_N \frac{\Omega(\rho,S)}{n} \exp\left[-\frac{Q_1}{k_B T_1}\right] \tag{5-21}$$

is considered to be valid between Equations (2-6) and (5-18) for the common $T = T_1$ and $Q = Q_1$.

As can be seem from Appendix A in the existing theory, the effect of a driving force on the diffusivity is incorporated into the difference of a jump frequency between nearest neighbor crystal plane 1 and plane 2. In relation to the situation, we discuss the difference between the present diffusivity expression and the existing one as follows.

The existence probability of a micro particle in a material depends on the Boltzmann factor of Equation (5-17). The potential energy Q_{df} as well as the activation energy Q gives motive energy and vibrational one to each micro particle in a material. In that case, the potential energy Q_{df} causes that the vibrational energy of micro particle on the crystal plane 1 is different from that of one on the plane 2, since the potential energy Q_{df} depends on the space coordinate. In other words, the existence probability of a micro particle on the plane 1 is different from that of one on the plane 2.

The difference of existence probabilities between a micro particle on the plane 1 and one on plane 2 corresponds to that of the frequencies between them. The effect of a driving force on diffusivity should be thus incorporated into the Boltzmann factor as a potential energy Q_{df} in the present case.

When the potential energy is not so much large, i.e., $Q_{df}/k_B T_1 \ll 1$, the diffusivity D of Equation (2-7) corresponds to

$$D = D_N \frac{\Omega(\rho,S)}{n} \left\{1 + \frac{Q_{df}}{k_B T_1}\right\} \exp\left[-\frac{Q_1}{k_B T_1}\right] = D_0 + D_0 \frac{Q_{df}}{k_B T_1}, \tag{5-22}$$

incorporating a driving force effect into the Boltzmann factor for $T = T_1$ and $Q = Q_1$. Equations (2-7) and (5-22) show that the relation of

$$D_{df} = D_0 \frac{Q_{df}}{k_B T_1} \tag{5-23}$$

is valid. Although the diffusivity gradient of

$$\frac{\partial D_{\mathrm{df}}}{\partial x} = D_0 \frac{1}{k_{\mathrm{B}} T_1} \frac{\partial Q_{\mathrm{df}}}{\partial x} = -D_0 \frac{F}{k_{\mathrm{B}} T_1} \tag{5-24}$$

has the velocity dimension, it is directly relevant to a driving force F because of the definition of mechanics $F = -\partial Q_{\mathrm{df}} / \partial x$. The left-hand side of Equation (5-24) is thus called the drift velocity in the existing diffusion field (See Appendix A). However, the diffusivity gradient relevant to a driving force is obviously not the velocity concept but the force one.

Substituting Equations (5-22)-(5-24) into the one dimension diffusion equation of the general nonlinear Fickian second law given by

$$\partial_t C = \partial_x \left\{ D \partial_x C \right\} \tag{5-25}$$

the diffusion equation is obtained as

$$\partial_t C = D_0 \partial_x \left\{ \left(1 + \frac{Q_{\mathrm{df}}}{k_{\mathrm{B}} T_1} \right) \partial_x C \right\} = D_0 \partial_x^2 C + D_0 \frac{S_{\mathrm{df}}}{k_{\mathrm{B}}} \partial_x^2 C - D_0 \frac{F}{k_{\mathrm{B}} T_1} \partial_x C, \tag{5-26}$$

where $S_{\mathrm{df}} = Q_{\mathrm{df}} / T_1$ is entropy caused by a driving force. It is considered that the second term and third term in the right-hand side of Equation (5-26) mean entropy effect and a frequency effect caused by a driving force F, respectively. The general diffusivity expression is thus obtained by rewriting Equation (5-18) into

$$D = D_{\mathrm{N}} \frac{\Omega(\rho, S)}{n} \exp\left[-\frac{Q - Q_{\mathrm{df}}}{k_{\mathrm{B}} T} \right]. \tag{5-27}$$

In view of the derivation process, the diffusivity expression of Equation (5-27) is applicable to every material in an arbitrary thermodynamic state.

5.4. Operators in Quantum Mechanics

In the quantum mechanics, energy E, momentum $|p\rangle$ and angular momentum $|L\rangle$ are expressed as operators yielding

$$E = i\hbar\frac{\partial}{\partial t}, \quad |p\rangle = -i\hbar|\nabla\rangle, \quad L_{r_3} = -i\hbar\frac{\partial}{\partial\theta_3}.$$

We cannot observe imaginary physical quantities. Therefore, the eigenvalues of their operators are meaningful in the quantum theory.

In a collision problem mentioned above, the concept of acceleration disappears in the quantum theory because of $\langle a|a\rangle \to \infty$. However, Equation (5-4) shows that the velocity is still a real physical quantity in the minute time and space region, since the momentum conservation law is valid in the collision process.

It is physically considered that the wave image of a micro particle disappears in the quantum theory when the corresponding wave packet exists in the minute space region $0 < r < 2\pi\Delta r$. Substituting $k_j \to ik_j$ and $\omega \to -i\omega$ into Equation (5-3) yields the relation of

$$C(t,r) = A\exp\left[-\langle k|r\rangle - \omega t + \alpha\right], \tag{5-28}$$

and it converges at zero in a minute time. For the wave image, the relations of $k_j \to ik_j$ and ω □ $-i\omega$ mathematically correspond to the relations of $r \to \xi = ir$ and $t \to \tau = -it$ then, compared with $\xi = ir$ and $\tau = it$ for the particle image.

In mathematics, except for physical constant quantities, physical variables with odd time or space dimensions should be thus accepted as physical operators in the minute time and space region.

The kinetic energy $E_k = \langle p|p\rangle/2m$ of Hamiltonian is acceptable as a physical quantity and/or a usual differential operator since it is a real variable for the sake of even time or space dimensions. On the other hand, the photon energy $E_p = h\upsilon\ (= \hbar\omega)$ expressed by using a frequency $\upsilon\ (\omega = 2\pi\upsilon)$ should be accepted as an operator, since h is the Planck constant and $\upsilon \to 1/\Delta\tau$ is an imaginary variable because of $\tau = -it$ in accordance with the wave image of υ. Therefore, the relation of $\upsilon \to i\partial/\partial t$ yields the operator $E_p \to i\hbar\partial/\partial t$ described above.

Using relations of $r_j \to ir_j$, $k_j = 2\pi/\lambda_j$ and $\lambda_j = 2\pi\Delta r_j$ for the wave packet in the time Δt and space Δr_j region, the relation of

$$p_j = \hbar k_j \tag{5-29}$$

yields the operator $p_j = -i\hbar\partial/\partial r_j$ described above. Further, using $|p\rangle = -i\hbar|\nabla\rangle$ for the definition of an angular moment $|L\rangle = |r \times P\rangle$, the above angular moment operator $L_{r_3} = -i\hbar\partial/\partial\theta_3$ is also obtained.

We investigate a characteristic of diffusivity in the minute time and space regions in the following. Equation (2-5) is rewritten as

$$D_0 = (\Delta\xi_j)^2/2\Delta\tau = i(\Delta r_j)^2/2\Delta t \tag{5-30}$$

for a particle image. Equation (5-4) is also rewritten as

$$m\Delta r_j/\Delta t = \hbar/\Delta r_j \tag{5-31}$$

because of $\lambda_j = 2\pi\Delta r_j$ and $v = \Delta r_j/\Delta t$. Substituting Equation (5-31) into Equation (5-30), the diffusivity expression of

$$D_0 = i\hbar/2m \tag{5-32}$$

is obtained as an operator.

Substituting Equation (5-32) into Equation (2-6), we have

$$\partial_t C = i\frac{\hbar}{2m}\langle\tilde{\nabla}|\nabla C\rangle \tag{5-33}$$

Multiplying both sides of Equation (5-33) by $i\hbar$ and rewriting C into ψ, the *S* Equation (5-14) is thus obtained again. In mathematics, the diffusion equation and the *S* equation are thus correlated only to the replacement of $D_0 \rightleftarrows i\hbar/2m$ between their equations in the real time and space.

6. RESULTS

In 1855, Fick applied the heat conduction equation proposed by Fourier in 1822 to diffusion phenomena as it had been. In general, however, the discussion of coordinate transformation of the general nonlinear F2 law is indispensable for understanding the diffusion phenomena. Nevertheless, the coordinate transformation of a diffusion equation had not been discussed for a long time until the recent works [14] - [16] [19] [20] [22].

Applying the divergence theorem to a diffusion equation, therefore, we investigated the problems of coordinate transformation in the diffusion systems. In relation to the investigation, a new diffusion flux of Equation (2-13) was defined in accordance with the mathematical physics. The new diffusion flux revealed that the F1 law is incomplete without the intrinsic diffusion flux.

The coordinate transformation theory shows also that the K effect is caused by a shift between the coordinate systems like the Doppler effect of the wave equation is caused by a shift between them. The theoretical Equation (3-15) relevant to the K effect was reasonably obtained in accordance with the empirical Equation (3-14) then. At the same time, it was also revealed that the concept of intrinsic diffusion is nonexistent in the diffusion theory, although we had accepted as an indispensable concept for understanding the K effect for a long time until recently.

In relation to the coordinate transformation theory, further, using the new diffusion flux applicable to various diffusion problems, the unified diffusion theory relevant to every material in an arbitrary thermodynamic state was established, including a self-diffusion theory and also an N elements system interdiffusion theory.

The self-diffusion theory has been indirectly discussed in the existing one, where the self-diffusion behavior of a pure material is visualized by any other diffusion particles in the pure material. In the present theory, the self-diffusion mechanism of a pure material itself is discussed using the intrinsic diffusion flux. In an N elements system interdiffusion, it was found that each of diffusion particles moves randomly with a common diffusivity D of Equation (3-32) in the diffusion field, as if the gas molecules in a given space move randomly, regardless of their species in accordance with the Boyle Charles law.

The essence of diffusion phenomena is shown in the parabolic law. Based on the concept, a new diffusion flux applicable to analyzing diffusion problems was derived as Equation (4-7) and/or (4-8) in the parabolic space. The analytical method in the parabolic space was thus established. It was then revealed that the analysis in the parabolic space is exceedingly superior in calculation to the existing analytical methods. For example, the solutions of nonlinear diffusion equation were, for the first time, obtained as the analytical expressions of Equations (4-48) and (4-50). The effect of a driving force on the diffusivity was then obtained as Equation (4-49) from the solution of Equation (4-48).

The definition of diffusivity indicates that the diffusivity should be essentially investigated as a few many bodies problem in the quantum mechanics. Based on the concept, the S equation was derived from applying the diffusion equation to a diffusion elementary process. In the analytical process, we found that the diffusivity corresponds to an angular momentum operator in the quantum mechanics. The new universal diffusivity expression was thus obtained as Equation (5-27) using an essential diffusion constant of Equation (5-19). It is essentially applicable to every material in an arbitrary thermodynamic state.

It was also found that the material wave equation proposed by de Broglie, which is the most fundamental relation in the materials science, was derived from the relation between diffusivity expressions of Equations (2-5) and (5-13) in the diffusion elementary process. This gives evidence for the validity for the present diffusion theory.

7. DISCUSSION

Even if a driving force exists in the diffusion system under the condition of no sink and source, the divergence theorem shows that a collective behavior of micro particles depends only on a diffusivity of the general nonlinear F2 law. Whenever the diffusivity of a micro particle in a material is given, the collective behavior of micro particles is thus determined only by solving the diffusion equation. The diffusivity depends on a material internal structure composed of micro particles. Therefore, the investigation of diffusivity is the most important subject in the diffusion study. Based on the concept discussed here, the present study was performed.

In relation to the Darken equation, even if a concentration profile is reasonably reproduced by using intrinsic diffusion coefficients as parameters, we cannot directly evaluate its validity of the used diffusivity values, since the diffusivity profile is not obtained by a direct observation. In other words, we can usually perform the numerical simulation to a high precision only by using fitting parameters for the given equation, regardless of a physical validity. As far as the Darken equation is not mathematically correct as shown in the Appendix B, it is thus necessary for us to reexamine a great many results of interdiffusion problems obtained by using the unsuitable Darken equation.

As revealed in the text and Appendix B, we believe that the F1 law and the intrinsic diffusion concept as well as the Darken equation are misunderstanding matters in the diffusion history. Nevertheless, they have been accepted as reasonable ones for such a long time in the concerned field.

If we consider serious influences of their misunderstanding matters on student education and young researcher training in the concerned field, they should be solved as soon as possible, since they have been still described as plausible ones in a lot of existing textbooks. One of the present work aims is thus to make them universally known in the concerned field.

Since the general solutions of nonlinear diffusion equation are analytically obtained in the parabolic space, the diffusion problems under the condition of no sink and source are essentially solved. The influence of a driving force on the diffusivity is thus reasonably obtained as Equation (4-49) from the analytical solutions of Equations (4-48) and (4-50). The discussions about diffusion fluxes are thus now essentially unnecessary for fundamental diffusion problems.

In the actual diffusion problems in detail, further investigation will be required for various difficult problems, for example, such a problem where the sink and source exists in the concerned diffusion system. Even in that case, however, it is no doubt that the fundamental theories discussed here are still useful for their fundamental investigations, just because of fundamental ones.

The investigations relevant to the many elements system interdiffusion problems, such as an alloy, a compound semiconductor, a multilayer thin film, and a microstructure material, are extremely important as an actual problem. Under the condition of Equation (3-27), corresponding Equations (4-48) and (4-50) to solutions of Equation (3-33) and using solutions of Equation (3-34) obtained numerically for a many elements system interdiffusion problems, their solutions are possible.

As mentioned at the beginning of this section, the foundation of diffusion problem is relevant to not the diffusion flux but the diffusion equation for the concerned diffusion system. The influence of a driving force on a diffusion system should be thus incorporated into the diffusivity of the concerned diffusion equation, e.g., Equation (5-25), although the problem relevant to a driving force is discussed in the existing theory using a drift velocity in the diffusion flux representation (see Appendix A).

In the present study, the fundamental diffusion problems are surveyed in accordance with the mathematical physics. As a result, the new useful findings for the diffusion study are reasonably derived from the essential discussions about the general nonlinear Fickian diffusion equation. In particular, the analytical method established in the parabolic space will be applicable not only to diffusion problems but also to Brown problems in various science fields as a useful method.

Finally, we hope that the new findings are useful for the material science researches and also that the present aims are accomplished.

REFERENCES

1. Markov, A.A. (1960) American Mathematical Society Translations Series, 15, 1-14.
2. Brown, R. (1828) Philosophical Magazine N. S., 4, 161-173.

3. Einstein, A. (1905) Annals of Physics, 18, 549-560. http://dx.doi.org/10.1002/andp.19053220806
4. Skvortsov, A.A., Orlov, A.M. and Muradov, V.E. (2009) Technical Physics Letters, 35, 606-609. http://dx.doi.org/10.1134/S1063785009070062
5. Ting, P.Y. (1994) IEEE Transactions on Neural Networks, 5, 622-638. http://dx.doi.org/10.1109/72.298232
6. You, Y.L., Xu, W.Y., Tannenbaum, A. and Kaveh, M. (1996) IEEE Transaction on Image Processing, 11, 1539-1553.
7. Das, I., Agrawal, N.R., Gupta Sanjeev, K., Gupta Sujeet, K. and Rastogi, R.P. (2009) Journal of Physical Chemistry A, 113, 5296-5301. http://dx.doi.org/10.1021/jp8064147
8. Raji, A., Kuceyeski, A. and Weiner, M. (2012) Neuron, 73, 1204-1215. http://dx.doi.org/10.1016/j.neuron.2011.12.040
9. Zhou, Q. (2015) International Journal of Multimedia and Ubiquitous Engineering, 10, 173-182. http://dx.doi.org/10.14257/ijmue.2015.10.1.16
10. Gauss, C.F. (1840) Res Beob Magn Vereins, 4, 1.
11. Fourier, J.B.J. (1822) Theorie analytique de la chaleur. Didot, Paris, 499-508.
12. Mehrer, H. (2007) Diffusion in Solid. Springer-Verlag, Berlin Heidelberg. http://dx.doi.org/10.1007/978-3-540-71488-0
13. Fick, A. (1855) Philosophical Magn J Sci, 10, 31-39.
14. Okino, T. (2012) Journal of Modern Physics, 3, 1388-1393. http://dx.doi.org/10.4236/jmp.2012.310175
15. Okino, T. (2013) Journal of Modern Physics, 4, 1495-1498. http://dx. doi.org/10.4236/jmp.2013.411180
16. Okino, T. (2014) Applied Physics Research, 6, 1-7. http://dx.doi.org/10.5539/apr.v6n2p1
17. Smigelskas, A.D. and Kirkendall, E.O. (1947) Transactions of the American Institute of Mining, Metallurgical and Petroleum Engineers, 171, 130-142.
18. Schrödinger, E. (1926) Annals of Physics, 79, 361-376. http://dx.doi.org/10.1002/andp.19263840404
19. Okino, T. (2011) Materials Transactions, 52, 2220-2227. http://dx.doi.org/10.2320/matertrans.M2011137
20. Okino, T. (2012) Journal of Modern Physics, 3, 255-259. http://dx.doi. org/10.4236/jmp.2012.33034
21. Boltzmann, L. (1894) Annual Review of Physical Chemistry, 53, 959-964. http://dx.doi.org/10.1002/andp.18942891315
22. Okino, T. (2013) Journal of Modern Physics, 4, 612-615. http://dx.doi.org/ 10.4236/jmp.2013.45088
23. de Broglie, L. (1923) Nature, 112, 540. http://dx.doi.org/10.1038/112540a0
24. Dirac, P.A.M. (1958) Principles of Quantum Mechanics. Clarendon Press, Oxford.
25. Manning, J.R. (1972) Theory of Diffusion. In Diffusion American Society for Metals, 1-23.
26. Matano, C. (1933) Japanese Journal of Physics, 8, 109-113.
27. Langevin, P. (1908) Comptes Rendus de l'Académie des Sciences (Paris), 146, 530-533.
28. Cornet, J.F. and Calais, D (1972) Journal of Physics and Chemistry of Solids, 33, 1675-1684. http://dx.doi.org/10.1016/S0022-3697(72)80463-3
29. Cornet, J.F. (1974) Journal of Physics and Chemistry of Solids, 35, 1247-1252. http://dx.doi.org/10.1016/S0022-3697(74)80148-4
30. Iijima, Y., Hirano, K. and Kikuchi, M. (1982) Transactions of the Japan Institutetute of Metals, 23, 19-23. http://dx.doi.org/10.2320/matertrans1960.23.19
31. Alizadeh, R., Mahmudi, R., Ngan, A.H.W. and Langdon, T.G. (2015) Journal of Materials Science, 50, 4940-4951. http://dx.doi.org/10.1007/s10853-015-9041-x
32. Saad, A., Fedotov, A., Velichko, O., Pachynin, V. and Davydko, A. (2006) Physica Status Solidi (b), 12, 2665-2671. http://dx.doi.org/10.1002/pssb.200541272
33. Wang, C.W., Yiu, P., Chu, J.P., Shek, C.H. and Hsueh, C.H. (2015) Journal of Materials Science, 50, 2085-2092. http://dx.doi.org/10.1007/s10853-014-8770-6
34. Gong, M.M., Castro, R.H.R. and Liu, F. (2015) Journal of Materials Science, 50, 4610-4621. http://dx.doi.org/10.1007/s10853-015-9010-4

35. Darken, L.S. (1948) Transactions of the American Institute of Mining, Metallurgical and Petroleum Engineers, 175, 184-201.
36. Xin, L., Schnell, S.K., Simon, J.M., Krgüer, P., Bedeaux, D., Kjelstrip, S., Bardow, A. and Vlugt, T.J.H. (2013) International Journal of Thermophysics, 34, 1169-1196. http://dx.doi.org/10.1007/s10765-013-1482-3
37. Meyer, A. and Kargl, F. (2013) Int J Microgravity Sci Appl, 30, 30-35.
38. Perrin, J. (1909) Annales de Chimie et de Physique, 18, 5-114.
39. Shewmon, P.G. (1963) Diffusion in Solids. Series in Materials Science and Engineering. McGraw-Hill, New York.
40. Onishi, M., Ikeda, T., Wakamatu, Y. and Shimozaki, T. (1988) Transactions of the Japan Institutetute of Metals, 29, 383-387. http://dx.doi.org/10.2320/matertrans1960.29.383
41. da Silva, L.C.C. and Mehl, R.F. (1951) Transactions of the American Institute of Mining, Metallurgical and Petroleum Engineers, 191, 155-157.
42. Hoshino, H., Iijima, Y. and Hirano, K. (1982) Transactions of the Japan Institutetute of Metals, 21, 674-682. http://dx.doi.org/10.2320/matertrans1960.21.674
43. Onishi, M. and Miura, H. (1977) Transactions of the Japan Institutetute of Metals, 18, 107-112. http://dx.doi.org/10.2320/matertrans1960.18.107
44. Boltzmann, L. (1872) Wiener Berichte, 66, 275-370.
45. Ohya, M. and Watanabe, N. (2010) Entropy, 12, 1194-1245. http://dx.doi.org/10.3390/e12051194
46. John, L. and Haller, J. (2013) Journal of Modern Physics, 4, 1393-1399. http://dx.doi.org/10.4236/jmp.2013.410167

APPENDIX A

In Ref. [25] , the influence caused by a driving force in the planar diffusion phenomena is investigated in relation to jump frequencies of diffusion particles between the nearest neighbor crystal planes.

The net diffusion flux J of diffusing particles across a plane between two nearest neighbor crystal planes 1 and 2 is expressed as

$$J = n_1\Gamma_{12} - n_2\Gamma_{21}, \tag{A-1}$$

where n_1 $(x-\Delta x)$ and n_2 $(x-\Delta x)$ are the atom concentrations per unit area on each plane, and $\Gamma_{12}(x-\Delta x)$ and $\Gamma_{21}(x+\Delta x)$ are the jump frequencies for an atom to jump from plane 1 to plane 2 and from plane 2 to plane 1. The coordinate axis x is perpendicular to each plane and equal to the diffusion direction.

The relation between concentration $C(x)$ per unit volume and concentration $n(x)$ per unit area is expressed as

$$n(x) = aC(x) \tag{A-2}$$

where a is a lattice constant and $a = 2\Delta x$ in the present case.

Here, the following relations are considered as;

$$\begin{cases} n_1 = n - \dfrac{a}{2}\dfrac{\mathrm{d}n}{\mathrm{d}x} \\ n_2 = n + \dfrac{a}{2}\dfrac{\mathrm{d}n}{\mathrm{d}x} \end{cases}, \tag{A-3}$$

Where $\mathrm{d}n/\mathrm{d}x = (n_2 - n_1)/a$.

Substituting Equations (A-2), (A-3) into Equation (A-1), the relation of

$$J = -\frac{a^2}{2}(\Gamma_{12} + \Gamma_{21})\frac{\mathrm{d}C}{\mathrm{d}x} + a(\Gamma_{12} - \Gamma_{21})C \tag{A-4}$$

is obtained. Using the notations given by

$$D^* = \frac{a^2}{2}(\Gamma_{12} + \Gamma_{21}) = a^2\Gamma \quad \text{and} \quad v_{\mathrm{F}} = a(\Gamma_{12} - \Gamma_{21}),$$

the diffusion flux of

$$J = -D^*\frac{\mathrm{d}C}{\mathrm{d}x} + v_{\mathrm{F}}C \tag{A-6}$$

is derived. Here, $D^*$is defined as a diffusivity of tracer diffusion. If $\Gamma_{12} = \Gamma_{21}$, D^* means the diffusivity when a driving force is nonexistent in the diffusion system.

On the other hand, if we use the Taylor expansion in the above calculations, the relation of

$$\begin{cases} n_1(x - \Delta x) = n(x) - \Delta x\dfrac{\mathrm{d}n}{\mathrm{d}x} \\ n_2(x + \Delta x) = n(x) + \Delta x\dfrac{\mathrm{d}n}{\mathrm{d}x} \end{cases} \tag{A-7}$$

and

$$\begin{cases} \Gamma_{12}(x-\Delta x)=\Gamma(x)-\Delta x\dfrac{\mathrm{d}\Gamma(x)}{\mathrm{d}x} \\ \Gamma_{21}(x+\Delta x)=\Gamma(x)+\Delta x\dfrac{\mathrm{d}\Gamma(x)}{\mathrm{d}x} \end{cases} \tag{A-8}$$

are valid. Using Equations (A-7), (A-8) for Equation (A-1), the relation similar to Equation (A-6) is obtained as

$$J=-D^{*}\frac{\mathrm{d}C}{\mathrm{d}x}-\frac{\mathrm{d}D^{*}}{\mathrm{d}x}C, \tag{A-9}$$

Where $\mathrm{d}D^{*}/\mathrm{d}x=\mathrm{d}(a^{2}\Gamma)/\mathrm{d}x$. Equations (A-6) and (A-9) show that v_F corresponds to the diffusivity gradient given by

$$v_{F}=-\frac{\mathrm{d}D^{*}}{\mathrm{d}x}. \tag{A-10}$$

Equation (A-9) should be essentially equivalent to Equation (2-7) as shown by

$$D_{0}\frac{\mathrm{d}C}{\mathrm{d}x}+D_{\mathrm{df}}\frac{\mathrm{d}C}{\mathrm{d}x}=D^{*}\frac{\mathrm{d}C}{\mathrm{d}x}+\frac{\mathrm{d}D^{*}}{\mathrm{d}x}C. \tag{A-11}$$

However, it seems that Equation (A-11) is not generally valid. The reason is as follows.

When a driving force exists in the diffusion system, it affects entropy and jump frequencies of diffusion particles. As a result, the diffusivity is affected by a driving force, since the diffusivity depends directly on entropy and a frequency. The problem of entropy is extremely important for the diffusion phenomena. However, the effect of entropy is not taken into account in the derivation process of Equation (A-6) or (A-9). Therefore, the diffusion flux discussed here is not generally applicable to the diffusion problems relevant to a driving force.

Equations (4-48) and (4-50) reveal that we can solve a nonlinear diffusion equation even if a driving force exists in the diffusion system. Further, the effect of a driving force on the diffusivity is obtained as Equation (4-49). The diffusion flux discussed here is thus unnecessary for solving a nonlinear diffusion equation.

Further, the divergence theorem shows that the diffusion equation is obtained from the diffusion flux of Equation (A-9) as

$$\frac{\partial C}{\partial t} = D^* \frac{\mathrm{d}^2 C}{\mathrm{d}x^2} + 2\frac{\mathrm{d}D^*}{\mathrm{d}x}\frac{\mathrm{d}C}{\mathrm{d}x} + \frac{\mathrm{d}^2 D^*}{\mathrm{d}x^2} C. \qquad \text{(A-12)}$$

As discussed in Section 2.1, even if a driving force exists in the diffusion system, the diffusion equation (2-6) of

$$\frac{\partial C}{\partial t} = D\frac{\mathrm{d}^2 C}{\mathrm{d}x^2} + \frac{\mathrm{d}D}{\mathrm{d}x}\frac{\mathrm{d}C}{\mathrm{d}x} \qquad \text{(A-13)}$$

must be valid for the diffusion particles, except such an external force as Coulomb force for an ion particle.

Equation (A-12) is obviously inconsistent with Equation (A-13). This is evidence that Equation (A-6) neglecting an influence of entropy caused by a driving force is incorrect. In relation to the Equation (5-24), the diffusivity gradient is directly relevant to a driving force. Although the last term in the right-hand side of Equation (A-12) is the derivation term of a driving force because of Equation (5-24), we cannot understand such an effect of $\partial F/\partial x$ relevant to the random movement of diffusion particles in the diffusion system inside. If the force F is an external one where it is deliberately introduced into the diffusion system from the diffusion system outside, such phenomena may be possible. In other words, such diffusion flux of Equation (A-6) is not suitable in the present discussion, since the influence of a driving force should be incorporated into the Boltzmann factor as a potential energy caused by a driving force, as discussed in Section 5.3.

For example, one can easily notice that a force moment of vector quantity is not relevant to a kinetic energy of scalar quantity, although they have the same dimension of [kg m^2 s^{-2}] each other. Even if the drift velocity has the dimension of [m^2 s^{-1}] in the present case, it is thus not relevant to a velocity itself, since it is relevant to a driving force F as can be seem from Equation (5-24).

In general, the diffusion problems should be discussed using not the diffusion flux but the original diffusion equation, since the mathematical theory shows that the diffusion flux should be originally understood after the diffusion equation has been solved.

APPENDIX B

In the text, the K-effect was reasonably understood as a shift between the coordinate systems. This indicates that the intrinsic diffusion concept is not only needless but also nonexistent in the diffusion phenomena.

Based on the intrinsic diffusion concept, Darken derived a relation between the interdiffusion coefficient and intrinsic diffusion coefficient [35]. However, he repeated misunderstandings inconsistent with the fundamental mathematics theory in the analyzing process. Nevertheless, the so-called Darken Equation (*D* equation) has been widely applied to interdiffusion problems for such a long time since 1948. In the following, we reveal that the D equation is incorrect in mathematics.

When a driving force exists in the diffusion system, the diffusion phenomena are expressed as

$$\partial_t C = \langle \tilde{\nabla} | D \nabla C \rangle. \tag{2-6}$$

In that case, using the Equation (2-17) transformed from Equation (2-6) for j = I, II, the relation of

$$\partial_\tau \left\{ C^{\mathrm{I}} + C^{\mathrm{II}} \right\} = \partial_\xi \left\{ D^{\mathrm{I}} \partial_\xi C^{\mathrm{I}} + D^{\mathrm{II}} \partial_\xi C^{\mathrm{II}} - v\left(C^{\mathrm{I}} + C^{\mathrm{II}} \right) \right\} \tag{B-1}$$

is valid, where the coordinate notation $(\tilde{t}, \tilde{x})$ and the velocity v_{QP} are, for convenience, rewritten as (τ, ξ) and v. Here, the velocity v (=v_{QP}) used in Equation (2-15) means the velocity of diffusion region space caused by the diffusion of solvent particles. At the same time, it is visualized by an inert marker set on the initial interface because of the inert characteristic of markers.

Equation (B-1) yields

$$\partial_\xi \left\{ \left(D^{\mathrm{I}} - D^{\mathrm{II}} \right) \partial_\xi C^{\mathrm{II}} + v \right\} = 0 \tag{B-2}$$

because of C^{I} + C^{II} = 1. Here, note that Equation (B-2) is obviously a partial differential equation of τ and ξ. Therefore, the mathematical theory shows that Equation (B-2) corresponds to

$$\left(D^{\mathrm{I}} - D^{\mathrm{II}} \right) \partial_\xi C^{\mathrm{II}} + v = \Gamma(\tau), \tag{B-3}$$

where $\Gamma(\tau)$ is an arbitrary function of τ.

Nevertheless, Darken solved Equation (B-2) as an ordinary differential equation with respect to ξ. In other words, using k as an integral constant value instead of the $\Gamma(\tau)$, he thought that the relation of

$$\left(D^{\mathrm{I}} - D^{\mathrm{II}}\right)\partial_{\xi}C^{\mathrm{II}} + v = k \tag{B-4}$$

is valid. Using an unsuitable initial value k = 0 for Equation (B-4), it was rewritten as

$$v = -D^{\mathrm{I}}\partial_{\xi}C^{\mathrm{II}} + D^{\mathrm{II}}\partial_{\xi}C^{\mathrm{II}}. \tag{B-5}$$

Here, the functional equation (B-5) is not valid, since the left-hand side depends only on τ and the right- hand side depends on ⍰ and ⍰. Thus, he failed again to notice the mistake in the fundamental mathematics.

Under the condition of J_{eq} = 0 in Equation (2-18) in accordance with the existing theory, the diffusion flux becomes

$$J_{\mathrm{Q}}^{\mathrm{I}} = -D^{\mathrm{I}}\partial_{\xi}C^{\mathrm{I}} + vC^{\mathrm{I}}. \tag{B-6}$$

In any case, if we substitute Equation (B-5) into Equation (B-6), the relation of

$$J_{\mathrm{Q}}^{\mathrm{I}} = -\left(D^{\mathrm{I}}C^{\mathrm{II}} + D^{\mathrm{II}}C^{\mathrm{I}}\right)\partial_{\xi}C^{\mathrm{I}} \tag{B-7}$$

is obtained. By comparing Equation (B-7) with the F1 law of $J_{\mathrm{F}}^{\mathrm{I}} = -\tilde{D}\partial_{\xi}C^{\mathrm{I}}$, the D equation is perfunctorily derived as

$$\tilde{D} = D^{\mathrm{I}}C^{\mathrm{II}} + D^{\mathrm{II}}C^{\mathrm{I}} \tag{B-8}$$

He also failed to notice the mistake relevant to the basic theory of physics, since we cannot directly compare $J_{\mathrm{F}}^{\mathrm{I}}$ with $J_{\mathrm{Q}}^{\mathrm{I}}$ expressed in the different coordinate system, as can be seen from the derivation of Equation (2-17).

It seems that D^{I} and D^{II} are independent of each other in the above analytical process. In other words, he thought that the K effect is caused by the difference between D^{I} and D^{II}. They are symbolically rewritten as $D_{\mathrm{int}}^{\mathrm{I}}$ and $D_{\mathrm{int}}^{\mathrm{II}}$ in Equation (B-8) and are named "intrinsic diffusion coefficients" then.

In addition, even if we accept Equation (A-6) in Appendix A, it is not suitable to use it for the movement of K interface in the present case, since the drift velocity v_F depends on a time and a space coordinate, while the experimental result of Equation (3-14) shows that the K effect depends on only a time. It is obvious that the problem of K effect is a one to be discussed using not the diffusion flux Equation (A-6) but the diffusion Equation (2-6) taking account of a driving force. Further, the present interdiffusion problem should be expressed by Equation (2-6). If it is so, the corresponding diffusion flux must be not Equation (A-6) but Equation (B-6) obtained from Equation (2-6).

Hereinbefore, it was revealed that the D equation is obviously inconsistent with the theory in mathematical physics. Equation (B-8) is thus entirely meaningless and at the same time the concept of intrinsic diffusion is an illusion in the diffusion history. That the D equation has been highly applied to various interdiffusion problems by a lot of researchers for a long time since 1948 is one of the extremely few cases of misunderstanding problems in the science history.

CHAPTER 2

Unifying decoherence and the Heisenberg Principle

Bas Janssens[1]

1 Max Planck Institute for Mathematics, Vivatsgasse 7, 53111 Bonn, Germany

ABSTRACT

We exhibit three inequalities involving quantum measurement, all of which are sharp and state independent. The first inequality bounds the performance of joint measurement. The second quantifies the trade-off between the measurement quality and the disturbance caused on the measured system. Finally, the third inequality provides a sharp lower bound on the amount of decoherence in terms of the measurement quality. This gives a unified description of both the Heisenberg uncertainty principle and the collapse of the wave function.

Keywords: Decoherence, CP-maps, Uncertainty, relations, Quantum measurement

1 INTRODUCTION

Initial Remark Modulo minor modifications, this manuscript has been on the ArXiv since 2006, under the identifier quant-ph/0606093. Since that time, several results have found their way into the literature [17, 18, 19, 20, 24, 25]. This convinced me to submit the manuscript for publication.

In quantum mechanics, observables are modelled by self-adjoint operators A on a Hilbert space H, and a quantum mechanical system is described by the von Neumann algebra A generated by its observables. A normal state $\rho \in S(A)$ on AA induces a probability measure on the spectrum Spec(A)Spec(A) of an observable A, and it is the objective of a *quantum measurement* to portray this probability measure as faithfully as possible.A

For the more detailed results of Sects. 5 and 6, we will assume that A is the algebra B(H) of all bounded operators on a separable Hilbert space H, an assumption which is justified for quantum systems consisting of finitely many particles. In this case, all normal states on A are of the form $\rho(A) := tr(rA)$, where $r \in T_1,+(H)$ is a positive normalized trace class operator on H, the *density operator*.

According to the uncertainty relation $\sigma_X \sigma_Y \geq \frac{1}{2}|\rho([X, Y])|$, (see [5, 14, 26]), there is an inherent variance in the quantum state. Furthermore, quantum theory puts severe restrictions on the performance of measurement. These restrictions, which come on top of the measurement restrictions implied by the above uncertainty relation, fall into three distinct classes.

1) The impossibility of perfect *joint measurement*. It is not possible to perform a simultaneous measurement of two noncommuting observables in such a way that both measurements have perfect quality.

2) The *Heisenberg Principle*, (see [5]). In this paper, this means that quantum information cannot be extracted from a system without disturbing that system. (There are various other formulations of the Heisenberg Principle in the literature, but throughout the paper, it will exclusively have the above meaning.)

3) The *collapse of the wave function*. When information is extracted from a quantum system, a so-called decoherence is experimentally known to occur on this system.

We will see that this collapse of the wave function is a mathematical consequence of information extraction. In the process, II and III will be clearly exhibited as two sides of the same coin.

The subject of uncertainty relations in quantum measurement is already endowed with an extensive literature. For example, the Heisenberg Principle and the impossibility of joint measurement are quantitatively illustrated in [2, 4, 8, 22].

However, the inequalities in these papers depend on the state ρ, which somewhat limits their practical use. Indeed, the bound on the measurement quality can only be calculated if the state ρ is known, in which case there is no need for a measurement in the first place.

Our state-independent figures of merit (Sects. 2, 3) will lead us quite naturally to state-independent bounds on the performance of measurement. In order to illustrate their practical use, we will give some applications. We investigate the beamsplitter, resonance fluorescence and nondestructive qubit measurement.

In Sect. 4, we will prove a sharp, state-independent bound on the performance of jointly unbiased measurement. This generalizes the impossibility of perfect joint measurement.

In Sect. 5, we will prove a sharp, state-independent bound on the performance of a measurement in terms of the maximal disturbance that it causes. This generalizes the Heisenberg Principle.

In contrast to the Heisenberg Principle and its abundance of inequalities, the phenomenon of decoherence has mainly been investigated in specific examples (see, e.g. [6, 11, 32]). Although there are some bounds on the remaining coherence in terms of the measurement quality (see [10, 27]), a sharp, information-theoretic inequality does not yet appear to exist.

We will provide such an inequality in Sect. 6, where we will prove a sharp upper bound on the amount of coherence which can survive information transfer. Not only does this generalize the collapse of the wave function, it also shows that no information can be extracted if all coherence is left perfectly intact. It is therefore a *unified* description of both the Heisenberg Principle and the collapse of the wave function.

2 INFORMATION TRANSFER

In quantum mechanics, a system is described by a von Neumann algebra A of bounded operators on a Hilbert space H. The space S(A) of *normal states* of the system A is formed by the linear functionals ρ: $A \to C$ which are positive (i.e. $\rho(A^\dagger A) \geq 0$ for $A \in$ AA$\in A$), normalized (i.e. $\rho(\mathbf{1}) = 1$ and normal (i.e. weakly continuous on the unit ball A1). If A is the algebra B(H) of all bounded operators on a separable Hilbert space H, then every normal state on A is of the form $\rho(A) := tr(r\,A)$, with $r \in T_1$, + (H) a positive normalized ($tr(r)$ = 1 trace class operator on H (cf. [13, Ch. 7]), the *density operator*. With the system in state $\rho \in S$(A), observation of an (Hermitean) observable $A \in$ A is postulated to yield the average value $\rho(A)$.

Definition 1 Let A and B be von Neumann algebras. A map T : B → A is called *Completely Positive* (or CP for short) if it is linear, normalized (i.e. $T(1)$ = 1, positive (i.e. $T(X^\dagger X) \geq 0$ for all $X \in B$) and if, moreover, the extension $id_n \otimes$T : $Mn \otimes$B → $M_n \otimes$A is positive for all $n \in$ N, where M_n is the algebra of complex $n \times n$-matrices. In this paper, we will require CP-maps to be weakly continuous on the unit ball B1 unless specified otherwise.

Its dual $T^* : S(A) \rightarrow S(B)$, defined by $T^*(\rho) := \rho \circ T$, has a direct physical interpretation as an operation between quantum systems. First of all, due to positivity and normalization of T, each state $\rho \in S(A)$ is again mapped to a state $T^*(\rho) \in S(B)$. Secondly, linearity implies that T^* satisfies

$$pT^*(\rho_1) + (1-p)T^*(\rho_2) = T^*(p\rho_1 + (1-p)\rho_2)$$

for all $p \in [0,1]$, ρ_1, $\rho_2 \in S(A)$. This expresses the *stochastic equivalence principle*: a system which is in state ρ_1 with probability p and in state ρ_2 with probability $(1 - p)$ cannot be distinguished from a system in state $p\rho_1 + (1 - p)\rho_2$. Finally, it is possible to extend the systems A and B under consideration with another system M_n, on which the operation acts trivially. Due to complete positivity, states in $S(M_n \otimes A)$ are once again mapped to states in $S(M_n \otimes B)$. Incidentally, any CP-map T automatically satisfies $T(X^\dagger) = T(X)^\dagger$ and $\|T(X)\| \leq \|X\|$ for all $X \in B$.

2.1 General, unbiased and perfect information transfer

Suppose that we are interested in the distribution of the observable $A \in A$, with the system A in some unknown state ρ. We perform the operation T^* : $S(A) \rightarrow S(B)$ and then observe the 'pointer' B in B in order to obtain information on A. One may (see [4]) take the position that *any* CP-map $T : B \rightarrow A$ is an information transfer from *any* observable $A \in A$ to *any*pointer $B \in B$. The following is a figure of demerit for the quality of such an information transfer.

Definition 2 Let $T : B \rightarrow A$ be a CP-map. Its *measurement infidelity* δ in transferring information from A to the pointer B is defined as

$$\delta := \sup_S \|\mathbf{1}_S(A) - T(\mathbf{1}_S(B))\|$$

where S runs over the Borel subsets of $\mathbb{R}$.

It measures how accurately probability distributions on the measured observable A are copied to the pointer B.

The initial state ρ defines a probability distribution $\mathbb{P}_i$ on the spectrum of A by $\mathbb{P}_i\ (S) := \rho(1S(A))$, where $1_S(A)$ denotes the spectral projection of A associated with the set S. Similarly, the final state $T^*(\rho)$ defines a probability distribution $\mathbb{P}_f$ on the spectrum of B. δ is now the maximum distance between $\mathbb{P}_i$ and $\mathbb{P}_f$, where the maximum is taken over all initial states ρ. That is, $\delta = \sup_\rho D(\mathbb{P}_i, \mathbb{P}_f)$.

The *trace distance* (a.k.a. variational distance or Kolmogorov distance) is defined as

$$D(\mathbb{P}_f, \mathbb{P}_i) := \sup_S \{|\mathbb{P}_i(S) - \mathbb{P}_f(S)|\},$$

the difference between the probability that the event S occurs in the distribution $\mathbb{P}_i$ and the probability that it occurs in the distribution $\mathbb{P}_f$, for the worst-case Borel set S. Writing out this definition, we see that indeed

$$\begin{aligned} \sup_\rho D(\mathbb{P}_i, \mathbb{P}_f) &= \sup_{\rho, S} |\rho(\mathbf{1}_S(A)) - \rho(T(\mathbf{1}_S(B)))| \\ &= \sup_S \|\mathbf{1}_S(A) - T(\mathbf{1}_S(B))\|, \end{aligned}$$

which equals δ. The measurement infidelity δ is thus precisely the worst-case difference between input and output probabilities.

In this paper, we will devote considerable attention to the class of unbiased information transfers.

Definition 3 A CP-map $T : \mathrm{B} \to \mathrm{A}$ is called an *unbiased information transfer* from the Hermitean observable $A \in \mathrm{A}$ to a Hermitean $B \in \mathrm{B}$ if $T(B) = A$.

Recall that we are interested in the distribution of A, with the system A in some unknown state $\rho \in S(\mathrm{A})$. We perform the operation $T^* : \mathrm{S}(A) \to \mathrm{S}(\mathrm{B})$ and then observe the 'pointer' Bin B. Since $T^*(\rho)(B) = \rho(T(B))$ by definition of the dual, and $\rho(T(B)) = \rho(A)$ by definition of unbiased information transfer, the expectation value of B in the final state $T^*(\rho)$ is the same as that of A in the initial state ρ. We conclude that the expectation of A was transferred to B.

Definition 4 An information transfer $T : \mathrm{B} \to \mathrm{A}$ from $A \in \mathrm{A}$ to $B \in \mathrm{B}$ is called *perfect* if $\delta = 0$. Equivalently, it is perfect if $T(B) = A$, and if the restriction of T to B'', the von Neumann algebra generated by B, is a $*$-homomorphism $B'' \to A''$.

The entire probability distribution of A is then transferred to B, rather than merely its average value. Indeed, for all moments $\rho(A^n)$, we have $T^*(\rho)(B^n) = \rho(T(B^n)) = \rho(T(B)^n) = \rho(A^n)$. Everything there is to know about A in the initial state ρ can be obtained by observing the 'pointer' B in the final state $T^*(\rho)$.

Between the different kinds of information transfer (IT), the following relations hold:

$$\{\text{General IT}\} \supset \{\text{Unbiased IT}\} \supset \{\text{Perfect IT}\}.$$

2.2 Example: von Neumann qubit measurement

Let $\Omega := \{+1, -1\}$. Denote by $C(\Omega)$ the (commutative) algebra of $\mathbb{C}$-valued random variables on Ω. A state on $C(\Omega)$ is precisely the expectation value $\mathbb{E}$ w.r.t. a probability distribution $\mathbb{P}$ on Ω, in which context $\rho(f)$ is denoted $E(f)$. Define the probability distributions $P\pm$ to assign probability 1 to $\pm1\pm1$.

The von Neumann measurement $T : M_2 \otimes C(\Omega) \to M_2$ on a qubit (described by the algebra M_2 of complex 2×2 matrices) is defined as

$$T(X \otimes f) := f(+1)P_+ X P_+ + f(-1)P_- X P_-,$$

with $P_+ = |\uparrow\rangle\langle\uparrow|$ and $P_- = |\downarrow\rangle\langle\downarrow|$. Then $T^* : S(M_2) \to S(M_2) \otimes S(C(\Omega))$ is given by

$$T^*(\rho) = \rho(P_+)|\uparrow\rangle\langle\uparrow| \otimes P_+ + \rho(P_-)|\downarrow\rangle\langle\downarrow| \otimes P_-.$$

In words: with probability $\rho(P_+)$, the output +1 occurs and the qubit is left in state $|\uparrow\rangle$. With probability $\rho(P_-)$, the output −1 occurs, leaving the qubit in state $|\downarrow\rangle$. The von Neumann measurement T is a perfect (and thus unbiased) information transfer from $\sigma_z \in M_2$ to the pointer $1 \otimes (\delta_{+1} - \delta_{-1}) \in M_2 \otimes C(\Omega)$.

2.3 CP-maps and POVMs

Quantum measurements are often (e.g. [4, 7]) modelled by positive operator-valued measures or POVMs. From the above CP-map T, we may distil the POVM $\mu : \Omega \to M_2$ by $\mu(\omega) := T(1 \otimes \delta_\omega)$, i.e. $\mu(+1) = P_+$ and $\mu(-1)$ P_-.

This procedure is fully general: given a CP-map $T : B \to A$ and a 'pointer' $B \in$ B, we obtain an A-valued POVM $\mu_{B,T}$ on Spec(B) by $\mu_{T,B}(S) := T(1_S)$μT,B(S):=T(1S). (This is why we require CP-maps to be weakly continuous on the unit ball B1.) Conversely, any A-valued POVMμ on Ω gives rise to the CP-map $T\mu : L\infty(\Omega) \to$ A by integration, $T\mu(f) = \int_\Omega f(\omega)\mu(d\omega)$.

A CP-map can thus be seen as an extension of a POVM that keeps track of the system output as well as the measurement output. Since we will be interested in disturbance of the system, it is imperative that we consider the full CP-map rather than merely its POVM.

3 MAXIMAL ADDED VARIANCE

For unbiased information transfer, there exists a figure of demerit more attractive than δ. Consider the variance $\mathrm{Var}(B,T^*(\rho))$ of the output, where the variance of X in the state ρ is defined as

$$\mathbf{Var}(X,\rho) := \rho(X^\dagger X) - \rho(X)^* \rho(X).$$

The output variance can be split into two parts. One part $\mathbf{Var}(A,\rho)$ is the variance of the input, which is intrinsic to the quantum state ρ. The other part $\mathbf{Var}(B,T^*(\rho)) - \mathbf{Var}(A,\rho) \geq 0$ is *added* by the measurement procedure. This second part determines how well the measurement performs.

The maximal added variance (where the maximum is taken over the input states ρ) will be our figure of demerit. For example, perfect information transfer from A to B satisfies $\mathbf{Var}(B,T^*(\rho)) = \mathbf{Var}(A,\rho)$, so that the maximal added variance is 0. There is uncertainty in the measurement outcome, but all uncertainty 'comes from' the quantum state, and none is added by the measurement procedure.

Definition 5 The *maximal added variance* of an unbiased information transfer T is defined as

$$\Sigma^2 := \sup_{\rho \in \mathscr{S}(\mathscr{A})} \mathbf{Var}(B,T^*(\rho)) - \mathbf{Var}(A,\rho)$$

It is straightforward to verify that $\Sigma^2 = \|T(B^\dagger B) - T(B)^\dagger T(B)$. This inspires the following definition.

Definition 6 Let $T : \mathrm{B} \to \mathrm{A}$ be a CP-map. We define the operator-valued sesquilinear form $(\cdot,\cdot) : \mathrm{B} \times \mathrm{B} \to A$ by

$$(X,Y) := T(X^\dagger Y) - T(X)^\dagger T(Y).$$

It satisfies $(X,Y)^\dagger = (Y,X)$ and is positive semi-definite: $(B,B) \geq 0$ for all $B \in \mathrm{B}$ (cf. also [12]). This 'length' has the physical interpretation $\|(B,B)\| = \Sigma^2$, and there is even a Cauchy–Schwarz inequality:

Lemma 1 (Cauchy–Schwarz) Let $T : \mathrm{B} \to \mathrm{A}$ be a CP-map, and $(X,Y) := T(X^\dagger Y) - T(\mathrm{i})^\dagger T(Y)$. Then for all $X, Y \in B$:

$$(X,Y)(Y,X) \leq \| (Y,Y) \| (X,X).$$

Proof By Stinespring's theorem (see [28]), we may assume without loss of generality that T is of the form $T(X) = V^\dagger XV$ for some contraction V. Writing this out, we obtain $(X, Y) = V^\dagger X^\dagger(1 - VV^\dagger)YV$. Defining $g(X) := \sqrt{1 - VV^\dagger}XV$, we have $(X, Y) = g(X)^\dagger g(Y)$. Hence $(X, Y)(Y, X) = g(X)^\dagger g(Y)g(Y)^\dagger g(X)$, which is less or equal than $\|g(Y)\|^2 g(X)^\dagger g(X) = \|(Y, Y)\|(X, X)$.

If an information transfer is perfect, then of course $\Sigma^2 = \|(B, B)\| = 0$. (No variance is added.) We will now show that the converse also holds: if $\Sigma^2 = 0$, then T is a $*$-homomorphism on B''. (Compare this with the fact that probability distributions of zero variance are concentrated in a single point.)

Theorem 1 Let $T : B \to$ A be a CP-map, and let $B \in$ BB∈B be Hermitean. Then among

1) $(B, B) = 0$.
2) The restriction of T to B'', the von Neumann algebra generated by B, is a $*$-homomorphism $B'' \to T(B)''$.
3) $(f(B), f(B)) = 0$ for all measurable functions f on the spectrum of B.
4) T maps the relative commutant $B' = \{X \in A; [X, B] = 0\}$ into $T(B)'$.

the following relations hold: (1) ⇔ (2) ⇔ (3) ⇒ (4).

Proof For (1) ⇒ (2), use Cauchy–Schwarz (Lemma 1) to find $T(B^n) - T(B)$ $T(B^{n-1}) \leq \|(B, B)\|(B^{n-1}, B^{n-1}) = 0$. By induction, we have $T(B^n) = T(B)^n$, and by linearity $T(f(B)) = f(T(B))$ for all polynomials f. Thus, T is a $*$-homomorphism from the algebra of polynomials on the spectrum of B to that on $T(B)$. Since T is positive, it is norm continuous, hence extends to a C$*$-algebra homomorphism between the algebras $C(\mathbf{Spec}(B))$ and $C(\mathbf{Spec}(A))$ of continuous functions on the spectra. Moreover, since we require CP-maps to be continuous in the weak operator topology on the unit ball and since every measurable function on the spectrum can be approximated weakly by a uniformly bounded sequence of continuous functions, the statement even extends to the algebras of measurable functions on the spectra of B and $T(B)$, isomorphic to B'' and $T(B)''$, respectively. For (2) ⇒ (3), note that $T(f(B)^2) = T(f(B))^2$. For (3) ⇒ (1), take $f(x) = x$. Finally, we prove (1) ⇒ (4). Suppose that $[A, B] = 0$. Then $[T(B), T(A)] = T([A , B]) - [T(A), T(B)] = (A^\dagger, B) - (B^\dagger, A)$ (B is Hermitean). By Cauchy–Schwarz, the last term equals zero if (B, B) does.

We see that the maximal added variance Σ^2 equals 0 if and only if T is a perfect information transfer. We shall take Σ to parametrize the imperfection of unbiased information transfer.

4 JOINT MEASUREMENT

In a jointly unbiased measurement, information on two observables A and $\tilde{A}$ is transferred to two *commuting* pointers B and $\tilde{B}$. If A and $\tilde{A}$ do not commute, then it is not possible for both information transfers to be perfect. (See [29, 30].) Indeed, the degree of imperfection is determined by the amount of noncommutativity.

Let $T : B \to A$ be a CP-map, and let $B, \tilde{B} \in B$ be Hermitean observables in BB. Set $A := T(B)$, $\tilde{A} := T(\tilde{B})$, $\Sigma^2_B := \|(B, B)\|$, and $\Sigma^2_{\tilde{B}} := \|(\tilde{B}, \tilde{B})\|$.

Theorem 2 If B and $\tilde{B}$ commute, then

$$\Sigma_B \Sigma_{\tilde{B}} \geq \tfrac{1}{2}\|[A, \tilde{A}]\|. \quad (1)$$

Proof Since $[B, \tilde{B}] = 0$, we have $[\tilde{A}, A] = T([B, \tilde{B}]) - [T(B), T(\tilde{B})] = (B, \tilde{B}) - (\tilde{B}, B)$. By Cauchy–Schwarz, the latter is at most $2\Sigma_B\Sigma\tilde{B}$ in norm.

Remark 1 The A-valued POVMs μ_i on $\Omega i \subseteq \mathbb{R}$ are called jointly measurable (e.g. [15, §2 and 7]) if there exists a POVM μ on Ω and measurable functions $B_i : \Omega \to \Omega_i$ such that $\mu_i = \mu \circ B_i^{-1}$. If the sets Ω_i are bounded, we can apply Theorem 2 to the CP-map $T\mu : L\infty(\Omega) \to A$ obtained by integrating against μ. Since $\Sigma^2_{B_i, T_\mu} = \Sigma^2_{x, T_{\mu_i}}$, we find $\Sigma_x, T_{\mu i} \cdot \Sigma_x, \quad T_{\mu j} \geq \frac{1}{2}\|$. Here, $\Sigma^2_{x, T_{\mu_i}} = \| \int_{\Omega_i} x^2 \mu_i(dx) - (\int_{\Omega_i} x\mu_i(dx))^2 \|$ is the maximal added variance of $T_{\mu i}$ as an unbiased measurement of $Ai := \int_{\Omega i} x\mu i(dx)$.

We now show that the bound (1) is sharp in the sense that for all $S, \tilde{S} > 0$, there exist $T, B, \tilde{B}$ such that (1) attains equality with $\Sigma_B = S, \Sigma\tilde{B} = \hat{S}$.

4.1 Application: The beamsplitter as a joint measurement

A *beamsplitter* is a device which takes two beams of light as input. A certain fraction of each incident beam is refracted and the rest is reflected, in such a way that the refracted part of the first beam coincides with the reflected part of the second and vice versa (cf. Fig. 1). We will show that the beamsplitter serves as an optimal joint unbiased measurement.

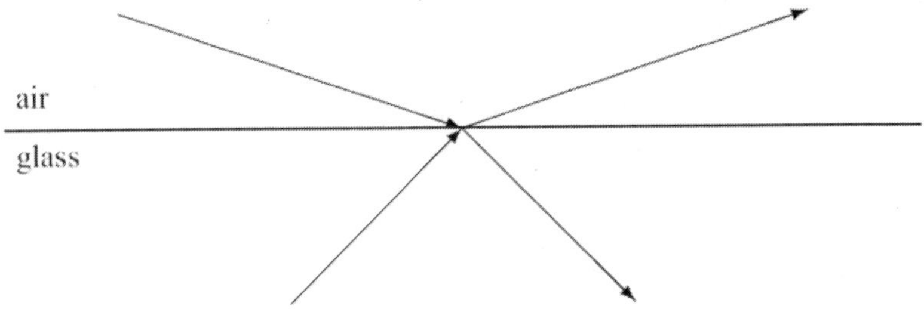

Figure. 1 Beamsplitter

In cavity QED, a single mode in the field is described by a Hilbert space H of a harmonic oscillator, with creation and annihilation operators $a^\dagger$ and a satisfying $[a, a^\dagger] = 1$, as well as $x = \frac{a+a^\dagger}{\sqrt{2}}$ and $p = \frac{a-a^\dagger}{\sqrt{2}i}$. The coherent states $|\alpha\rangle = e^{-|\alpha|^2/2}\sum_{n=0}^{\infty}\frac{\alpha^n}{\sqrt{n!}}|n\rangle$ are dense in H, and satisfy $a|\alpha\rangle = \alpha|\alpha\rangle$.

Quantummechanically, a beamsplitter is described by the unitary operator U on H⊗ H, given by $U = \exp(\theta(a^\dagger \otimes a - a \otimes a^\dagger))$. In terms of the coherent vectors, we have $U|\alpha\rangle\otimes|\beta\rangle = |\alpha\cos(\theta) + \beta\sin(\theta)\rangle\otimes| -\alpha\sin(\theta) + \beta\cos(\theta)\rangle$. Note that $U^\dagger a\otimes 1U = \cos(\theta)a\otimes 1 + \sin(\theta)1\otimes a$ and that $U^\dagger 1\otimes aU = -\sin(\theta)a\otimes 1 + \cos(\theta)1\otimes a$. (This can be seen by sandwiching both sides between coherent vectors.) Since the map $Y \mapsto U^\dagger YU$ respects $+,\cdot$ and †, we readily calculate

$$
\begin{aligned}
U^\dagger x\otimes \mathbf{1}U &= \cos(\theta)x\otimes\mathbf{1} + \sin(\theta)\mathbf{1}\otimes x,\\
U^\dagger x^2\otimes \mathbf{1}U &= \cos^2(\theta)x^2\otimes\mathbf{1} + 2\sin(\theta)\cos(\theta)x\otimes x + \sin^2(\theta)\mathbf{1}\otimes x^2,\\
U^\dagger \mathbf{1}\otimes pU &= -\sin(\theta)p\otimes\mathbf{1} + \cos(\theta)\mathbf{1}\otimes p,\\
U^\dagger \mathbf{1}\otimes p^2U &= \sin^2(\theta)p^2\otimes\mathbf{1} - 2\cos(\theta)\sin(\theta)p\otimes p + \cos^2(\theta)\mathbf{1}\otimes p^2.
\end{aligned}
$$

Since the von Neumann algebra describing a single mode in the field is B(H), we identify the normal state space $S(B(H))$ with the space $T_{1,+}(H)$ of positive normalized trace class operators, and we are interested in the map $r \mapsto Ur\otimes|0\rangle\langle 0|U^\dagger$ from $T_{1,+}(H)$ to $T_{1,+}(H)\otimes T_{1,+}(H)$. In other words, we feed the beamsplitter only one beam of light in a state ρ_r described by the density operator r, the other input being the vacuum. The dual of this is the CP-map $T: B(H)\otimes B(H) \to B(H)$ defined by $T(Y) := \mathrm{id}\otimes\phi_0(U^\dagger YU)$, with ϕ_0 the vacuum state $\phi_0(X) = \langle 0|X|0\rangle$.

Take $B = \cos^{-1}(\theta)x\otimes 1$ for instance. Then $T(B) = x\langle 0|1|0\rangle + \tan(\theta)1\langle 0|x|0\rangle = x$. Similarly, with $\tilde{B} = -\sin^{-1}(\theta)1\otimes p$, we have $T(\tilde{B}) = p$. Apparently, splitting a beam of light in two parts, measuring $x\otimes 1$ in the first beam and $1\otimes p$ in the second, and then compensating for the loss of intensity provides a *simultaneous unbiased measurement* of x and p in the original beam. Since $[x, p] = i$, we must[1] have $\Sigma_B \Sigma_{\tilde{B}} \geq \frac{1}{2}$.

We now calculate Σ_B and $\Sigma\tilde{B}$ explicitly. From $\langle 0|x^2|0\rangle = \frac{1}{2}$, we see that $T(B^2) = x^2 + \frac{1}{2}\tan^2(\theta)1$. Thus, $\Sigma_B^2 = \|(B, B)\| = \frac{1}{2}\tan^2(\theta)$. Similarly, $\Sigma_{\tilde{B}}^2 = \frac{1}{2}\tan^{-2}(\theta)$.

We see that $\Sigma_B \Sigma_{\tilde{B}} = \frac{1}{2}$, so that the beamsplitter is indeed an optimal jointly unbiased measurement.

By scaling B, optimal joint measurements can be found for arbitrary values of Σ_B and $\Sigma_{\tilde{B}}$, which shows the bound in Theorem 2 to be sharp. It may therefore be used to evaluate joint measurement procedures. For example, it was shown in [9] that homodyne detection of the spontaneous decay of a two-level atom constitutes a joint measurement with $\Sigma\Sigma' = 1.056$, slightly above the bound $\Sigma\Sigma' \geq 1$ provided by Theorem 2.

The beamsplitter is an optimal joint measurement in the sense that it minimizes $\Sigma\Sigma'$. It also performs well with other figures of merit. For example, if the quality of joint measurement is judged by the state-dependent cost $R(T) := \mathrm{Var}(B,T^*(\rho)) + \mathrm{Var}(\tilde{B}, T^*(\rho))$, then at least for Gaussian ρ, the optimal measurement is again the above beamsplitter with $\theta = \pi/4$. (See [7].)

5 THE HEISENBERG PRINCIPLE

The Heisenberg Principle may be stated as follows:

If all states are left intact, no quantum information can be extracted from a system.

This alludes to an information transfer from an initial system A to a final system consisting of two parts: the system A and an ancilla B, containing the pointer B. We thus have an information transfer $T : A\otimes B \rightarrow$ from A to $1\otimes B$.

An initial state $\rho \in S(A)$ gives rise to a final state $T^*(\rho) \in S(A\otimes B)$. Restricting this final state to the system $A \simeq A \otimes 1 \subseteq A \otimes B$ (this is called *taking the partial trace over* B) yields a 'residual' state $R^*(\rho) \in S(A)$, whereas taking the partial trace over A yields the final state $Q^*(\rho) \in S(B)$ of the ancilla. We define the CP-maps $R : A \rightarrow A$ by $R(A) := T(A\otimes 1)$ and $Q : B \rightarrow A$ by $Q(B) := T(1\otimes B)$. The map R describes what happens to A if we forget about the ancilla B, and Q describes the ancilla, neglecting the original system A.

We wish to find a quantitative version of the Heisenberg Principle, i.e. we want to relate the imperfection of the extracted quantum information to the amount of state disturbance.

Definition 7 The *maximal disturbance* Δ of a map $R : \mathcal{A} \to \mathcal{A}$ is given by

$$\Delta := \sup\{\|R(P) - P\|; P \in A, P^2 = P^\dagger = P\}.$$

The trace distance (or Kolmogorov distance) $D(\tau, \rho)$ is the maximal difference between the probability $\tau(P)$ that an event P occurs in the state τ, and the probability $\rho(P)$ that it occurs in the state ρ, for the worst-case event (projection operator) P. For short, $D(\tau, \rho) := \sup_P\{|\tau(P) - \rho(P)|\}$. If τ and ρ correspond to density operators t and r, then $D(\tau, \rho) = \frac{1}{2}\mathbf{tr}(|t - r|)$

(see, e.g. [21]).

The maximal disturbance Δ is now the worst-case distance between the input ρ and the output $R^*(\rho)$, i.e. $\Delta = \sup\{D(\rho, R^*(\rho)); \rho \in S(\mathcal{A})\}$. Indeed,

$$\sup_{\rho}\{D(\rho, R^*(\rho))\} = \sup_{\rho, P}\{\rho(P) - \rho(R(P))\}$$

which equals $\sup_P\{\|R(P) - P\|\} = \Delta$.

5.1 Heisenberg Principle for unbiased information transfer

We first turn our attention to unbiased information transfer. The imperfection of the information is then captured in the maximal added variance Σ^2.

The Heisenberg Principle only holds for quantum information. Classical observables are contained in the centre $\mathcal{Z} = \{A \in A; [A, X] = 0 \ \forall X \in \mathcal{A}\}$, whereas quantum observables are not. The degree in which an observable A is 'quantum' is given by its distance to the centre $d(A, \mathcal{Z}) = \inf_{Z \in \mathcal{Z}}\|A - Z\|$.

In the rest of the paper, we will take the algebra of observables to be $B(\mathcal{H})$ *for some separable Hilbert space* $\mathcal{H}$.

The centre is then simply $\mathbb{C}1$. Since normal states $\rho \in S(B(\mathcal{H}))$ are given by density operators $r \in T_{1,+}$ as $\rho(A) = \mathbf{tr}(rA)$, the dual $T^* : S(B(\mathcal{H})) \to S(B(\mathcal{H}'))$ of a CP-map $T : B(\mathcal{H}') \to B(\mathcal{H})$ yields a map $T_{1,+}(\mathcal{H}) \to T_{1,+}(\mathcal{H}')$, which we denote by T_*. We then have

$$T^*(\rho)(A) = \mathbf{tr}(rT(A)) = \mathbf{tr}(T_*(r)A).$$

Theorem 3 Let $T : B(H) \otimes B \to B(H)$ be a CP-map, let $B \in B$ be Hermitean. Define $A := T(\mathbf{1} \otimes B)$, and $\Sigma_2 := \|(\mathbf{1} \otimes B, 1 \otimes B)\|$. Further, define $\Delta := \sup_P\{\|R(P) - P\|\}$, with R the restriction of $\boldsymbol{T}$ to $\mathbf{B}(H) \otimes \mathbf{1}$. Then

$$\Sigma \geq d(A, \mathscr{Z})\frac{\frac{1}{2} - \Delta}{\sqrt{\Delta(1 - \Delta)}}. \quad (2)$$

This bound is sharp in the sense that for all $\Delta \in [0, \frac{1}{2}]$, there exist T and A for which (2) attains equality.

Proof For the sharpness, see Sect. 6.5. As for the bound, we may assume $\Delta < \frac{1}{2}$, since inequality (2) is trivially satisfied otherwise. Denote the spectrum of A by $\mathbf{Spec}(A)$. Let $x := \sup(\mathbf{Spec}(A))$ and $y := \inf(\mathbf{Spec}(A))$, so that $d(A, \mathscr{Z}) = \frac{x-y}{2}$. Without loss of generality, assume that there exist normalized eigenvectors ψ_x and ψ_y satisfying $A\psi_x = x\psi_x$ and $A\psi_y = y\psi_y$. (If this is not the case, choose x' and y' in $\mathbf{Spec}(A)$ arbitrarily close to x and y and complete the proof using approximate eigenvectors.) Define $\psi_- := \frac{1}{\sqrt{2}}(\psi_x + \psi_y)$, $\tilde{B} := |\psi_-\rangle\langle\psi_-|$, $\tilde{A} := T(\tilde{B} \otimes \mathbf{1})$.

We thus have $\|[A, \tilde{B}]\| = d(A, Z)\|$. Since $\|\tilde{A} - \tilde{B}\| \leq \Delta$, we have $\|[A, \tilde{A} - \tilde{B}]\| \leq 2\Delta d(A, Z)\|$. Then by the triangle inequality,

$$\|[T(\mathbf{1} \otimes B), T(\tilde{B} \otimes \mathbf{1})]\| = \|[A, \tilde{B}] + [A, \tilde{A} - \tilde{B}]\| \geq d(A, \mathscr{Z})(1 - 2\Delta),$$

which brings us in a position to apply Theorem 2 to the commuting pointers $\tilde{B} \otimes \mathbf{1}$ and $1 \otimes B$. This yields

$$2\Sigma\sqrt{\|(\tilde{B} \otimes \mathbf{1}, \tilde{B} \otimes \mathbf{1})\|} \geq d(A, \mathscr{Z})(1 - 2\Delta). \quad (3)$$

In order to estimate $\|(\tilde{B} \otimes 1, \tilde{B} \otimes 1)\|$, we first prove that $\mathbf{Spec}(\tilde{A}) \subseteq [0, \Delta]$? $[(1 - \Delta),1]$. Let $a \in \mathbf{Spec}(\tilde{A})$. Since T is a contraction and $0 \leq \tilde{B} \leq 1$, we have $0 \leq a \leq 1$. Without loss of generality, assume that there exists a normalized eigenvector ψ_a such that $\tilde{A}\psi_a = a\psi_a$. (Again, if this is not the case, one may use approximate eigenvectors.) Decompose ψ_a over the eigenspaces of $\tilde{B}$, i.e., write $\psi_a = \chi_1 + \chi_0$, with $\chi_1 \perp \chi_0$, $\tilde{B}\chi1=\chi1B\sim\chi1=\chi1$ and $\tilde{B}\chi0 = 0$. Then, we have

$$(\tilde{A} - \tilde{B})\psi_a = (a - 1)\chi_1 + a\chi_0.$$

Since $\|\chi 1\|^2 + \|\chi 0\|^2 = 1$, the inequality

$$\Delta^2 \geq \|(\tilde{A} - \tilde{B})\psi_a\|^2 = (a - 1)^2\|\chi_1\|^2 + a^2\|\chi_0\|^2$$

implies that either $|1 - a| \leq \Delta$ or $a \leq \Delta$. Thus $\mathbf{Spec}(\tilde{A}) \subseteq [0, \Delta] \cup [(1 - \Delta), 1]$, as desired. From this, it follows that $\mathbf{Spec}(\tilde{A} - \tilde{A}^2) \subseteq [0, \Delta(1 - \Delta)]$. Since $\tilde{B}^2 = \tilde{B}$, we may estimate

$$\|(\tilde{B} \otimes \mathbf{1}, \tilde{B} \otimes \mathbf{1})\| = \|T(\tilde{B} \otimes \mathbf{1}) - T(\tilde{B} \otimes \mathbf{1})^2\| = \|\tilde{A} - \tilde{A}^2\| \leq \Delta(1 - \Delta).$$

Combining this with inequality (3) yields $2\Sigma\sqrt{\Delta(1 - \Delta)} \geq d(A, \mathscr{Z})(1 - 2\Delta)$, which was to be demonstrated.

In the case of no disturbance, Δ = 0, we see that $\Sigma \to \infty$. No information transfer from A is allowed if all states on A are left intact. This is Werner's (see [31]) formulation of the Heisenberg Principle. In the opposite case of perfect information transfer, $\Sigma = 0$, inequality 2shows that Δ must equal at least one half. We shall see in Sect. 6 that this corresponds with a so-called collapse of the wave function.

These two extreme situations are connected by Theorem 3 in a continuous fashion, as indicated in Fig. 2. The upper left corner of the curve illustrates the Heisenberg Principle. In Sect. 6, we will see that the lower right corner represents the collapse of the wave function.

5.2 Heisenberg Principle for general information transfer

We now prove a version of the Heisenberg Principle for general information transfer. Let $T : B(\mathrm{H}) \otimes \mathrm{B} \to B(\mathrm{H})$ be a CP-map, and let $A \in B(\mathrm{H})$ and $B \in \mathrm{B}$ be Hermitean, with $A \notin \mathrm{Z} = \mathbb{C}1$. Define $\Delta := \sup_P\{\|R(P) - P\|\}$, with R the restriction of T to $B(\mathrm{H}) \otimes 1$. Define $\delta := \sup_S\{\|T(1 \otimes 1_S(B)) - 1_S(A)\|\}$. In this setting, we find:

Corollary 1 For δ and Δ in [0,12][0,12], we have

$$\left(\tfrac{1}{2} - \delta\right)^2 + \left(\tfrac{1}{2} - \Delta\right)^2 \leq \tfrac{1}{4} \tag{4}$$

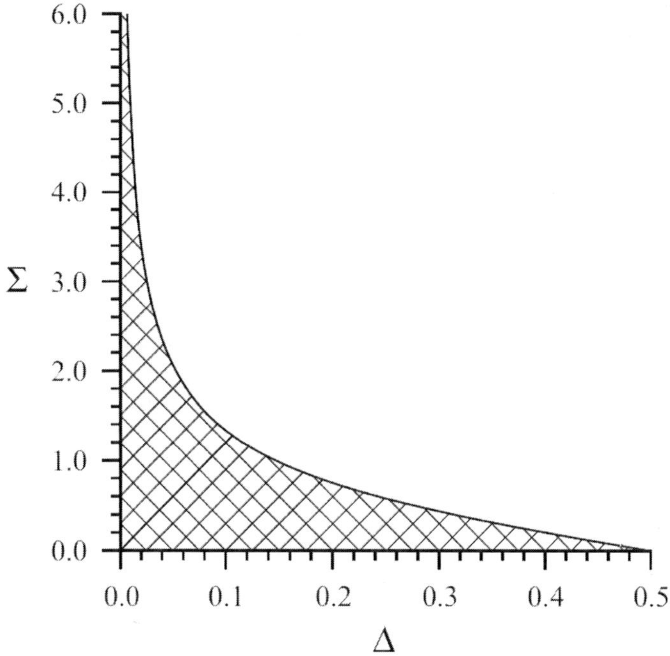

Figure. 2 Combinations (Δ, Σ) below the curve are forbidden, those above are allowed. (With $d(A, Z) = 1$.)

This bound is sharp; for all $\Delta \in [0,\frac{1}{2}]$, there exists a T for which (4) attains equality.

Proof Choose a nontrivial measurable subset S of **Spec**(A), and put $P := 1 \otimes 1_S(B)$. Since $\|T(P) - 1_S(A)\| \le \delta\|$ and **Spec**($1_S(A)$) = {0,1}, we find that **Spec**($T(P)$) $\subseteq [0, \delta] \cup [1 - \delta, 1]$ (cf. the proof of Theorem 3). Thus $\Sigma^2 = \|T(P) - T(P)\ 2\| \le \delta(1 - \delta)$. Similarly, $d(T(P), Z) \ge \frac{1}{2} - \delta$ since **Spec**($T(P)$) contains points in both $[0, \delta]$ and $[1 - \delta, 1]$. Applying Theorem 3 to the pointer P, we obtain

$$\sqrt{\delta(1-\delta)} \ge \left(\tfrac{1}{2} - \delta\right)\left(\tfrac{1}{2} - \Delta\right)/\sqrt{\Delta(1-\Delta)}$$

or equivalently $\left(\frac{1}{2} - \delta\right)^2 + \left(\frac{1}{2} - \Delta\right)^2 \le \frac{1}{4}$. For sharpness, see Sect. 6.5.

A measurement that does not disturb any state ($\Delta = 0$) cannot yield information ($\delta \ge 12$). This is the Heisenberg Principle. On the other hand, perfect information ($\delta = 0$) implies full disturbance ($\Delta \geq 12$), corresponding to the collapse of the wave function. Both extremes are connected in a continuous fashion, as depicted in Fig. 3.

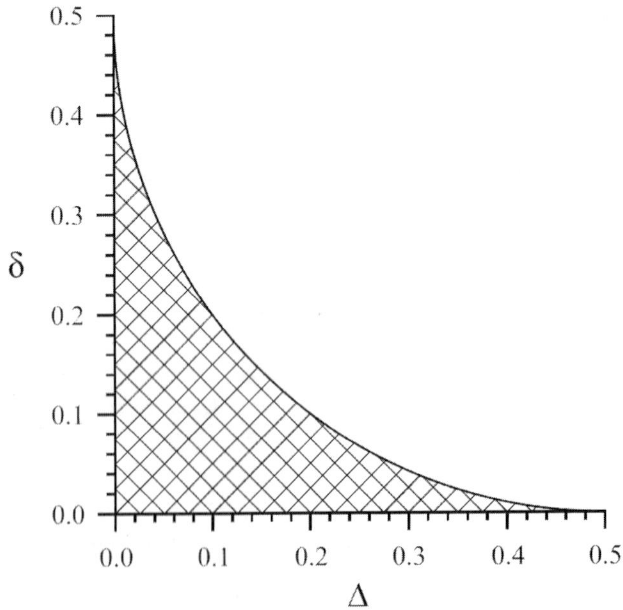

Figure. 3 Combinations (Δ,δ)(Δ,δ) below the curve are forbidden, those above are allowed

5.3 Application: Resonance fluorescence

Corollary 1 may be used to determine the minimum amount of disturbance if the quality of the measurement is known. Alternatively, if the system is only mildly disturbed, one may find a bound on the attainable measurement quality. Let us concentrate on the latter option.

We investigate the radiation emission of a laser-driven two-level atom. The emitted EM radiation yields information on the atom. A two-level atom (i.e. a qubit) only has three independent observables: σ_x, σ_y and σ_z. There are various ways to probe the EM field: photon counting, homodyne detection, heterodyne detection, etc. For a strong ($\Omega \gg 1$) resonant ($\omega_{\text{laser}} = \omega_{\text{atom}}$) laser, we will use Corollary 1 to prove that *any* EM measurement of σxσx, σyσy or σzσzwill have a measurement infidelity of at least

$$\delta \geq \frac{1}{2} - \frac{1}{2}\sqrt{1 - e^{-\frac{3}{2}\lambda^2 t}}$$

with λ the coupling constant. For a measurement with two outcomes, δ is the maximal probability of getting the wrong outcome. (Cf. Fig. 4.)

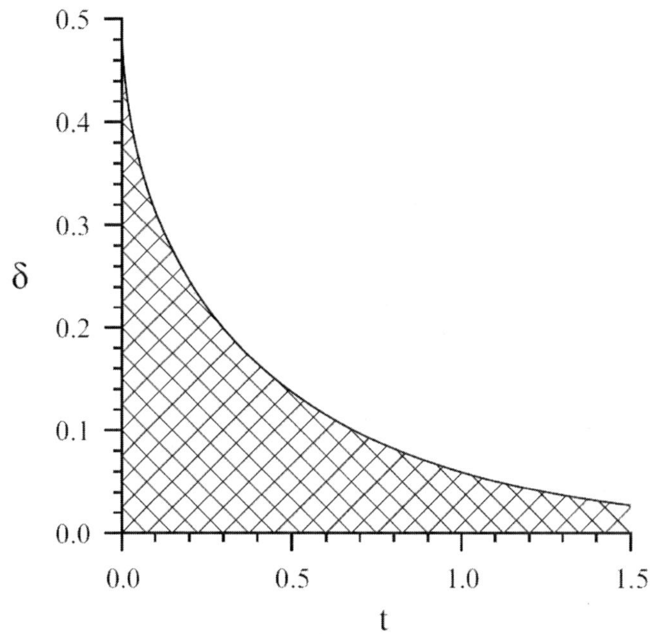

Figure. 4 Lower bound on δ in terms of t (in units of λ^{-2})

5.3.1 Unitary evolution on the closed system

The atom is modelled by the Hilbert space $\mathbb{C}^2$ (only two energy levels are deemed relevant). In the field, we discern a *forward* and a *side* channel, each described by a bosonic Fock space F. The laser is put on the forward channel, which is thus initially in the coherent state with frequency ω and strength Ω, described by the density matrix ϕ_Ω. (The field strength is parametrized by the frequency of the induced Rabi oscillations). The side channel starts in the vacuum state, described by the density matrix ϕ_0. If the two-level atom starts in the state described by the density matrix r, then the state at time t is given by the density matrix

$$T_{t*}(r) = U(t)(r \otimes \phi_\Omega \otimes \phi_0)U^\dagger(t)$$

with time evolution

$$\frac{d}{dt}U_t = -i(H_S + H_F + \lambda H_I)U_t$$

where $H_S \in B(\mathbb{C}^2)$ is the Hamiltonian of the two-level atom, $H_F \in B(F \otimes F)$ that of the field, and $\lambda H_I \in B(\mathbb{C}^2) \otimes B(F \otimes F)$ is the interaction Hamiltonian. Define the interaction picture time evolution by

$$\hat{T}_{t*}(r) := U_1(t)^\dagger U_2(t)^\dagger T_{t*}(r) U_2(t) U_1(t),$$

where $U_1(t) := e^{-iH_S t}$ and $U_2(t) := e^{-iH_F t}$ form the 'unperturbed' time evolution.

We now investigate $\hat{T}t$ instead of T_t. Indeed, we are looking for a bound on the measurement infidelity $\delta = \sup_S\{\|T(1 \otimes 1_S(B)) - 1_S(A)\|\}$ of T. Yet if $\hat{B} := U_2^\dagger B U_2$, then $\hat{T}(1_S(\hat{B})) = T(1_S(B))$, so that

$$\hat{\delta} = \sup_S\{\|\hat{T}(\mathbf{1} \otimes \mathbf{1}_S(\hat{B})) - \mathbf{1}_S(A)\|\} = \delta$$

If we find the interaction picture disturbance $\hat{\Delta}$, Corollary 1 will yield a bound on $\hat{\delta}$ and thus on δ.

In the weak coupling limit $\lambda \downarrow 0$, $\hat{T}_t$ is given by

$$\hat{T}_{t*}(\rho) = \hat{U}(t/\lambda^2)(\rho \otimes \phi_\Omega \otimes \phi_0)\hat{U}^\dagger(t/\lambda^2),$$

where the evolution of the unitary cocycle $t \mapsto \hat{U}_t$ is described (see [1]) by a quantum stochastic differential equation or QSDE. Explicitly calculating the maximal added variances Σ^2 by solving the QSDE is in general rather nontrivial, if possible at all. (See [9] for the case of spontaneous decay, i.e. $\Omega = 0$, with the map $\hat{T}t$ restricted to the commutative algebra of homodyne measurement results.)

5.3.2 Master equation for the open system

Fortunately, in contrast to the somewhat complicated time evolution $\hat{T}t$ of the combined system, the evolution *restricted to the two-level system* is both well known and uncomplicated. If we use λ^{-2} as a unit of time, then the restricted evolution $\hat{\bar{R}}t_*(r) := \mathrm{tr}_{\ F} \otimes_{\ F} \hat{\bar{T}}t_*(r)$ of the two-level system is known (see [3]) to satisfy the Master equation

$$\frac{d}{dt}\hat{R}_{t*}(r) = L(\hat{R}_{t*}(r)), \tag{5}$$

with the Liouvillian $L(r) := \frac{1}{2}i\Omega[e^{-i(\omega-E)t}V + e^{i(\omega-E)t}V^\dagger, r] - \frac{1}{2}\{V^\dagger V, r\} + VrV^\dagger$. In this expression, E is the energy spacing of the two-level atom and $V^\dagger = \sigma_+$, $V = \sigma_-$ are its raising and lowering operators. In the case $\omega = E$ of *resonance fluorescence*, we obtain

$$L(r) = \frac{1}{2} i\Omega[V + V^\dagger, r] - \frac{1}{2}\{V^\dagger V, r\} + VrV^\dagger.$$

If we parametrize a state by its Bloch vector $\hat{R}_t^*(r) = \frac{1}{2}(\mathbf{1} + x\sigma_x + y\sigma_y + z\sigma_z)$, then Eq. (5) is simply the following differential equation on $\mathbb{R}^3$:

$$\frac{d}{dt}\begin{pmatrix} x \\ y \\ z \end{pmatrix} = \begin{pmatrix} -\frac{1}{2} & 0 & 0 \\ 0 & -\frac{1}{2} & \Omega \\ 0 & -\Omega & -1 \end{pmatrix}\begin{pmatrix} x \\ y \\ z \end{pmatrix} - \begin{pmatrix} 0 \\ 0 \\ 1 \end{pmatrix}.$$

This can be solved explicitly. For $\Omega \gg 1$, the solution approaches

$$\begin{pmatrix} x \\ y \\ z \end{pmatrix} = \begin{pmatrix} e^{-\frac{1}{2}t} & 0 & 0 \\ 0 & e^{-\frac{3}{4}t}\cos(\Omega t) & e^{-\frac{3}{4}t}\sin(\Omega t) \\ 0 & -e^{-\frac{3}{4}t}\sin(\Omega t) & e^{-\frac{3}{4}t}\cos(\Omega t) \end{pmatrix}\begin{pmatrix} x_0 \\ y_0 \\ z_0 \end{pmatrix}.$$

If we move to the interaction picture once more to counteract the Rabi oscillations, i.e. with $U_1(t) = e^{\frac{i}{2}\Omega t\sigma_x}$ and $U_2 = 1$, we see that the time evolution is transformed to

$$\begin{pmatrix} x \\ y \\ z \end{pmatrix} = \begin{pmatrix} e^{-\frac{1}{2}t} & 0 & 0 \\ 0 & e^{-\frac{3}{4}t} & 0 \\ 0 & 0 & e^{-\frac{3}{4}t} \end{pmatrix}\begin{pmatrix} x_0 \\ y_0 \\ z_0 \end{pmatrix}$$

Since the trace distance $D(\rho, \tau)$ is exactly half the Euclidean distance between the Bloch vectors of ρ and τ, (see [21]), we see that $\Delta = \frac{1}{2}(1 - e^{-3t/4})$. For any measurement of σ_x, σ_y or σ_z, we therefore have $\delta \geq \frac{1}{2} - \frac{1}{2}\sqrt{1 - e^{-3t/2}}$ by Corollary 1 (remember that t is in units of λ^{-2}).

6 COLLAPSE OF THE WAVE FUNCTION

The 'collapse of the wave function' may be seen as the flip side of the Heisenberg Principle. It states that if information is extracted from a system, then its states undergo a very specific kind of perturbation, called decoherence.

6.1 Collapse for unbiased information transfer

We start out by investigating unbiased information transfer. We prove a sharp upper bound on the amount of remaining coherence in terms of the measurement quality. Let $T : \mathrm{A} \otimes \mathrm{B} \to \mathrm{A}$ be a CP-map. Let $B \in \mathrm{B}$ be Hermitian, and define $A := T(1 \otimes B)$.

Theorem 4 Suppose that ψ_x and ψ_y are eigenvectors of A with different eigenvalues x and y, respectively. Define $R : A \to \mathrm{A}$ to be the restriction of T to $\mathrm{A} \otimes 1$, and put $\Sigma^2 := \|(B, B)\|$. Then for all $\alpha, \beta \in \mathbb{C}$ with $|\alpha|^2 + |\beta|^2 = 1$, we have

$$D\Big(R_*\big(|\alpha\psi_x + \beta\psi_y\rangle\langle\alpha\psi_x + \beta\psi_y|\big), R_*\big(|\alpha|^2|\psi_x\rangle\langle\psi_x| + |\beta|^2|\psi_y\rangle\langle\psi_y|\big)\Big) \leq \frac{\Sigma/|x-y|}{\sqrt{1 + 4(\Sigma/|x-y|)^2}}. \tag{6}$$

This bound is sharp in the sense that for all values of $\Sigma/|x - y|$, there exist T, ψ_x, ψ_y, α and β for which (6) attains equality.

Proof For sharpness of the bound, see Sect. 6.5. To prove (6), note that the l.h.s. equals

$$\sup\{\bar{\alpha}\beta\langle\psi_x, R(P)\psi_y\rangle + \mathbf{c.c.} \mid P \in \mathscr{A}, P^2 = P^\dagger = P\}$$

Furthermore, $2|\alpha||\beta| \leq 1$, so that it suffices to bound the 'coherence' $\langle\psi_x, R(P)\psi_y\rangle$ on all projections P. Now $(x-y)\langle\psi_x, R(P)\psi_y\rangle = \langle\psi_x, [A, R(P)]\psi_y\rangle$, and $[A, R(P)] = (P \otimes 1, 1 \otimes B) - (1 \otimes B, P \otimes 1)$. Thus,

$$(x-y)\langle\psi_x, R(P)\psi_y\rangle = \langle\psi_x, (P \otimes \mathbf{1}, \mathbf{1} \otimes B)\psi_y\rangle - \langle\psi_x, (\mathbf{1} \otimes B, P \otimes \mathbf{1})\psi_y\rangle \quad (7)$$

and we will bound these last two terms. In the notation of Lemma 1, we have $\Sigma = \|g(1 \otimes B)\|$. Therefore,

$$\begin{aligned}\langle\psi_x, (P \otimes \mathbf{1}, \mathbf{1} \otimes B)\psi_y\rangle &= \langle g(P \otimes \mathbf{1})\psi_x, g(\mathbf{1} \otimes B)\psi_y\rangle \\ &\leq \|g(P \otimes \mathbf{1})\psi_x\|\|g(\mathbf{1} \otimes B)\psi_y\| \\ &\leq \Sigma\sqrt{\langle\psi_x, (P \otimes \mathbf{1}, P \otimes \mathbf{1})\psi_x\rangle}.\end{aligned}$$

We will bound $\langle\psi_x, (T(P^2 \otimes 1) - T(P \otimes 1)^2)\psi_x\rangle = \langle\psi_x, (R(P) - R(P)^2)\psi x\rangle$ in terms of the coherence. For brevity, denote $X_{xx'} := \langle\psi_x, X_{\psi x'}\rangle$. Since $\psi_x \perp \psi_y$, we have

$$(R(P)^2)_{xx} \geq |R(P)_{xx}|^2 + |R(P)_{xy}|^2,$$

so that

$$(R(P) - R(P)^2)_{xx} \leq R(P)_{xx}(1 - R(P)_{xx}) - |R(P)_{xy}|^2.$$

Since $x(1-x) \leq \frac{1}{4}$ for all $x \in R$, this is at most $\frac{1}{4} - |R(P)_{xy}|^2$. All in all, we have obtained

$$(P \otimes \mathbf{1}, \mathbf{1} \otimes B)_{xy} \leq \Sigma\sqrt{\tfrac{1}{4} - |R(P)_{xy}|^2}$$

and of course the same for $x \leftrightarrow y$. Plugging these into Eq. (7) yields

$$|x-y||R(P)_{xy}| \leq 2\Sigma\sqrt{\tfrac{1}{4} - |R(P)_{xy}|^2}$$

or equivalently $|R(P)_{xy}| \leq \frac{\Sigma/|x-y|}{\sqrt{1+4(\Sigma/|x-y|)^2}}$, which was to be proven.

Consider the ideal case of perfect ($\Sigma = 0$) information transfer. Suppose that the system A is initially in the coherent state $|\alpha\psi_x + \beta\psi_y\rangle\langle\alpha\psi_x + \beta\psi_y|$. Then, Theorem 4 says that, after the information transfer to the ancilla B, the system A cannot be distinguished from one that started out in the 'incoherent' state

$$|\alpha|^2|\psi_x\rangle\langle\psi_x| + |\beta|^2|\psi_y\rangle\langle\psi_y|.$$

As far as the behaviour of A is concerned, it is therefore completely harmless to assume that a collapse

$$|\alpha\psi_x + \beta\psi_y\rangle\langle\alpha\psi_x + \beta\psi_y| \mapsto |\alpha|^2|\psi_x\rangle\langle\psi_x| + |\beta|^2|\psi_y\rangle\langle\psi_y|$$

has occurred at the start of the procedure.

Now consider the other extreme of a measurement which leaves all states intact, i.e. $R^*(\rho) = \rho$ for all ρ. Then there exist states for which the l.h.s. of Eq. (6) equals $\frac{1}{2}$, forcing $\Sigma \to \infty$; no information can be obtained. This is Werner's formulation of the Heisenberg Principle [31].

Theorem 4 thus unifies the Heisenberg Principle and the collapse of the wave function. For $\Sigma = 0$, we have a full decoherence, whereas if all states are left intact, we have $\Sigma \to \infty$. For all intermediate cases, the bound (6) on the remaining coherence is an increasing function of $\Sigma/|x - y|$ (Fig. 5).

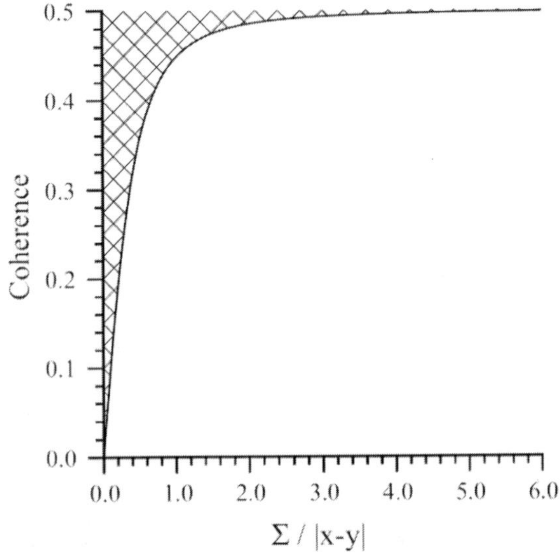

Figure. 5 Bound on the coherence as a function of $\Sigma/|x - y|$. All points above this curve are forbidden; all points below are allowed

This agrees with physical intuition: decoherence between ψ_x and ψ_y is expected to occur in case the information transfer is able to distinguish between the two. This is the case if the variance is small w.r.t. the differences in mean.

6.2 Application: Perfect qubit measurement

In Sect. 2, we have encountered the von Neumann Qubit measurement. Now consider any perfect measurement T of σ_z with pointer $1 \otimes (\delta_+ - \delta_-)$ which leaves $|\uparrow\rangle\langle\uparrow|$ and $|\downarrow\rangle\langle\downarrow|$ in place, i.e. $R_*(|\uparrow\rangle\langle\uparrow|) = |\uparrow\rangle\langle\uparrow|$ and $R_*(|\downarrow\rangle\langle\downarrow|) = |\downarrow\rangle\langle\downarrow|$. (Such a measurement is called nondestructive.) Theorem 4 then reads

$$R^*(|\alpha \uparrow +\beta \downarrow\rangle\langle\alpha \uparrow +\beta \downarrow|) = |\alpha|^2 |\uparrow\rangle\langle\uparrow| + |\beta|^2 |\downarrow\rangle\langle\downarrow|,$$

illustrated in Fig. 6.

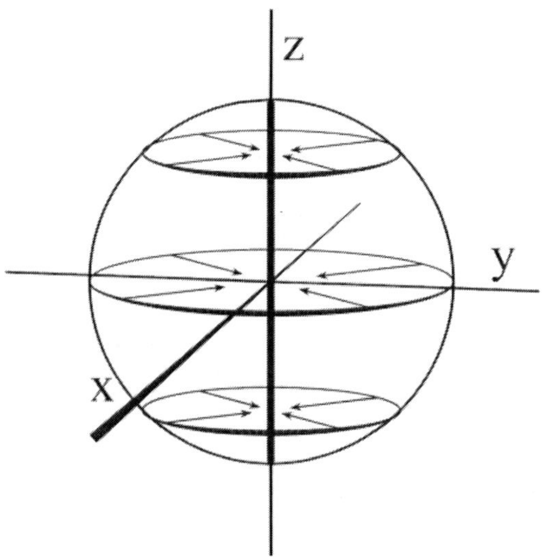

Figure. 6 Collapse on the Bloch sphere for perfect measurement

Incidentally, the trace distance between the centre of the Bloch sphere and its surface is $\frac{1}{2}$, so that we read off $\Delta = \sup\{D(R^*(\rho),\rho);\rho \in S(M_2)\} = \frac{1}{2}$. This was predicted by Theorem 3.

6.3 Collapse of the wave function for general measurement

We will prove a sharp bound on the remaining coherence in general information transfer. For technical convenience, we will focus attention on nondestructive measurements. A measurement of A is called 'nondestructive' (or 'conserving' or 'quantum nondemolition') if it leaves the eigenstates of A intact, so that repetition of the measurement will yield the same result. For example, the measurement in Sect. 2.2 is nondestructive, the one in Sect. 5.3 is destructive. Restriction to nondestructive measurements is quite common in quantum measurement theory (see [23]).

Let $T : B(\mathrm{H}) \otimes \mathrm{B} \to B(\mathrm{H})$ be a CP-map, let $A \in B(\mathrm{H})$ and $B \in \mathrm{B}$ be Hermitean and suppose that $\{\psi_i\}$ is an orthogonal basis of eigenvectors of A, with eigenvalues aiai. Define the measurement infidelity $\delta := \sup_S\{\| T(1 \otimes 1_S(B)) - 1_S(A) \|\}$.

Corollary 2 Suppose that T is nondestructive, i.e. $R_*(|\psi_i\rangle\langle\psi_i|) = |\psi_i\rangle\langle\psi_i|$ for all ψ_i, with R the restriction of T to $B(\mathrm{H}) \otimes 1$. Then if $\delta \in [0, 12]$, and $a_i \neq a_j$,

$$D\Big(R_*(|\alpha\psi_i + \beta\psi_j\rangle\langle\alpha\psi_i + \beta\psi_j|), (|\alpha|^2|\psi_i\rangle\langle\psi_i| + |\beta|^2|\psi_j\rangle\langle\psi_j|)\Big) \leq \sqrt{\delta(1-\delta)}. \tag{8}$$

This bound is sharp in the sense that for all $\delta \in [0, 12]$, there exist T, ψ_i, ψ_j, α and β for which (8) attains equality.

Proof For sharpness, see Sect. 6.5. Choose a set S such that $a_i \in S$ and $a_j \notin S$. T is an unbiased measurement of $T(1 \otimes 1_S(B))$ with pointer $1_S(B)$ and maximal added variance $\Sigma^2 \leq \delta(1 - \delta)$ (cf. the proof of Corollary 1). We will prove that ψ_i and ψ_j are eigenvectors of $T(1 \otimes 1_S(B))$ with eigenvalues x and y which differ at least $1^{-2}\delta$.

Let $P_i := |\psi_i\rangle\langle\psi_i|$. Since T is nondestructive, we have $\langle\psi_j, R(P_i)\psi_j\rangle = \langle\psi_j, P_i\psi_j\rangle$ for all j. Apparently, $R(P_i)$ has only one nonzero diagonal element, a 1 at position (i, i). Since $R(P_i) \geq 0$, this implies $R(P_i) = P_i$. Then $(P_i \otimes 1, P_i \otimes 1) = 0$, so that by Cauchy–Schwarz $(P_i \otimes 1, 1 \otimes 1_S(B)) = 0$. Since $P_i = T(P_i \otimes 1)$, we have

$$[T(\mathbf{1} \otimes \mathbf{1}_S(B)), P_i] = (P_i \otimes \mathbf{1}, \mathbf{1} \otimes \mathbf{1}_S(B)) - (\mathbf{1} \otimes \mathbf{1}_S(B), P_i \otimes \mathbf{1}) = 0.$$

Therefore, ψ_i is an eigenvector of $T(1 \otimes 1_S(B))$, with eigenvalue x, say. By a similar reasoning, ψ_j is also an eigenvector, denote its eigenvalue by y.

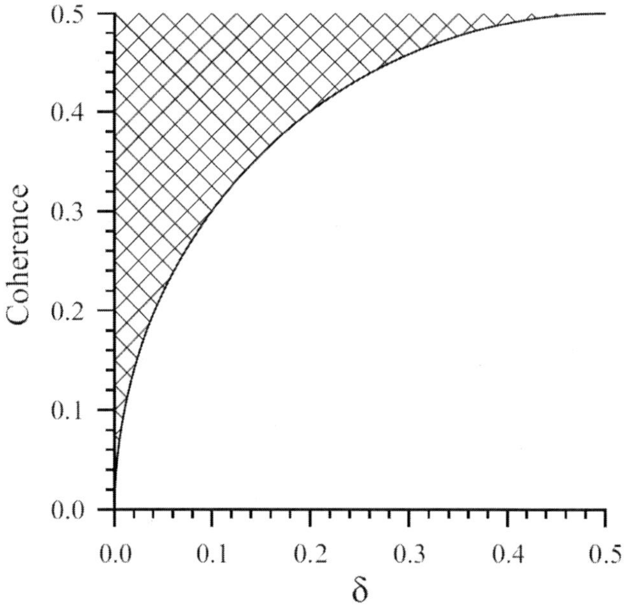

Figure. 7 Bound on the coherence in terms of δ. All points above the curve are forbidden; all points below are allowed

Since $\|T(1 \otimes 1_S(B)) - 1_S(A)\| \leq \delta$, we have in particular

$$\|(T(\mathbf{1} \otimes \mathbf{1}_S(B)) - \mathbf{1}_S(A))\psi_i\| = |x - 1| \leq \delta,$$
$$\|(T(\mathbf{1} \otimes \mathbf{1}_S(B)) - \mathbf{1}_S(A))\psi_j\| = |y| \leq \delta,$$

so that $|x - y| \geq 1 - 2\delta$. We can now apply Theorem 4. On the l.h.s. of the bound (6), we may substitute

$$R_*(|\alpha|^2|\psi_i\rangle\langle\psi_i| + |\beta|^2|\psi_j\rangle\langle\psi_j|) = |\alpha|^2|\psi_i\rangle\langle\psi_i| + |\beta|^2|\psi_j\rangle\langle\psi_j|$$

on account of T being nondestructive. On the r.h.s., we substitute $\Sigma = \sqrt{\delta(1-\delta)}$ and $|x - y| = (1 - 2\delta)$. Strikingly enough, this yields the bound

$$\left(\sqrt{\delta(1-\delta)}/(1-2\delta)\right)/\sqrt{1+4\left(\delta(1-\delta)/(1-2\delta)^2\right)} = \sqrt{\delta(1-\delta)}.$$

For perfect measurement ($\delta = 0$), this yields

$$R_*(|\alpha\psi_i + \beta\psi_j\rangle\langle\alpha\psi_i + \beta\psi_j|) = |\alpha|^2|\psi_i\rangle\langle\psi_i| + |\beta|^2|\psi_j\rangle\langle\psi_j|,$$

all coherence between ψ_i and ψ_j must vanish. This collapse of the wave function is illustrated in the lower left corner of Fig. 7. On the other hand, if all states are left intact so that $R_* = \mathrm{id}$, then we must have $\delta = 12$; no information can be gained. This is illustrated in the upper right corner of Fig. 7. Corollary 2 is a unified description of the Heisenberg Principle and the collapse of the wave function.

6.4 Application: Nondestructive qubit measurement

In quantum information theory, a σ_z-measurement is often taken to yield output +1 or −1, according to whether the input was $|\uparrow\rangle|\uparrow\rangle$ or $|\downarrow\rangle|\downarrow\rangle$. It is nondestructive if it leaves the states $|\uparrow\rangle|\uparrow\rangle$and $|\downarrow\rangle|\downarrow\rangle$ intact, yet it is only unbiased if it is perfect. Corollary 2 shows that in the nondestructive case, the Bloch sphere collapses to the cigar-shaped region depicted in Fig. 8.

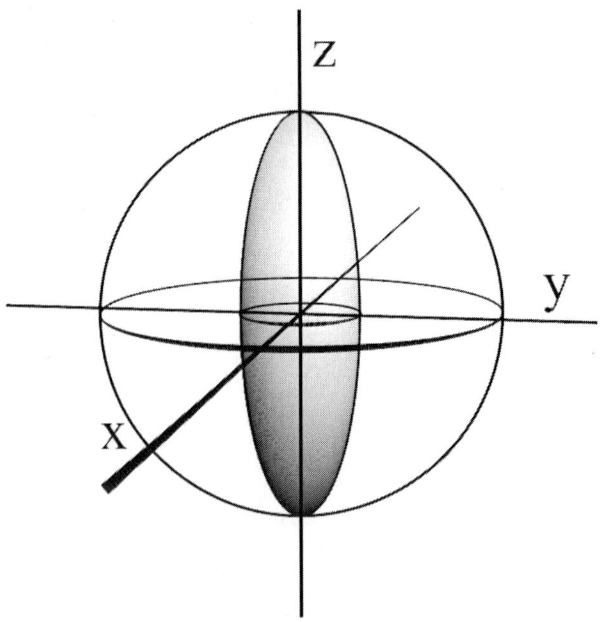

Figure. 8 Collapse on the Bloch sphere with $\delta = 0.01$

Current single-qubit read-out technology is now in the regime where the bound (8) becomes significant; in [16], a nondestructive measurement of a SQUID-qubit was described, with experimentally determined measurement infidelity $\delta = 0.13$. The bound then equals 0.336.

6.5 Sharpness of the bounds

We have yet to prove sharpness of all bounds. Let

$$V_+ := \begin{pmatrix} \sqrt{1-p} & 0 \\ 0 & \sqrt{p} \end{pmatrix}, \quad V_- := \begin{pmatrix} \sqrt{p} & 0 \\ 0 & \sqrt{1-p} \end{pmatrix}$$

and define $T : M_2 \otimes C(\Omega) \to M_2$ by $T(X \otimes f) := f(+1)V_+XV_+ + f(-1)V_-XV_-$. For $p = 0$, this is the von Neumann measurement. As a measurement of σ_z with pointer $B := (\delta_+ - \delta_-)/(1-2p)$, we have $\delta = p$. This yields bounds on the disturbance and on the coherence. Corollary 1 and Theorem 3 yield $\Delta \geq \frac{1}{2} - \sqrt{p(1-p)}$, corollary 2 and Theorem 4yield

$$D\big(R_*(|\alpha \uparrow + \beta \downarrow\rangle\langle\alpha \uparrow + \beta \downarrow|), (|\alpha|^2 |\uparrow\rangle\langle\uparrow| + |\beta|^2 |\downarrow\rangle\langle\downarrow|)\big) \leq \sqrt{p(1-p)}$$

We now explicitly calculate the restriction of T to M_2 and find

$$R_*(r) = \begin{pmatrix} r_{11} & 2\sqrt{p(1-p)}r_{12} \\ 2\sqrt{p(1-p)}r_{21} & r_{22} \end{pmatrix}$$

The maximal remaining coherence occurs for $\alpha = \beta = 1/\sqrt{2}$, for which it equals $\sqrt{p(1-p)}$. The maximal disturbance equals $\Delta = \frac{1}{2} - \sqrt{p(1-p)}$. This shows all bounds to be sharp.

7 CONCLUSION

Our investigation of joint measurement, the Heisenberg Principle and decoherence has yielded the following results.

1. Theorem 2 provides a sharp, state-independent bound on the performance of unbiased joint measurement of noncommuting observables. In the case of perfect ($\Sigma=0\Sigma=0$) measurement of one observable, it implies that no information whatsoever ($\Sigma'=\infty\Sigma'=\infty$) can be gained on the other.

2. Theorem 3 (for unbiased information transfer) and Corollary 1 (for general information transfer) provide a sharp, state-independent bound on the performance of a measurement in terms of the maximal disturbance that it causes. In the case of zero disturbance, when all states are left intact, it follows that no information can be obtained. This is the Heisenberg Principle.
3. Theorem 4 (for unbiased information transfer) and Corollary 2 (for general information transfer) provide a sharp upper bound on the amount of coherence which can survive information transfer. For perfect information transfer, all coherence vanishes. This clearly proves that decoherence on a system is a mathematical consequence of information transfer out of this system. If, on the other hand, all states are left intact, then it follows that no information can be obtained. This is the Heisenberg Principle. Theorem 4 and Corollary 2 connect these two extremes in a continuous fashion; they form a unified description of the Heisenberg Principle and the collapse of the wave function.

FOOTNOTES

1 We neglect the technical complication of x and p being unbounded operators.

ACKNOWLEDGEMENTS

Open access funding provided by Max Planck Society. I would like to thank Hans Maassen for his invaluable guidance and advice.

REFERENCES

1. Accardi, L., Frigerio, A., Lu, Y.G.: The weak coupling limit as a quantum functional central limit. Commun. Math. Phys. **131**, 537–570 (1990)
2. Arthurs, E., Kelly, J.: On simultaneous measurement on a pair of conjugate observables. Bell Syst. Tech. J. **44**, 725 (1965)
3. Bouten, L., Guţă, M., Maassen, H.: Stochastic Schrödinger equations. J. Phys. A **37**, 3189–3209 (2004)
4. Hall, M.: Prior information: how to circumvent the standard joint-measurement uncertainty relation. Phys. Rev. A **69**, 052113 (2004)
5. Heisenberg, W.: Über den anschaulichen Inhalt der quantentheoretischen Kinematik und Mechanik. Z. Phys. **43**, 172–198 (1927)

6. Hepp, K.: Quantum theory of measurement and macroscopic observables. Helv. Phys. Acta **45**, 237–248 (1972)
7. Holevo, A.S.: Probabilistic and Statistical Aspects of Quantum Theory. North Holland Publishing Company, Amsterdam (1982)
8. Ishikawa, S.: Uncertainty relations in simultaneous measurements for arbitrary observables. Rep. Math. Phys. **29**, 257–273 (1991)
9. Janssens, B., Bouten, L.: Optimal pointers for joint measurement of σxσx and σzσz via homodyne detection. J. Phys. A **39**, 2773–2790 (2006)
10. Janssens, B., Maassen, H.: Information transfer implies state collapse. J. Phys. A: Math. Gen. **39**, 9845–9860 (2006)
11. Joos, E., Zeh, H.: The emergence of classical properties through interaction with the environment. Z. Phys. B **59**, 223–243 (1985)
12. Kadison, R.V.: A generalized Schwarz inequality and algebraic invariants for operator algebras. Ann. Math. **2**(56), 494–503 (1952)
13. Kadison, R.V., Ringrose, J.R.: Fundamentals of the Theory of Operator Algebras, Volume II, Advanced Theory. Academic Press, Cambridge (1986)
14. Kennard, E.: Zur Quantenmechanik einfacher Bewegungstypen. Z. Phys. **44**, 326–325 (1927)
15. Lahti, P.: Coexistence and joint measurability in quantum mechanics. Int. J. Theor. Phys. **42**(5), 893–906 (2003)
16. Lupaşcu, A.E.A.: High-contrast dispersive readout of a superconducting flux qubit using a nonlinear resonator. Phys. Rev. Lett. **96**, 127003 (2006)
17. Majgier, K., Maassen, H., Życzkowski, K.: Protected subspaces in quantum information. Quantum Inf. Process. **9**(3), 343–367 (2010)
18. Miyadera, T., Imai, H.: Heisenberg's uncertainty principle for simultaneous measurement of positive operator-valued measures. Phys. Rev. A **78**, 052119 (2008)
19. Miyadera, T.: Relation between strength of interaction and accuracy of measurement for a quantum measurement. Phys. Rev. A **83**, 052119 (2011)
20. Miyadera, T., Loveridge, L., Busch, P.: Approximating relational observables by absolute quantities: a quantum accuracy-size trade-off. J. Phys. A: Math. Theor. **49**, 185301 (2016)
21. Nielsen, M., Chuang, I.: Quantum Computation and Quantum Information. Cambridge University Press, Cambridge (2000)
22. Ozawa, M.: Universally valid reformulation of the Heisenberg uncertainty principle on noise and disturbance in measurement. Phys. Rev. A **67**, 042105 (2003)
23. Peres, A.: Quantum Theory: Concepts and Methods. Kluwer Academic Publishers, Dordrecht (1993)
24. Polterovich, L.: Quantum unsharpness and symplectic rigidity. Lett. Math. Phys. **102**(3), 245–264 (2012)
25. Polterovich, L.: Symplectic geometry of quantum noise. Commun. Math. Phys. **327**(2), 481–519 (2014)
26. Robertson, H.: The uncertainty principle. Phys. Rev. **34**, 163–164 (1929)
27. Sewell, G.: On the mathematical structure of quantum measurement theory. Rep. Math. Phys. **56**, 271–290 (2005)
28. Takesaki, M.: Theory of Operator Algebras I. Springer, New York (1979)
29. von Neumann, J.: Mathematische Grundlagen der Quantenmechanik. Springer, Berlin (1932)
30. Werner, R.: Optimal Cloning of Pure States. Phys. Rev. A **58**, 1827–1832 (1998)
31. Werner, R.: Quantum information theory—an invitation. Quantum Inf. **173**, 14–57 (2001)
32. Zurek, W.: Environment-Induced Superselection Rules. Phys. Rev. D **26**, 1862–1880 (1982)

CHAPTER 3

A Multi-MW Proton/Electron Facility at KEK

R. Belusevic

IPNS, High Energy Accelerator Research Organization (KEK), Ibaraki, Japan

ABSTRACT

The main "bottleneck" limiting the beam power in circular machines is caused by space charge effects that produce beam instabilities. To increase maximally the beam power of a "proton driver", it is proposed to build a facility consisting solely of a 2.5 GeV injector linac (PI) and a 20 GeV pulsed superconducting linac (SCL). Such a facility could be constructed using the existing KEK accelerator infrastructure. The PI, based on the European Spallation Source (ESS) linac, would serve both as an injector to the SCL and a source of proton beams that could be used to copiously produce, e.g., muons and "cold" neutrons. Protons accelerated by the SCL would be transferred through the KEK Tristan ring in order to create neutrino, kaon and muon beams for fixed-target experiments. At a later stage, a 70 GeV proton synchrotron could be installed inside the Tristan ring. The SCL, comprising 1.3 GHz ILC-type rf cavities, could also accelerate polarized or unpolarized electron beams. After acceleration, electrons could be used to produce polarized positrons, or may traverse an XFEL undulator.

Keywords: Proton Driver, Superconducting Linac, Neutrino Oscillations, Rare Kaon Decays, XFEL

1. INTRODUCTION

The Standard Model (SM) of particle physics is supported by two theoretical "pillars": the gauge principle and the Higgs mechanism for particle mass generation. In this theoretical model, the mass of a particle depends on its interaction with the Higgs field, a medium that permeates the universe. The SM predicts the existence of a neutral spin-0 particle (the Higgs boson) associated with the Higgs field, but it does not predict its mass. Whereas the gauge principle has been firmly established through precision electroweak measurements, the Higgs mechanism is yet to be fully tested.

The Higgs-boson mass, m_H, affects the values of electroweak observables through radiative corrections. Many of the electroweak measurements obtained so far may be combined to provide a global test of consistency with the SM. The best constraint

on m_H is obtained by making a global fit to the electroweak data. Such a fit is consistent with the value of the Higgs mass measured at the Large Hadron Collider (LHC) [1] [2

To discover a new particle (such as the Higgs boson), or to search for physics beyond the SM, usually requires the use of high-energy hadron or electron-po- sitron colliders. However, many important discoveries in particle physics have been made using proton beams with relatively low energies but high intensities (flavor mixing in quarks and in neutrinos are noteworthy examples). Experiments with high-intensity neutrino beams, e.g., are designed primarily to explore the mass spectrum of the neutrinos and their properties under the CP symmetry.

Some of the most important discoveries emerged from high-precision studies of K mesons ("kaons"), in particular neutral kaons. A deeper insight into CP violation is expected to be gained from measurements of ultra-rare kaon decays such as $K_L^0 \to \pi^0 \nu\overline{\nu}$ and $K_L^{+} \to \pi^{+} \nu\overline{\nu}$. These decays provide important information on higher-order effects in electroweak interactions, and therefore can serve as a probe of new phenomena not predicted by the Standard Model.

The physics programs envisaged at the proposed facility are, to a large extent, complementary to each other. For instance, neutrino oscillation experiments and searches for permanent electric dipole moments both look for new sources of CP violation, a phenomenon which reflects the fundamental difference between matter and antimatter.

A unique feature of the proposed facility is the use of superconducting ILC-type cavities to accelerate both protons and electrons, which considerably increase its physics potential. Polarized electrons and positrons can be used to study the structure of composite particles and the dynamics of strong interactions, as well as to search for new physics beyond the Standard Model.

2. THE PROPOSED PROTON/ELECTRON FACILITY AT KEK

The main "bottleneck" limiting the beam power in circular machines is caused by space charge effects that produce beam instabilities. Such a "bottleneck" exists at the J-PARC proton synchrotron complex, and is also intrinsic to the "proton drivers" envisaged at CERN and Fermilab. To increase maximally the beam power of a "proton driver", it is proposed to build a facility consisting solely of a low-energy injector linac and a high-energy pulsed superconducting linac. Pulsed operation is preferred over the CW mode (continuous wave, 100% duty) mainly because the former allows the use of rf cavities with high accelerating gradients. This would considerably reduce the overall length of the machine, which is limited by the size of the KEK site.

The layout of the proposed proton/electron facility at KEK is shown in Figure 1. A 2.5 GeV proton linac (PI) serves both as an injector to a superconducting linac (SCL) and a source of proton beams that can be used to copiously produce neutrons and muons. Protons accelerated by the SCL to 20 GeV are transferred through the KEK Tristan ring in order to create beams for various fixed-target experiments. At a later stage, a 70 GeV proton synchrotron could be installed inside the Tristan ring. The SCL, comprising 1.3 GHz superconducting ILC-type rf cavities, can also accelerate polarized or unpolarized electron bunches. After acceleration, electrons may be used to produce polarized positrons. An SCL-based XFEL and a synchrotron light source for applications in materials science and medicine are also envisaged.

The proposed facility could be constructed using the existing KEK accelerator infrastructure. As shown in Figure 2, the present KEK linac tunnel and klystron gallery could be extended to increase the length, and hence the maximum energy, of each linac in Figure 1. The cryomodules, RF sources and cryogenic plant units of the proposed linac complex would be installed inside these extended structures (see Figure 3). The Tristan ring (TR) would initially be equipped only with magnets capable of steering the SCL proton bunches to various fixed targets. The four 200 m-long straight sections of the TR, each with an experimental hall in the middle (see Figure 2), would house beam lines and detectors.

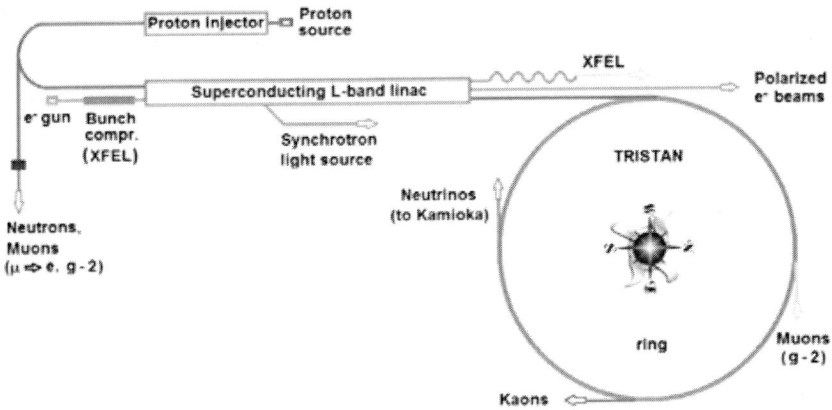

Figure 1. The layout of the proposed multi-purpose proton/electron facility at KEK. The Tristan ring would initially be equipped only with magnets capable of steering the proton bunches, accelerated by the superconducting linac, to various fixed targets.

The beam power of a pulsed linear accelerator is given by the expression

$$P_b\left[\mathrm{MW}\right] = \mathrm{E}_b\left[\mathrm{MV}\right] \times I\left[\mathrm{A}\right] \times \tau_p\left[\mathrm{s}\right] \times R\left[\mathrm{Hz}\right] \tag{1}$$

where P_b is the beam power, $E_b$is the beam energy, I is the average current per pulse, τ_p is the beam pulse length, and R is the repetition rate. The duty cycle of a pulsed linac is $D = \tau_p R$. Using the values from Table 1, and assuming E_b = 20GeV, one obtains D = 0.024 and

$$P_b = 20000\ \mathrm{MV} \times 31\ \mathrm{mA} \times 1.2\ \mathrm{ms} \times 20\ \mathrm{s}^{-1} \approx 15\ \mathrm{MW} \tag{2}$$

The beam parameters in Table 1 are mutually constrained by the following relations: The number of protons per second $N = P_b/E_b$ and the number of protons per pulse $N_p = N/R$; the average current per pulse is

$I = (N_p \times 1.6 \times 10^{-19}\mathrm{C})/\tau_p$. The klystron pulse length is the sum of the rf cavity fill time (current dependent) and the beam pulse length: $\tau = \tau_f + \tau_p$. For ILC-type cavities and $I \sim$ 30mA, $\tau_f \approx$ 0.3ms, Since τ =1.5ms, the beam pulse length $\tau_p \approx$ 1.2ms

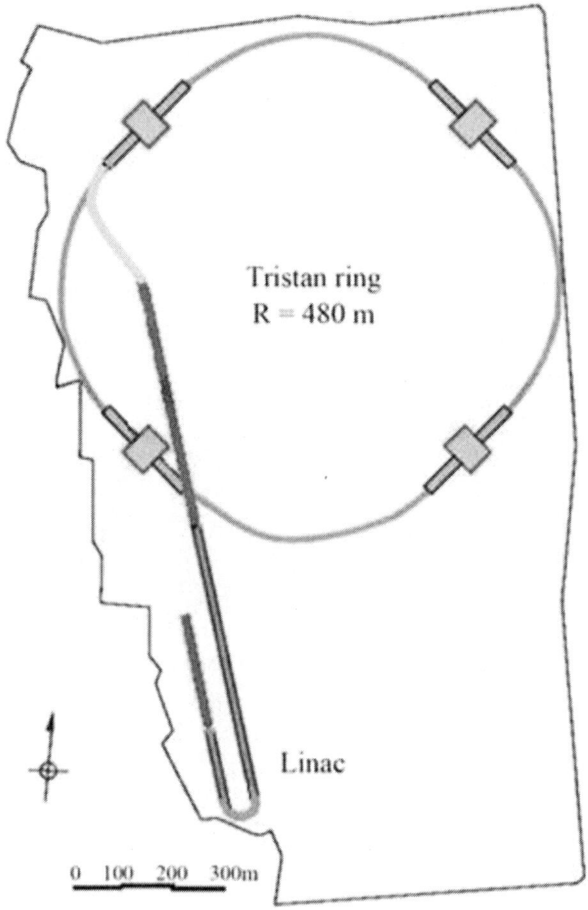

Figure 2. A sketch of the KEK site showing the Tristan ring and the existing electron linac (in blue). The lines drawn in red indicate possible extensions of the present linac tunnels. Alternatively, a new underground linac tunnel could be excavated at a greater depth in case a larger SCL beam energy (up to 25 GeV) is required.

2.1. Main Characteristics of an ILC-Type Linac

The main characteristics of a linear accelerator are determined by the properties of its rf source (klystrons) and accelerating cavities. For a pulsed ILC-type superconducting linac, one of the currently available rf sources is the Toshiba E3736 Multi-Beam Klystron [3]. This source has the following well-tested specifications: rf frequency—1.3 GHz; peak rf power—10 MW; average power—150 kW; efficiency—65%; pulse length—1.5 ms; repetition rate—10 Hz. If the repetition rate of the Toshiba klystron is increased by a factor of two, while its peak power is reduced by the same factor (thus keeping the average power constant) one obtains the klystron specifications presented in Table 1. For such a klystron, a suitable 20 Hz pulse modulator has to be developed.

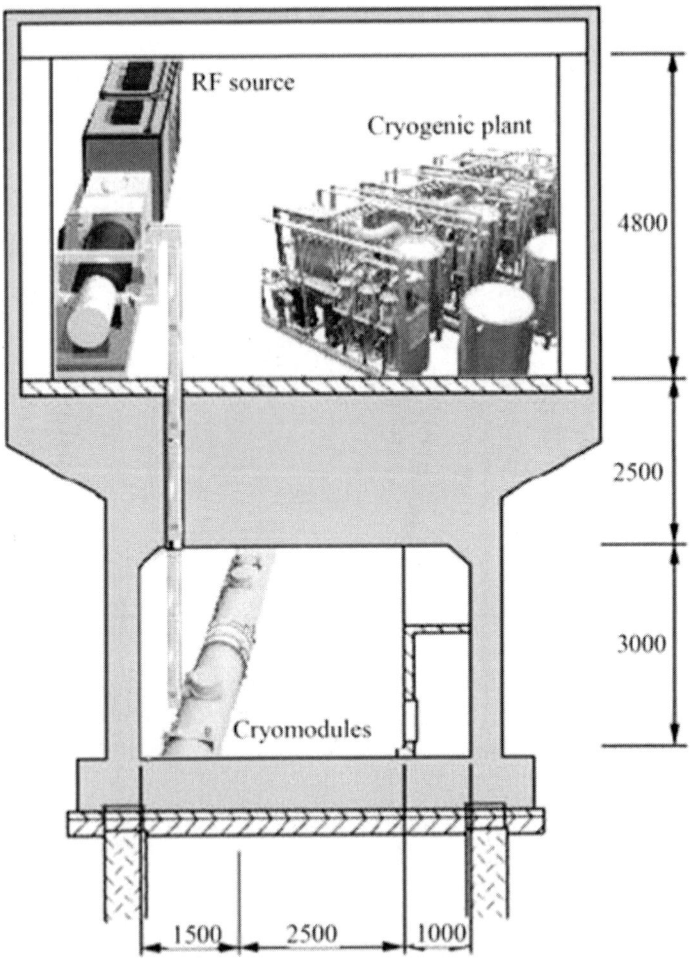

Figure 3. Front view of the linac tunnel and the klystron gallery housing the cryomodules, RF sources and cryogenic plant units of the proposed superconducting linac complex.

Table 1. Parameters of the SCL.

Beam energy	20 GeV
Beam power	15 MW
Repetition rate	20 Hz
Protons per pulse	2.3×10^{14}
Beam pulse length	1.2 ms
Average current per pulse	31 mA
Duty cycle	2.4%
RF frequency	1.3 GHz
Klystron average power	150 KW
Klystron peak power	5 MW
Klystron pulse length	1.5 ms
Effective accelerating gradient	20 MV/m
Peak power per coupler	460 kW

A very important parameter that determines, to a large extent, the power conversion efficiency of a klystron is its perveance, defined by

$$K \equiv \frac{I_0}{U^{3/2}} \tag{3}$$

In this expression, $I_0$is the beam current and U is the anode voltage. Since there is an upper limit to the applied voltage, low perveance can be only obtained by operating with low currents. For single-beam klystrons, this requirement is not compatible with the need for high output power. With this in mind, multi-beam klystrons (MBK) were originally developed in the 1960s [4]. An MBK is a parallel assembly of low-current (low-perveance) beamlets within a common rf structure, which efficiently generates high output power. Using Equation (3), the output rf power of an MBK can be expressed as

$$P_k = \eta I_0 U = \eta K U^{5/2} \tag{4}$$

where n is the klystron efficiency and $I_0 = N_b I_b$ is the total beam current; $N_b$is the number of beamlets and I_b is the current carried by each beamlet. For Toshiba's E3736 MBK, n = 65%, U = 116kV and I_0 = 134A. Hence, K =3.4×10^{-6} A/V$^{3/2}$, klystron's peak power P_k = 10MW and its average power

$\overline{P}_k \equiv P_k \times \tau R = 150\ \mathrm{kW}$ for τ = 1.5ms and R=10Hz. Since the klystron has six beamlets, I_b = 22.3A.

In order to increase the beam power of the SCL, R could be increased to 20 Hz (see Equation (1)). As already mentioned, the peak power of the klystron would then have to be reduced to 5 MW, so that its average power is still 150 kW for the klystron pulse length τ =1.5ms (see Table 1).

The basic properties of a 1.3 GHz superconducting ILC-type cavity are presented, e.g., in [5]. There are two important parameters that characterize rf cavities: the accelerating gradient $E_{acc}$and the unloaded quality factor Q_0. The former is a measure of how much the energy of a particle is increased over a given length of the linac (typically expressed in units of MV/m), while the latter specifies how well the cavity can sustain the stored rf power. A higher value of Q_0 implies a lower rate of power loss realtive to the stored energy[1]. ILC-type cavities must have a nominal Q_0greater than 1×10^{10} (a dimensionless parameter) at E_{acc} = 31.5 MV/m.

Each ILC-type cryomodule for the proposed SCL would contain eight niobium 9-cell cavities and a quadrupole magnet at its centre. Other major components of such a cryomodule are the vacuum vessel, thermal and magnetic shields, cryogenic piping, interconnections, etc. The inactive spaces between cavities or cryomodules (the "packing fraction") are responsible for a substantial reduction in the average accelerating gradient of the linac. The beam physics and the lattice design of the superconducting L-band linac described in [6] are applicable to the SCL.

The average usable accelerating gradient in ILC-type cavities is $\overline{E}_{acc} = 29.3 \pm 5.1\ \mathrm{MV/m}$ [7]. Taking into account an estimated linac "packing fraction" of about 70%, the effective accelerating gradient of the SCL is $E_{ff} \approx$ 20MV/m. Hence, the total length of a 20 GeV linac is ~1000 m.

Since the length of an ILC 9-cell cavity is 1 m, a linac with E_b = 20GeV would require about N_{cav}= (20000NeV)/(29MeV) = 690 cavities. The average input rf power per cavity is thus $\overline{P}_{cav} = P_b / N_{cav} \approx 22\ \mathrm{kW}$, and the corresponding peak power $P_{cav} \equiv \overline{P}_{cav} / D = 916\ \mathrm{kW}$. Although this value is acceptable for a pulsed linac with D ~2%, it would be prudent to use two rf couplers per cavity. In that case the peak rf power per coupler would be about 460 kW.

For E_{acc} =30MV/m, ohmic losses in an ILC 9-cell cavity amount to P_c = 100W in the CW mode of operation, but only P_c = $(100 \times D)$W = 2.4W (plus static loss) in the pulsed mode with a duty factor D = 0.024. Because of large ohmic losses, which scale with the square of the accelerating gradient, $E_{acc}$is generally lower in the CW mode of operation than in the pulsed mode.

2.2. Proton Injector (PI)

A typical ~1 GeV proton linear accelerator consists of three main sections:

- Front end, comprising a proton source and a radiofrequency quadrupole accelerator;
- Medium-velocity linac, which accelerates proton beams to ~100 MeV;
- High-velocity linac, which accelerates protons to energies exceeding 1 GeV.

The most complex part of a proton linac is the low-energy (low-b) section, situated between the proton (or ion) source and the first drift-tube-based accelerating section. The continuous beam of protons coming from an electron cyclotron resonance (ECR) source [8] has to be focused, bunched and accelerated in the first rf structure. These three essential functions are nowadays successfully performed by radio frequency quadrupoles (RFQ) [9]. However, the beam has to be shrunk before it can be fed into an RFQ. This is accomplished within a low-energy beam transport (LEBT) section by means of cylindrical magnets (solenoids).

As soon as the beam is bunched---which is essential for further acceleration— it enters a medium-energy beam transport (MEBT) section, where it is collimated and steered from the RFQ into the medium-velocity linac (MVL). The MEBT may also contain a number of buncher cavities. Inside the MVL, the beam is accelerated to about 100 MeV ($\beta\sim$ 0.1 to 0.5). The MVL usually contains normal-conducting drift-tube linac (DTL) and cell-coupled drift tube linac (CCDTL) structures. A DTL incorporates accelerating components of increasing length in order to match precisely the increase in beam velocity, while quadrupole magnets provide strong focusing. The main advantage of using CCDTL structures is that they provide longitudinal field stability.

High-velocity linac (HVL) structures accelerate the beam to energies around 1 GeV. They consist either of normal-conducting side-coupled linac (SCL) structures[2] or superconducting elliptical cavities. The latter offer some advantages over the former, such as higher accelerating gradients and lower operating costs. The superconducting HVL can also feature spoke resonators, characterized by their simplicity, high mechanical stability and compact size [10].

The European Spallation Source (ESS) is an example of a typical ~1 GeV proton linear accelerator [11] (see Figure 4). The transverse beam size along the linac varies in the range 1 - 4 mm, and the bunch length decreases from 1.2 cm to 3 mm towards the end of the linac.

One of the main concerns in the design of a high-power proton linac is to restrict beam losses. A careful beam dynamics study is therefore needed in order to avoid halo formation, a major source of beam loss. Another important issue is the preservation of beam emittance [6].

High-power proton linear accelerators have a wide range of applications including spallation neutron sources, nuclear waste transmutation, production of radioisotopes for medical use, etc. A number of laboratories worldwide have expressed interest in building "proton drivers" that would primarily deliver high- intensity neutrino, kaon and muon beams [12] [13].

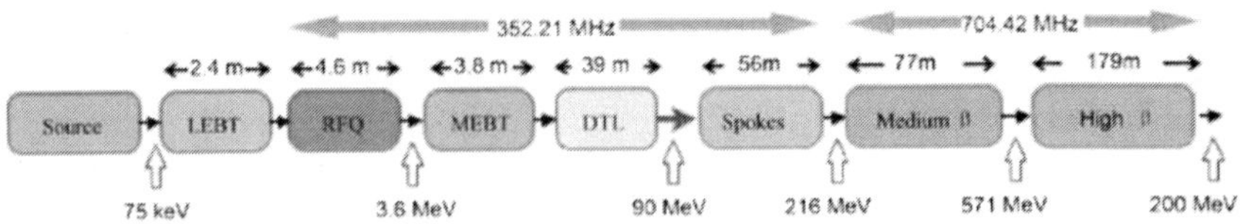

Figure 4. Block diagram of the ESS linac [11]. The RFQ and DTL structures are normal- conducting, while the spoke resonator and elliptical cavities are superconducting.

3. PHYSICS AT THE PROPOSED FACILITY

The physics potential of a multi-MW "proton driver" is extensively discussed, for instance, in [14]. This note is mainly concerned with the application of a high-intensity proton source to investigate the properties of long-baseline neutrino oscillations. A unique feature of the proposed facility is the use of a superconducting linac to accelerate both protons and electrons, which considerably increases its physics potential. As described in [15] [16] , polarized electron and positron beams can be used to study the structure of composite particles and the dynamics of strong interactions, as well as to search for new physics beyond the Standard Model.

3.1. Neutrino Flavor Oscillations and Leptonic CP Violation

The observed transformation of one neutrino flavor into another ("neutrino oscillation") indicates that these particles have finite masses. Cosmological data suggest that the combined mass of all three neutrino species is a million times smaller than that of the next-lightest particle, the electron [17]. The phenomenon of neutrino oscillations implies not only the existence of neutrino mass, but also of neutrino mixing. That is, the neutrinos of definite flavor are not particles of definite mass (mass eigenstates), but coherent quantum-mechanical superpositions of such states. Converesely, each neutrino of definite mass is a superposition of neutrinos of definite flavor. Neutrino mixing is large, in striking contrast to quark mixing. Whatever the origin of the observed neutrino masses and mixings, it implies a profound modification of the Standard Model.

Mathematically, the phenomenon of neutrino mixing can be expressed as a unitary transformation relating the flavor and mass eigenstates. The neutrino oscillation rate depends, in part, on (1) the difference between neutrino masses and (2) the three parameters in the transformation matrix known as mixing angles. The complex phase factors in the transformation matrix (also called mixing matrix) are associated with the violation of CP symmetry in the lepton sector. The size of the CP violation is determined both by the phases and the mixing angles.

Experiments with high-intensity neutrino beams are designed primarily to explore the mass spectrum of the neutrinos and their properties under the CP symmetry, and thus provide a deeper insight into the nature of these elusive particles and their role in the universe. For instance, if there is experimental evidence for CP violation in neutrino oscillations, it could be used to explain the observed asymmetry between matter and antimatter [18].

To search for CP violation in neutrino oscillations, a 100 kiloton water Cherenkov detector could be built at Okinoshima, located at a distance $L_2 \approx$ 650 km from KEK. Using the proposed KEK "proton driver", the detector at Okinoshima and the existing 0.022 Mt Super-Kamiokande detector (situated at a distance $L_1 \approx$ 300km from KEK), the neutrino mass hierarchy could be determined either by comparing the ν_e appearance probabilities measured at the two vastly different baseline lengths L_1 and L_2, or by measuring at L_1 and L_2 the neutrino energy of the first oscillation maximum. Once the mass hierarchy is determined, the CP-violating phase in the mixing matrix can be measured with a precision of $^{\pm 20}$, assuming that 2.5×10^{21} protons are delivered on target for both ν_e and $\overline{\nu_e}$ beams [19].

Proton Target and Magnetic Horn

The main challenge in the design of a multi-MW neutrino beam facility is to build a proton target that could dissipate large amounts of deposited energy, withstand the strong pressure waves created by short beam pulses, and survive long-term effects of radiation damage. Simulation studies of the pion production and energy deposition in different targets (liquid mercury jet, tungsten powder jet, solid tungsten bars and gallium liquid jet) are presented in [20]. Those studies also provide estimates of the amount of concrete shielding needed to protect the environment from the high radiation generated by each target. A proof-of- principle demonstration of a

4 MW target station comprising a liquid mercury jet inside a 20 T solenoidal magnetic field is described in [21]. Alternatively, one could use a rotating, gas-cooled tungsten target that would require the least amount of development effort, and would also have good thermal and mechanical properties [11]. A 15 MW proton beam could be separated by a series of magnets into four beam lines. Each of the four beams would be focused by a series of quadrupoles and correctors to an assembly consisting of four targets and the same number of magnetic horns (see, e.g., [22]).

To maximize the discovery potential of a neutrino beam facility, it is important to properly design the magnetic horn that focuses the charged particles produced in the proton target. For proton beam pulses lasting 1 ms, a DC horn has been designed at KEK by Yukihide Kamiya [23]. The toroidal magnetic field of the horn, characterized by B(*r*) =const is generated by hollow aluminium conductors containing water. The strength of the magnetic field B = 0.2T, and its length $\ell = 5\,\text{m}$; hence, $\text{B} \cdot \ell = 1\,\text{T} \cdot \text{m}$. The radius of the magnet, r, is determined by $r = L\tan(\theta) + \ell\tan(\theta/2)$, where $\theta \approx 0.03+0.3/p$ is the initial angle a charged pion makes with respect to the proton beam direction, L is the distance from the target to the horn, and p is the pion momentum. For example, if L = 5m then $r \approx$5m. The total power generated in the conductors is about 10 MW.

3.2. Physics with Polarized Electrons and Positrons

Electron and positron beams, polarized and/or unpolarized, can be used to study the structure of composite particles and the dynamics of strong interactions, as well as to search for new physics beyond the Standard Model. A detailed description of the physics potential of a facility that can provide such beams (e.g., the upgraded CEBAF facility at Jefferson Lab or the proposed KEK superconducting linac) is presented in [15] [16].

Polarized positrons are created in a conversion target by circularly polarized photons, which themselves are produced when polarized laser light is Compton- backscattered on a high-energy electron beam [24]. Circularly polarized photons can also be produced by bremsstrahlung from polarized electrons [25]. Using polarized electrons and positrons, the nucleon electromagnetic form factors and generalized parton distributions can be determined in a model-independent way [16].

Among the physics topics discussed in [15] , parity violation in electron-elec- tron (Mø ller) scattering is of particular interest. Mø ller scattering is a purely leptonic process that allows high-precision tests of the Standard Model. At four-momentum transfers much smaller than the mass of the Z boson $(q^2 \ll m_Z^2)$, the parity-violating asymmetry, A, is dominated by the interference between the electromagnetic and neutral weak amplitudes [26]. By definition,

$$A \equiv \frac{\mathrm{d}\sigma_R - \mathrm{d}\sigma_L}{\mathrm{d}\sigma_R + \mathrm{d}\sigma_L} \approx \frac{f_Z^R - f_Z^L}{f_\gamma} \tag{5}$$

In this expression, $\mathrm{d}\sigma_R$ ($\mathrm{d}\sigma_L$) is the differential cross-section for right-handed (left-handed) electron scattering on an unpolarized target:

$$\mathrm{d}\sigma_{R,L} \propto \left|f_\gamma + f_Z^{R,L}\right|^2 \approx \left|f_\gamma\right|^2 + 2f_\gamma f_Z^{R,L} \tag{6}$$

where f_Y and $f_Z^{R,L}$ are the scattering amplitudes with γ and Z exchange, respectively. From the four Feynman diagrams in Figure 6 of [27] , one can readily obtain the Born amplitudes for Mø ller scattering mediated by photons and Z bosons. The weak neutral current amplitudes are functions of the weak mixing (or Wienberg) angle θ_w, which relates the weak coupling constants g_w and to g_z the electromagnetic coupling constant. As shown in [27] , the polarization asymmetry for polarized electron scattering on an unpolarized target is given by

$$A^{\mathrm{Born}} = m_e \mathrm{E} \frac{G_{\mathrm{F}} Q_{\mathrm{W}}^e}{\sqrt{2}\pi\alpha} F(\theta) \tag{7}$$

where m_e is the mass of the electron, E is the incident beam energy, $G_F$is the Fermi coupling constant characterizing the strength of the weak interaction, a is the fine structure constant, and $F(\theta)$ is a function of the scattering angle in the center-of-mass frame. The weak charge of the electron, $Q_{\mathrm{w}}^e = 1 - 4\sin^2\theta_{\mathrm{w}}$, is proportional to the product of the electron's vector and axial-vector couplings to the Z boson.

Since the value of sin $^2\theta_w$ is close to 1/4, there is an enhanced sensitivity of A to small changes in the Weinberg angle. The value of θ_w varies as a function of the four-momentum transfer, q, at which it is measured. This variation, or "running", is a key prediction of the Standard Model. The one-loop electroweak radiative corrections to A^{Bom}(calculated once the renormalized parameters in (7) are properly defined) reduce its Born value by ~40%. This effect can be attributed to an increase of sin $^2\theta_w$ (q^2) by 3% as the four-momentum transfer "runs" from $q^2 = m_Z^2$ to $q^2 \approx 0$ [28].The precision with which sin $^2\theta_w$ can be measured in the MOLLER experiment

[29] (where a 11 GeV longitudinally polarized electron beam scatters on atomic electrons in a liquid hydrogen target) is shown in Figure 5.2 of [15].

Many of the electroweak meaurements obtained so far may be combined to provide a global test of consistency with the SM. Since the Higgs-boson mass affects the values of electroweak observables through radiative corrections, it is of fundamental importance to test the agreement between the directly measured value of m_H and that inferred from the measurements of electroweak parameters m_w, m_{top} , and $\sin^2\theta_w$ (see Figure 5.2 in [15]). High-precision electroweak measurements, therefore, represent a natural complement to direct studies of the Higgs sector.

Apart from providing a comprehensive test of the SM, precision measurements of weak neutral current interactions at $q^2 \ll m_Z^2$ also allow indirect access to new physics phenomena beyond the TeV energy scale. For instance, such measurements can be used to look for hypothetical Z' bosons, 4-fermion contact interactions, or very weekly coupled low-mass "dark bosons" [30].

3.3. Rare Kaon Decays

CP violation was introduced in the SM by increasing the number of quark and lepton families to at least three (M. Kobayashi and T. Maskawa, 1973). This idea became very attractive with the subsequent discovery (in 1977) of the bottom quark, which forms, together with the top quark (discovered in 1995), a third family of quarks. It is a remarkable property of the Kobayashi-Maskawa model that quark mixing and CP violation are intimately related [31].

A deeper insight into CP violation is expected to be gained from precision measurements of rare kaon decays such as $K_L^0 \to \pi^0\nu\bar{\nu}$ and $K^+ \to \pi^+\nu\bar{\nu}$. Both decays are theoretically 'clean' because hadronic transition amplitudes are matrix elements of quark currents between mesonic states, which can be extracted from the leading semileptonic decays using isospin symmetry. Since photons do not couple to neutrinos, $K \to \pi\nu\bar{\nu}$ decays are entirely due to second-order weak processes determined by Z-penguin and W-box diagrams [31].

The process $K_L^0 \to \pi^0\nu\bar{\nu}$ proceeds almost entirely through direct CP violation, and is completely determined by "short-distance" one-loop diagrams with top quark exchange. The Standard Model predicts its branching ratio to be [32]

$$B\left(K_L^0 \to \pi^0\nu\bar{\nu}\right) = (2.43 \pm 0.39) \times 10^{-11} \quad \text{theoretical value} \tag{8}$$

This decay is an important source of information on higher-order effects in electroweak interactions, and thus can serve as a probe of new physics (see [33] and references therein).

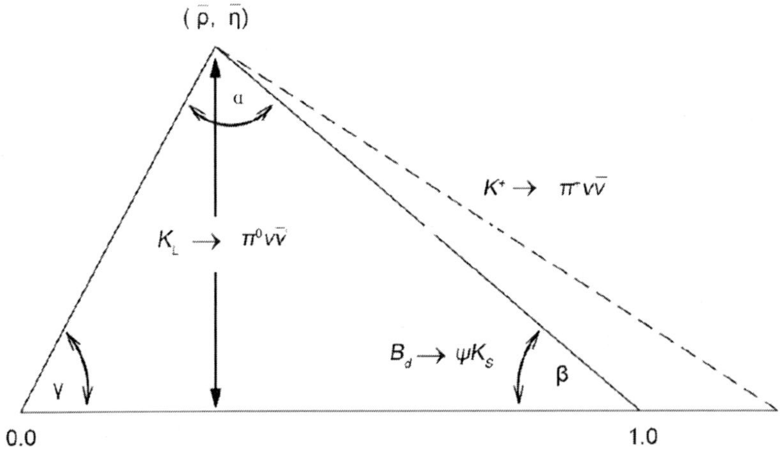

Figure 5. Unitarity triangle from $K \rightarrow \pi\nu\bar{\nu}$ decays. The displacement of the bottom- right vertex is due to the charm-quark contribution to $K^+ \rightarrow \pi^+\nu\bar{\nu}$.

The decay $K^+ \rightarrow \pi^+\nu\bar{\nu}$ receives both CP-conserving and CP-violating contributions. It has theoretical uncertainties that are somewhat larger than those in the process $K_L^0 \rightarrow \pi^0\nu\bar{\nu}$. Since both decays involve one-loop Feynman diagrams with top quark exchange, they can yield valuable measurements of the CKM matrix elements $|V_{td}|$ and $|V_{ts}|$. The quantity Im$V_{ts}^*V_{td}$, which can be obtained from $K_L^0 \rightarrow \pi^0\nu\bar{\nu}$ alone, plays a central role in the phenomenology of CP violation in K decays; this quantity is related to the Jarlskog parameter, the invariant measure of CP violation in the Standard model [31] [34].

By measuring the branching ratios of both $K \rightarrow \pi\nu\bar{\nu}$ decay modes, the unitarity triangle of the CKM matrix can be completely determined (see Figure 5), provided the matrix element V_{cb} and the top quark mass are known [34]. Of particular interest is the unitarity triangle parameter sin 2β, which can also be determined from the decay $B_d \rightarrow \Psi K_s$. Both determinations of this parameter have to coincide if the Standard Model is valid [33].

The current branching ratio measurement of the charged decay mode,

$$B\left(K^+ \to \pi^+ \nu\bar{\nu}\right) = \left(17.3^{+11.5}_{-10.5}\right) \times 10^{-11} \qquad 9$$

is based on the seven candidate events observed by the experiment E787/E949 at Brookhaven [35]. This result is consistent with $(7.8 \pm 0.80) \times 10^{-11}$, the value predicted by the SM [32].

As shown in Figure 11 of [33] , the kaon yield rises rapidly as a function of the incident proton momentum. From the figure one infers that the minimum energy of the proton beam should be about 20 GeV, for otherwise the kaon yield would be severely reduced. At the proposed KEK facility, the total proton beam energy would be (2.5+20) GeV. A 70 GeV proton synchrotron could be installed, at a later stage, inside the Tristan ring in order to increase the proton beam energy -albeit at the cost of a considerably lower beam power. For a given kaon yield, the required beam power would be lowest at E_b = 30 - 100 GeV [33].

3.4. A Novel g-2 Experiment with Ultra-Slow Muons

A charged elementary fermion has a magnetic dipole moment μ = $g_s(q/2m)s$ aligned with its spin s. The proportionality constant g_s is the Landé g-factor, q is the charge of the particle and m is its mass. Dirac's theory of the electron predicts that g_s = 2. For the electron (e), muon (μ) and tau lepton (τ), this prediction differs from the observed value by a small fraction of a percent. The difference is the anomalous magnetic moment; the anomaly is defined by $a \equiv (g_s - 2)/2 \sim 10^{-3}$. In the Standard Model, three distinct classes of Feynman diagrams contribute to the value of the anomaly for each lepton species: (1) the dominant QED terms that contain only leptons and photons; (2) terms that involve hadrons; and (3) electroweak terms containing the Higgs, W and Z bosons. The muon anomaly a_μ is about $(m_\mu/\ m_e)$ ~ 43000 mtimes more sensitive to the existence of yet unknown heavy particles than the electron anomaly a_e. The value of a_μ (a_e) is sensitive to new physics at the scale of a few hundred GeV (MeV) [36].

The current experimental uncertainty on a_μ is ± 0.54 ppm. In a novel g-2 experiment [37] , the main aim of which is to reduce this uncertainty to ± 0.1 ppm, 3 GeV protons impinge on a graphite target and produce pions that are stopped in the target. Some of the positive pions are brought to rest near the surface of the target, where they decay into positive muons with momenta p_μ = 30MeV/c and 100% spin polarization. The muons are collected using a large-aperture solenoid and transported to a silica-aerogel target in which they form muonium (electron—μ^+) atoms. As the atoms slowly diffuse from the target, they are ionized by a pulsed laser to produce 50% polarized muons with very low momenta[3]. Those "ultra-slow" muons ($4\times10^{4/\text{pulse}}$)) are then accelerated to p_μ = 300MeV/c by two linacs, and injected into a magnetic storage ring that contains a 3 T solenoid with a diameter of 66 cm. After injection, the muons circulate orthogonal to the magnetic field B. An orbiting muon decays within 6.6 ms into a positron, a neutrino and an antineutrino: $\mu^+ \rightarrow e^+ + \nu_e + \bar{\nu}_\mu$. The experiment is also designed to measure the electric dipole moment $d_\mu = \eta(qh/2m)s$ of the muon (see below).

The highest-energy positrons, preferentially emitted parallel to the muon spin direction in the μ^+ rest frame, are Lorentz-boosted to become the highest-energy positrons in the lab frame. Hence, the angular distribution of those positrons has its maximum in the direction of the muon spin [17]. By measuring the energy and time distributions of positrons one can determine the average spin direction. The time spectrum will show the muon lifetime modulated by the spin precession frequency. The relative precession of the spin with respect to the direction of the particle velocity u is given by

$$\omega_a + \omega_\eta \propto a_\mu \mathbf{B} - (\eta/2)(\beta \times \mathbf{B}) \tag{10}$$

where ω_η and ω_η arise from a_μ and d_μ, respectively, and $\beta \equiv u/c$. Since the rotation axes due to a_μ and d_μ are orthogonal, the corresponding signals can be separated [37]. In the case of μ_μ, the anomalous precession period is 2.2 ms, about 300 times the cyclotron period. Assuming that muons are 100% polarized, 1.5×10^{12}positrons have to be detected for a measurement precision of 0.1 ppm [37].

4. AN XFEL BASED ON THE PROPOSED SUPERCONDUCTING LINAC

To record the dynamics of atoms requires a probe with Ångstrom (10^{-10} m) wavelength and femtosecond temporal duration (10^{-15} s). Such probes have recently become available with the advent of X-ray free-electron lasers (XFELs). The ultrashort pulse duration of an XFEL matches the timescale of non-equilibrium microscopic processes such as electron transfer in molecules, evolution of che- mical reactions, vibration dynamics in solid state systems, etc. Optical lasers are also capable of producing pulses of femtosecond duration, but lack the required spatial resolution.

The spectral brightness (or brilliance), B, of a radiation field can be expressed, in practical units, as the number of photons per second passing through a given cross-sectional area and within a given solid angle and spectral bandwidth (BW). This quantity determines how much monochromatic radiation can be focused onto a tiny spot on the target. The peak spectral brightness—the brightness measured during the very short duration of an FEL pulse—of the two presently most powerful XFEL facilities (LCLS at SLAC in the United States and SACLA at SPring-8 in Japan) is billion times higher than that of any synchrotron radiation source.

Owing to the high intensity of XFEL radiation, laser-irradiated atoms, molecules and atomic clusters can be excited into previously unknown states. Although high-intensity pulses may also destroy molecular structures, they can still be used to produce high-resolution X-ray diffraction patterns, from which real-space images of the atomic positions in molecules can be reconstructed. In a typical "pump-probe" experiment, the evolution of a chemical (or biochemical) reaction, initiated by an optical or IR laser pulse, is observed by a time-delayed X-ray pulse. By varying the delay, such stroboscopic measurements result in femtosecond "movies" of the evolving system. Note that, due to their long pulse duration, X-rays from synchrotron light sources can be used to image atomic structures only in static measurements.

Nanometre-scale molecular imaging is made possible by the high degree of coherence of the XFEL radiation. The coherence quality of a light source is best described by the degeneracy parameter D, defined as the number of emitted photons per coherent phase-space volume per coherence time:

$$D \equiv \frac{B\lambda^3}{4c} \approx 8.3\times10^{-25}\,B\lambda^3 \tag{11}$$

Here B is the brilliance and λ [0.1 nm] the wavelength of the source.

Free-electron lasing is achieved by a single-pass, high-gain FEL amplifier operating in the so-called self-amplified spontaneous emission (SASE) mode. An FEL consists of an electron linear accelerator and an undulator, a long periodic array of magnets with period λ_u. The wavelength of the first harmonic of the observed FEL radiation is given by [38] [39]

$$\lambda = \frac{\lambda_u}{2\gamma^2}\left(1+\frac{K^2}{2}+\gamma^2\theta^2\right) \tag{12}$$

where the dimensionless quantity

$$K = \frac{eB_0\lambda_u}{2\pi m_e c} = 0.934B_0[\text{Tesla}]\lambda_u[\text{cm}] \tag{13}$$

γ is the Lorentz factor of the electron beam. Typically, $\gamma_u \approx 3$cm and $\gamma \approx 10^4$; hence,$\gamma \approx 0.1$nm.

The emission of radiation in an undulator does not occur at one wavelength, but in a wavelength band of width $\Delta\lambda$ around the central value given by Equation (12). Each electron propagating through the undulator emits a wave train consisting of a number of wavelengths equal to the number of undulator periods,N_u. The time duration Δt of this pulse is the pulse length $L_p \equiv N_u\lambda$ divided by the speed of light: $\Delta t = N_u\lambda/c$. A pulse of duration $\Delta\lambda$ has a frequency bandwidth $\Delta\nu \sim 1/\Delta t$. Hence, $\Delta\nu \sim c/N_u\lambda = \nu/N_u$, because $\lambda = c/\nu$. Thus,

$$\frac{\Delta\nu}{\nu} = \frac{\Delta\lambda}{\lambda} \sim \frac{1}{N_u} \approx 10^{-3} \tag{14}$$

The wave train is not monochromatic due to its finite length. For typical values $N_u \approx 10_3$ and $\lambda \approx 0.1$nm, one obtains $\Delta t \approx 0.33$ fs. Since the electrons are distributed throughout a bunch, the pulse duration is increased to $\tau_p \sim (\sigma_z/L_p)\Delta t \approx 80$ fs, where $\sigma \approx 25\mu$m is the bunch length and $L_p \approx 0.1\mu$m. An 80 fs pulse, therefore, consists of many micropulses of 0.33 fs duration.

The 'shot noise' in an electron beam, the origin of which is the random emission of the electrons from a photocathode, causes random fluctuations of the beam density. The radiation produced by such a beam has amplitudes and phases that are random in both space and time. For this reason, SASE X-ray FELs lack longitudinal (or temporal) coherence, characterized by the coherence length $L_{\text{coh}} = \lambda^2/\Delta\lambda \approx 0.1\ \mu\text{m}$. This quantity is defined as the distance of propagation over which radiation with spectral width $\Delta\lambda$ becomes 180° out of phase. The coherence time, defined by $t_{\text{coh}} \equiv L_{\text{coh}}/c \sim 1/\Delta\nu$, is much shorter than the pulse duration:$t_{\text{coh}} \approx 0.3$fs.

In order to increase the coherence length in the hard X-ray regime (photons with 0.1 nm wavelength), a "self-seeding" method was tested at LCLS [40]. FEL pulses, generated in the first modular section of the LCLS undulator, are spectrally "purified" by a crystal filter (a diamond monochromator). Since a typical monochromator delays X-rays, the electron bunches exiting the first modular section are appropriately delayed after being diverted around the crystal by a compact magnetic chicane (see Figure 1 in [40]). The crystal selects a very narrow part of the spectrum, which is further amplified in the second undulator section where the FEL radiation reaches saturation. At LCLS, "self-seeding" generated X-ray pulses with $\Delta\nu = 0.4 - 0.5$ eV at ν =8-9keV, which represents a factor of 40 - 50 bandwidth reduction with respect to SASE [40].

4.1 The European XFEL as a Prototype for the Proposed X-Ray FEL

The European XFEL, currently under construction at DESY (Germany), could serve as a prototype for the proposed X-Ray FEL at KEK. The XFEL at DESY is a free-electron laser based on self-amplified spontaneous emission (SASE) in the X-ray regime. The FEL consists of a 17.5 GeV superconducting electron linear accelerator and a set of undulators that can generate both SASE FEL X-rays and incoherent radiation. A schematic layout of the European XFEL is shown in [41]. Electron bunches, each with a charge of 1 nC, are extracted from a photocathode by short ultraviolet laser pulses and then focused and accelerated inside a radio-frequency cavity ("RF gun") to an energy of 120 MeV. In order to produce 5 kA peak currents necessary for lasing, the bunches are further accelerated and longitudinally compressed down to 25 mm using two magnetic chicanes (at 0.5 and 2.0 GeV). After traversing the main linac, where their energy is increased to 17.5 GeV, the bunches are sent through a number of FEL undulators.

A superconducting linac may accelerate 10 "bunch trains" per second, each train consisting of up to 2700 electron bunches. This results in 27,000 ultrashort X-ray flashes per second. The higher the number of electron bunches, the more scientific instruments can be operated simultaneously. The European XFEL facility will generate ultra-short pulses (≤ 100 fs) of spatially and temporally coherent X-rays with wavelengths in the range ~0.1 - 5 nm; its peak brilliance is expected to be

$$B \approx 5\times10^{33}/\text{photons/second/mm2/mrad2/0.1\%BW} \qquad (15)$$

The coherent superposition of the radiation fields from all microbunches within an electron bunch in the linac is responsible for the nearly monochromatic spectrum and small divergence of the radiation emitted in the forward direction [38]. Recall also that "self-seeding" can substantially improve longitudinal (temporal) coherence of SASE XFEL radiation. Thus, the radiation from an X-ray FEL has a narrow bandwidth, is transversely and longitudinally coherent, and is fully polarized. The coherently emitted XFEL spectral lines appear in addition to the spontaneously emitted undulator spectrum that extends into the MeV energy region (see, e.g., [42]).

The micropulses that form an FEL pulse give rise to "spikes" in the spectrum. The amplitudes of the micropulses vary greatly as a consequence of the amplified stochastic variations in the electron density. Within a micropulse, the radiation is both transversely and longitudinally coherent. The duration of a micropulse is roughly t_{coh}, the coherence time. In the SASE 1 and SASE 2 undulators at the European XFEL, t_{coh} = 0.2 - 0.38fs [41]. The number of "spikes" in a pulse is given by the ratio of the bunch length to the coherence length: $\sigma_z / L_{coh} = (25\,\mu\text{m})/(0.1\,\mu\text{m}) \approx 250$.

The spectrum of undulator radiation is sharply peaked around odd harmonics[4] (see, e.g., [42]). The photon energy that corresponds to the nth harmonic is given by

$$E_n[\text{keV}] = 0.9496 \frac{n E_e^2 [\text{GeV}]}{\lambda_u [\text{cm}] \left(1 + K^2/2 + \gamma^2 \theta^2\right)} \tag{16}$$

where E_e is the electron beam energy and θ is the radiation detection angle (with respect to the forward direction). For $\theta = 0$ and the SASE 2 undulator parameters λ_u = 4.8cm and K = 6.1, for example, Equation (16) yields E_1 = 12.2KeV [41].

At the exit of the SASE 1 undulator, the photon beam divergence is $\sigma_\theta \sim 1/\gamma \sqrt{N_u} \approx 1\ \mu\text{rad}$ [38] and the beam size is 70μm×70 μm, the diameter of a fine needle. This beam can be focused to an area of 0.1μm×0.1 μm (the size of a virus) at an experimental station located a couple of hundred meters from the undulator exit [43]. Through variable focusing, the flux density of an XFEL beam can therefore be tuned by a factor of about one million. The SASE 1 undulator will deliver 10^{12} photons in an ultra-short pulse of 100 fs duration (the timescale of molecular vibrations), yielding a peak power of about 20 GW at a photon energy E ≈ 12KeV (the photon wavelength is $\lambda = hc/E \approx 0.1$nm.

Recall that all photons in a single micropulse are completely coherent. Since each pulse contains ~10^{12} photons and a few hundred micropulses, there are 10^9 indistinguishable ("degenerate") photons in the coherence volume. In comparison, the degeneracy parameter $D \approx 0.03$ at a synchrotron source with λ = 0.1nm. Because of the large transverse coherence area of 70×0μm^2 and the large number of coherent photons per pulse, an interference ("speckle") pattern can be recorded with a single XFEL pulse [44]. Therefore, X-ray FELs can be used not only to probe the structure of matter down to the size of an atom, but also to take 'snapshots' of the motion of atoms and molecules [45].

5. SUMMARY

The main "bottleneck" limiting the beam power in circular machines is caused by space charge effects that produce beam instabilities. Such a "bottleneck" exists at the J-PARC proton synchrotron complex, and is also intrinsic to the "proton drivers" envisaged at CERN and Fermilab. In order to maximally increase the beam power of a "proton driver", it is proposed to build a facility consisting of a low-energy injector linac (PI) and a high-energy pulsed superconducting L-band linac (SCL). The 2.5 GeV PI, based on the European Spallation Source (ESS) linac, would serve both as an injector to the SCL and a source of proton beams that could be used to copiously produce muons and "cold" neutrons. Protons accelerated by the SCL to 20 GeV would be transferred through the KEK Tristan ring in order to create neutrino, kaon and muon beams for fixed-target experiments. At a later stage, a 70 GeV proton synchrotron could be installed inside the Tristan ring. The proposed facility would be constructed using the existing KEK accelerator infrastructure and the most advanced linac technologies currently available.

Experiments with high-intensity neutrino beams are designed primarily to explore the mass spectrum of the neutrinos and their properties under the CP symmetry. To search for CP violation in neutrino oscillations, a 100 kt water Cherenkov detector could be built at Okinoshima, located at a distance $L_2 \approx 650$ km from KEK. Using the proposed KEK "proton driver", the detector at Okinoshima and the existing 0.022 Mt Super-Kamiokande detector (placed $L_1 \approx 300$ km away from KEK), the neutrino mass hierarchy could be determined either by comparing the ν_e appearance probabilities measured at the two vastly different baseline lengths L_1 and L_2, or by measuring at L_1 and L_2 the neutrino energy of the first oscillation maximum. Once the mass hierarchy is determined, the CP-violating phase in the mixing matrix can be measured with a precision of ± 20, assuming that 2.5×10^{21} protons are delivered on target for both ν_e and $\overline{\nu}_e$ beams.

Some of the most important discoveries in particle physics emerged from high-precision studies of K mesons ("kaons"), in particular neutral kaons. A deeper insight into CP violation is expected to be gained from measurements of ultra-rare kaon decays such as $K_L^0 \to \pi^0 \nu\overline{\nu}$ and $K_L^+ \to \pi^+ \nu\overline{\nu}$. These decays provide important information on higher-order effects in electroweak interactions, and therefore can serve as a probe of new phenomena not predicted by the Standard Model.

A unique feature of the proposed facility is the use of superconducting ILC-type cavities to accelerate both protons and electrons, which considerably increases its physics potential. Polarized electrons and positrons can be employed to study the structure of composite particles and the dynamics of strong interactions, as well as to search for new physics beyond the Standard Model.

An SCL-based X-ray free-electron laser (XFEL) and a synchrotron light source for applications in materials science and medicine are also envisaged. The ultrashort pulse duration of an XFEL matches the timescale of non-equilibrium microscopic processes, allowing the dynamics of atoms and molecules to be recorded in the form of femtosecond "movies".

ACKNOWLEDGEMENTS

For useful comments and suggestions concerning various aspects of this work, I am grateful to K. Fujii, K. Hagiwara, T. Higo, E. Kako, Y. Kamiya, K. Oide, K. Takayama and K. Tokoshuku. I wish to express my special gratitude to Kaoru Yokoya for his invaluable help and encouragement.

This paper is an abridged version of KEK Preprint 2014-35 (2014), which was subsequently published as an arXiv e-print (arXiv:1411.4874).

REFERENCES

1. Aad, G., et al. (2012) Observation of a New Particle in the Search for the Standard Model Higgs Boson with the ATLAS Detector at the LHC. Physics Letters B, 716, 1-29. https://doi.org/10.1016/j.physletb.2012.08.020
2. Chatrchyan, S., et al. (CMS Collab.) (2012) Observation of a New Boson at a Mass of 125 GeV with the CMS Experiment at the LHC. Physics Letters B, 716, 30-61. https://doi.org/10.1016/j.physletb.2012.08.021
3. 3. Yano, A., et al. (2004) The Toshiba E3736 Multi-Beam Klystron. Proceedings of LINAC 2004, Lubeck, Germany, 706-708.
4. Gelvich, E., et al. (1993) The New Generation of High-Power Multiple-Beam Klystrons. IEEE Transactions on Microwave Theory and Techniques, 41, 15-19https:// doi.org/ 10.1109/22.210224
5. Yamamoto, A. (2009) Superconducting RF Cavity Development for the International Linear Collider. IEEE Trans. Appl. Supercond., 19, 1387-1393
6. Ostroumov, P.N. (2006) Physics Design of the 8 GeV H-Minus Linac. New Journal of Physics, 8, 281. https://doi.org/10.1088/1367-2630/8/11/281
7. Reschke, D. (2013) Infrastructure, Methods and Test Results for the Testing of 800 Series Cavities for the European XFEL. Proceedings of SRF 2013, Paris, France, 812-815.
8. Faircloth, D. (2013) Ion Sources for High-Power Hadron Accelerators. rXiv:1302.3745
9. Kapchinsky, I.M. and Tepliakov, V.A. (1970) Linear Ion Accelerator with Spatially Homogenous Focusing. Pribory i Tekhnika Eksperimenta, 2, 19-22.

10. Vretenar, M. (2012) Low-Beta Structures. CERN Yellow Report 2011-007, 319-340. arXiv:1201.2593
11. Danared, H., Lindroos, M. and Theroine, C. (2014) ESS: Neutron Beams at the High-Intensity Frontier. CERN Courier, June 2014, 21-24.
12. Garoby, R., et al. (2013) Proton Drivers for Neutrino Beams and Other High Intensity Applications. Journal Physics: Conference Series, 408, 012016. https://doi.org/ 10.1088/1742-6596/408/1/012016
13. Baussan, E., et al. (2014) A very Intense Neutrino Super Beam Experiment for Leptonic CP Violation Discovery Based on the European Spallation Source Linac. Nuclear Physics B, 885, 127-149. https://doi.org/10.1016/j.nuclphysb.2014.05.016
14. Al-Binni, U., et al. (2013) Project-X, Physics Opportunities. arXiv:1306.5009v2
15. Dudek, J., et al. (2012) Physics Opportunities with the 12 GeV Upgrade at Jefferson Lab. arXiv:1208.1244v2
16. Voutier, E. (2014) Physics Perspectives at JLab with a Polarized Positron Beam. rXiv:1402.2814v1
17. Belusevic, R. (2008) Relativity, Astrophysics and Cosmology. Wiley-VCH Verlag., Weinheim
18. Shaposhnikov, M. (2009) Baryon Asymmetry of the Universe and Neutrinos. Progress of Theoretical Physics, 122, 185-203. https://doi.org/10.1143/PTP.122.185
19. Hagiwara, K., et al. (2012) Physics Potential of Neutrino Oscillation Experiment with a Far Detector in Oki Island along the T2K Baseline, KEK-TH-1568. arXiv:1209.2763v1
20. Back, J., et al. (2013) Particle Production and Energy Deposition Studies for the Neutrino Factory Target Station. Physical Review Accelerators and Beams, 16, 021001. https://doi.org/10.1103/PhysRevSTAB.16.021001
21. McDonald, K.T., et al. (2010) The MERIT High-Power Target Experiment at the CERN PS. Proceedings of First International Particle Accelerator Conference, Kyoto, Japan, 3527-3529.
22. Baussan, E., et al. (2013) Target, Magnetic Horn and Safety Studies for the CERN to Frejus Super Beam. Journal of Physics: Conference Series, 408, 012061. https://doi.org/10.1088/1742-6596/408/1/012061
23. Kamiya, Y. DC Neutrino Horn. The manuscript (in Japanese) has not been published.
24. Omori, T., et al. (2006) Efficient Propagation of Polarization from Laser Photons to Positrons through Compton Scattering and Electron-Positron Pair Creation. Physical Review Letters, 96, 114801. https://doi.org/10.1103/PhysRevLett.96.114801
25. Grames, J., et al. (2013) PEPPo: Using a Polarized Electron Beam to Produce Polarized Positrons. 15th International Workshop on Polarized Sources, Targets and Polarimetry, Charlottesville, Virginia, USA.
26. Zel'dovich, Y.B. (1959) Parity Nonconservation in the First Order in the Weak-Interaction Constant in Electron Scattering and Other Effects. Journal of Experimental and Theoretical Physics (JETP), 36, 964-966.
27. Derman, E. and Marciano, W.J. (1979) Parity Violating Asymmetries in Polarized Electron Scattering. Annals of Physics, 121, 147-180.
28. Czarnecki, A. and Marciano, W.J. (1996) Electroweak Radiative Corrections to Polarized Moller Scattering Asymmetries. Physical Review D, 53, 1066-1072. https://doi.org/10.1103/PhysRevD.53.1066
29. Mammei, J. (2012) The MOLLER Experiment. Nuovo Cimento, 35, 203-208.
30. Kumar, K.S., et al. (2013) Low-Energy Measurements of the Weak Mixing Angle. Annual Review of Nuclear and Particle Science, 63, 237-267. https://doi.org/10.1146/annurev-nucl-102212-170556
31. Belusevic, R. (1999) Neutral Kaons. Springer-Verlag, Berlin.
32. Brod, J., Gorbahn, M. and Stamou, E. (2011) Two-Loop Electroweak Corrections for the Decays. Physical Review D, 83, 034030. https://doi.org/10.1103/PhysRevD.83.034030
33. Belyaev, A., et al. (2001) Kaon Physics with a High Intensity Proton Driver, CERN-TH/2001-175. arXiv:hep-ph/0107046v2
34. Buras, A. (1999) CP Violation and Rare Decays of K and B Mesons. arXiv:hep-ph/9905437

35. Artamonov, A., et al. (2009) Study of the Decay in the Momentum Region 140 < Pπ < 199 MeV/c. Physical Review D, 79, 092004. https://doi.org/10.1103/PhysRevD.79.092004
36. Miller, J., de Rafael, E. and Roberts, B.L. (2007) Muon (g-2): Experiment and Theory. Reports on Progress in Physics, 70, 795-881. https://doi.org/10.1088/0034-4885/70/5/R03
37. Mibe, T. (2010) New g-2 Experiment at J-PARC. Chinese Physics C, 34, 745-748.
38. Schmuser, P., Dohlus, M. and Rossbach, J. (2008) Ultraviolet and Soft X-Ray Free-Electron Lasers. Springer-Verlag, Berlin.
39. Huang, Zh. and Kim, K. (2007) Review of X-Ray Free-Electron Laser Theory. Physical Review Accelerators and Beams, 10, 034801. https://doi.org/10. 1103/ PhysRevSTAB.10.034801
40. Amann, J., et al. (2012) Demonstration of Self-Seeding in a Hard-X-Ray Free-Electron Laser. Nature Photonics, 6, 693-698. https://doi.org/10.1038/nphoton.2012.180
41. Abela, R., et al. (2007) The European X-Ray Free-Electron Laser. Technical Design Report, DESY 2006-097.
42. Arthur, J. (1999) Prospects for an X-Ray FEL Light Source and Some Possible Scientific Applications, SLAC-PUB-8276.
43. Agapov, I., et al. (2014) SASE Characteristics from Baseline European XFEL Undulators in the Tapering Regime. Proceedings of FEL 2014, Basel, Switzerland, 159-163.
44. Pellegrini, C. and Stohr, J. (2003) X-Ray Free-Electron Lasers—Principles, Properties and Applications. Nuclear Instruments and Methods in Physics Research Section A: Accelerators, Spectrometers, Detectors and Associated Equipment, 500, 33-40. https://doi.org/10.1016/S0168-9002(03)00739-3
45. Barty, A. (2010) Time-Resolved Imaging Using X-Ray Free Electron Lasers. Journal of Physics B: Atomic, Molecular and Optical Physics, 43, 194014. https://doi.org/ 10.1088/0953-4075/43/19/194014

NOTES

[1]The Q factor of an rf cavity is defined as $Q = 2\pi\pi'$ (energy stored/energy dissipated per cycle). For large values of Q, the Q factor is approximately the number of oscillations required for the energy of a freely oscillating system to fall off to $e^{-2\pi}$, or 0.2%, of its original value.

[2]The main reason for using these π/2-mode structures is that long chains of coupled cavities are often required for an efficient use of high-power rf sources [10].

[3]A much higher level of polarization can be obtained by using a magnetic field to align the particle spins [37].

[4]The occurrence of higher harmonics is explained in [38]. In the forward region ($\theta = 0$) of a planar undulator, only the odd higher harmonics are observed, while the off-axis radiation contains also the even harmonics. For the European XFEL, simulations predict that the relative contribution to the total radiation power of the 3rd and the 5th harmonic is about 1% and 0.03%, respectively [41]

CHAPTER 4

Remarks on global sub-Lorentzian geometry

Grochowski, M. Anal.Math.Phys.

Faculty of Mathematics and Natural Sciences, Cardinal Stefan Wyszy´nski University, ul. Dewajtis 5, 01-815 Warsaw, Poland

ABSTRACT

This paper aims at being a starting point for the investigation of the global sub-Lorentzian (or more generally sub-semi-Riemannian) geometry, which is a subject completely not known. We present an adaptation to the fiber bundles setting (and state its elementary proof in the real analytic case) of the decomposition result for indefinite metrics. Using it and the result by Sussmann on global spanning of distributions, we state and prove some global facts on sub-Lorentzian manifolds. We also construct a global control system describing nonspacelike curves and study its controllability properties in case of compact manifolds.

Keywords: Vector bundles, Indefinite metrics on vector, bundles Distributions

1 INTRODUCTION

The motivation of this note is the lack of global theorems in the sub-Lorentzian or more generally sub-semi-Riemannian geometry. A sub-semi-Riemannian manifold is, by definition, a triplet (M, H, g) where M is a smooth (smooth means of class C in this paper) connected and paracompact manifold, H is a smooth bracket generating vector distribution of constant rank on M, rank $H < \dim M$, and g is a semiRiemannian metric on H (H is bracket generating if for every $x \in M$ there exists a basis $X_0, \dots, X_k$ of H defined near x such that the fields $X_0, \dots, X_k$ and all their successive commutators evaluated at x span the whole tangent space $T_x M$). A couple (H, g) is called a sub-semi-Riemannian metric or structure on M. The index of a metric g at a point x is the maximal dimension of all linear subspaces in H_x on which $gx = g|H_x \times H_x$ is negative definite. As in the semi-Riemannian geometry [23] the index is independent of a base point x.If the index of g is equal to 1 then g is a Lorentzian metric, (H, g) is called a sub-Lorentzian metric on M, and (M, H, g) is a sub-Lorentzian manifold. In the sub-semi-Riemannian geometry we can ask similar question and try to solve similar problems as in the classical semi-Riemannian geometry. There have appeared lately several papers concerning the local sub-Lorentzian geometry— see e.g. [12–14], showing some significant differences between the Lorentzian and sub-Lorentzian case. There are also papers treating global problems but only in flat cases— see [8,17,18]. The problem is the little understanding of global phenomena that occur in the sub-geometries on general manifolds. For instance, as far as the author of this paper is concerned, there is only one known global fact from the general sub Lorentzian geometry proved in [10], and saying that any two causally related points on a time-oriented globally hyperbolic sub-Lorentzian manifold can be joined by a length maximizing nonspacelike future directed curve.

The main problem in proving global theorems is the fact that the appropriate tools are missing. The aim of this paper is twofold. First is to draw the reader's attention to two results (Theorems 1.1, 1.2) which can serve as a starting point for global studies. Examples of global facts that can be obtained in this way may be found in Sect. 3. Before we state the mentioned theorems let us fix some notation. If $\pi: E \to M$ is a vector bundle and x ⍰ M then by E_x we denote the fiber of E over x, i.e. $E_x = \pi^{-1}(x)$

Theorem 1.1 Let $\pi : E \to M$ be a smooth vector bundle. Suppose that there exists a smooth fiber metric g of index l on E. Then E can be represented in the form of the Whitney sum $E = E^- \oplus E^+$ of smooth subbundles, where E^- is of rank l and the restriction of g to E^- is negative definite, while the restriction of g to E^+ is positive definite. Moreover the sum can be chosen to be orthogonal, i.e. for each $x \in M$ the subspaces E_x^- and E_x^+ x are gx - orthogonal.

The decomposition $E = E^- \oplus E^+$ is not unique. The converse statement to Theorem 1.1 is of course true. If a (smooth) vector bundle admits a decomposition $E = E^- \oplus E^+$, where E^- is of rank l, then clearly there exists a metric of index l on E which is negative definite on E^-.

If E is equipped with an indefinite fiber metric g, then a vector v $\in E_x$ is called timelike if $g_x(v, v) < 0$ and is called spacelike if $g_x(v, v) > 0$.

Theorem 1.2 Let $\pi : E^{-}\rightarrow M$ be a smooth vector bundle of rank k over an ndimensional manifold M. Suppose that there exists a fiber metric g of index l on E. Then there exist global sections $\sigma_1,...,\sigma_d$, $\tau_1,...,\tau_e : M \rightarrow E$, where $d \leq l(n + 1)$, $e \leq (k - l)(n + 1)$, and such that for every $x \in M$

1) each of the vectors $\sigma_1(x), \ldots, \sigma_d(x)$ is either timelike or equal to zero,
2) each of the vectors $\tau_1(x), \ldots, \tau_e(x)$ is either spacelike or equal to zero,
3) $gx(\sigma_i(x), \tau_r(x)) = 0$, $i = 1,..., d$, $r = 1,...,$ e, and
4) $E_x = \text{Span}\{\sigma_1(x), \ldots, \sigma_d(x), \tau_1(x), \ldots, \tau_e(x)\}$.

This theorem is a corollary of Sussmann's theorem on global spanning of distributions [25] and the decomposition result from Theorem 1.1.

Suppose that (M, H, g) is a sub-Lorentzian manifold. A curve γ : [a, b] $\rightarrow M$ is called horizontal if it is absolutely continuous and $\gamma(t) \in H_{\gamma}(t)$ a.e. on $[a, b]$. By a time orientation of (H, g) we mean a continuous timelike vector field on M. Suppose that (M, H, g) is time-oriented by a vector field X. A horizontal curve $\gamma : [a, b] \rightarrow M$ is said to be nonspacelike future directed if $g(\dot{\gamma}(t), \dot{\gamma}(t)) \leq 0$ and $g(X(\dot{\gamma}(t)), \dot{\gamma}(t)) < 0$ a.e. on $[a, b]$. Nonspacelike curves on a given (sub-) Lorentzian manifold are responsible for its causal structure, so it is of highest importance to have as precise description of nonspacelike curves as possible. On certain time-oriented sub-Lorentzian manifolds (for instance on all compact), Theorem 1.2 facilitates this task, namely it allows one to construct a global affine control system having as its trajectories a class of nonspacelike future directed curves. This system can be written as

$$\dot{x} = \sigma + \sum_{r=1}^{e} u_r \tau_r \quad , \tag{1.1}$$

where σ and τ_r's are taken from Theorem 1.2, σ is a time orientation, the set of control parameters is equal to

$$\mathcal{U} = \left\{ u \in \mathbb{R}^e : \sum_{r=1}^{e} u_r^2 \leq \frac{1}{R} \right\}$$

$R = \max \{|c^{+}_{rs}(x)| :$ $x \in M$, $r,s = 1,..., e\}$, and $c^{+}_{rs}(x) = g_x(\tau_r(x), \tau_s(x))$, $r,s = 1,..., e$— all details are contained in Sect. 3 below. As we said, the set of all trajectories of the system (1.1) is contained in the class of all nonspacelike future directed curves on (M, H, g), and in some particular cases the two sets of curves are equal. The second aim of this note is to prove some results concerning controllability of the system (1.1). In order to state the result we need more definitions. So, if M is a manifold and H is a distribution of constant rank on M, then by $H^{\perp}$ we denote the annihilator of H i.e. $H^{\perp} = \{\lambda \in T^*M : \lambda, \xi = 0$ for every $\xi \in H_{\pi(\lambda)}\}$; here $\pi : T^*M \to M$ is the canonical projection. Now, a horizontal curve $\gamma : [a, b] \to M$ is called abnormal if there exists an absolutely continuous lift to the cotangent bundle (an abnormal lift) $\lambda : [a, b] \to T^*M$ (i.e. $\gamma = \pi \circ \lambda$) such that $\lambda([a, b]) \subset H^{\perp}$, $\lambda(t) = 0$ for every $t \in [a, b]$, and finally $\Omega_{\lambda(t)} (\lambda(t), \zeta) = 0$ for almost every $t \in [a, b]$ and every $\zeta \in T_{\lambda(t)}H^{\perp}$, where Ω is the restriction to $H^{\perp}$ of the standard symplectic structure on T^*M. An abnormal curve γ: $[a, b] \to M$ is a Goh curve if it has an abnormal lift $\lambda(t)$that additionally satisfies $\lambda([a, b]) \subset (H^2)^{\perp}$ (this is the so-called Goh condition), where H^2 is the distribution spanned by the fields tangent to H and their commutators of first order. Goh condition is a second-order necessary condition for optimality. Goh curves appear in the sub-Riemannian and sub-Lorentzian geometry, and often their presence is accompanied by some singular phenomena (cf. e.g. [1,14,20]). The second definition that we need is as follows. Let U be a subset of R^k . A control system (Σ) $\dot{x} = f(x, u)$, $x \in M$, $u \in U$, is called controllable if for every $x, y \in M$ there exists a trajectory of (Σ) joining x to y. (Σ) is controllable at a point x if there exists a neighbourhood U of x such that (Σ) is controllable on U; such a neighbourhood is called *a* controllable neighbourhood.

Theorem 1.3 Suppose that (M, H, g) is a compact time-oriented sub-Lorentzian manifold. Then the induced control system (1.1) has closed trajectories, among which there are closed timelike curves. Moreover, at every point belonging to any closed timelike curve which is a trajectory of (1.1) and which is not a Goh curve with respect to H, the system is locally controllable. More precisely, every closed timelike trajectory of (1.1), that is not a Goh curve, possesses a controllable neighbourhood for (1.1).

The analogous reasoning leads to the following similar but a little different in its character result.

Theorem 1.4 Suppose that (M, H, g) is a compact time-oriented sub-Lorentzian manifold. Then there exist closed timelike curves on M. Any such curve that is not *a* Goh curve (with respect to H) possesses a neighbourhood U with the following property: for every $x, y \in U$ there exists a timelike future directed curve contained in U and joining x to y.

Closed timelike curves from Theorem 1.4 need not be trajectories of (1.1). The method of the proofs of Theorems 1.3, 1.4 presented in this paper slightly generalizes considerations from [22], as it will be explained at the end of this paper. Theorem 1.4 has an immediate corollary:

Corollary 1.1 Let (M, H, g) be a time-oriented sub-Lorentzian manifold of step two, i.e. $H_2 = T M$. Then any closed timelike curve possesses a neighbourhood described in the assertion of Theorem 1.4.

The paper is organized as follows. In Sect. 2 we discuss proofs of Theorems 1.1 and 1.2. In Sect. 3 we recall basic notions from the sub-Lorentzian geometry. Then we present some results concerning global sub-Lorentzian geometry; part of them are corollaries of the mentioned theorems. In Sect. 4 we prove Theorems 1.3, 1.4 and show their possible generalizations.

2 PROOFS OF THEOREMS

1.1, 1.2 In this section all objects are smooth or real analytic. Let $\pi : E \to$ M be a rank k bundle. By a fiber metric on E we mean a section g of the bundle $E^{*}\otimes E^{*} \to M$ such that $g_x : E_x \times E_x \to$ R is a nondegenerate symmetric bilinear form for every $x \in M$. If the index of g, ind(g), is 0 then g is called a Riemannian metric, if ind(g) = 1, g is called Lorentzian. Generally, g is called semi-Riemannian if the index is greater than 0 and less than rank E. It is well known that Riemannian metrics exist on every vector bundle. In a smooth case it follows from the existence of partitions of unity, in a real analytic case from the imbedding theorem—see [9]. The existence of semiRiemannian metrics is a more subtle thing. Adapting for instance arguments from [24] to the fiber bundles setting, one can prove that: there exists a metric of index l on a vector bundle E $\to M$ if and only if E possesses a rank l subbundle $L \to M$. This theorem however does not say anything about the behaviour of the metric on L, Global sub-Lorentzian geometry 299 and from the point of view of applications one needs a more precise statement—such as Theorem 1.1.

Theorem 1.1 is known for Lorentzian metrics on manifolds, i.e. when E is a tangent bundle TM and ind(g) = 1—see e.g. [5]. It is also known for general semiRiemannian manifolds. The proof may be found in [3] and relies on the following observation. The presence on M of a metric of index l is equivalent to the existence of $O(l, n)$-structure in the bundle LM of linear frames over M.In turn, the structure group $O(l, n)$ can be reduced to its maximal compact subgroup $O(l)\times O(n-l)$(reduction to a maximal compact subgroup is always possible). Denote by $(P, M, O(l) \times O(n - l))$ the reduced bundle. Now we use properties of universal bundles. If ω_G denotes the universal principle G-bundle $EG \to BG$, then it is known that the universal principle $O(l) \times O(n - l)$-bundle $\omega_{O(l)\times O(n-l)}$ is isomorphic to the product bundle $\omega_{O(l)} \times \omega_{O(n-l)}$. Now, if $f : M \to B\,O(l) \times B\,O(0 - l)$ is the classifying map for P, then $P = P_1 \oplus P_2$, where $P_1 = (pr_1 \circ f)^{*}\omega_{O(l)}$, $P_2 = (pr_2 \circ f)^{*}w_{o}(n-l)$. The decomposition of P results in the decomposition of the tangent bundle $TM = \xi_1 \oplus \xi_2$sa, where $\xi_1 = P_1 \times O(l)$ R^l and $\xi_2 = P_2 \times_{O(n-l)} R^{n-l}$ (i.e. for instance ξ_1 is a vector bundle with typical fiber R^l , associated to P_1). It remains to check that the metric g is negative on ξ_1 and positive on ξ_2.

In the fiber bundle setting the proof might go along similar lines. Instead of this we will present elementary and fully constructive proof of real analytic version of Theorem 1.1.

2.1 Proof of Theorem 1.1 in a real analytic case

During the proof all objects are supposed to be real analytic. Let $\pi : E \to M$ be a rank k vector bundle equipped with a fiber metric g of index l. Introduce a Riemannian fiber metric h on E. For each point $x \in M$ we can define a linear mapping $f_x : E_x \to$ Ex by formula

$$g_x(v, w) = h_x(f_x(v), w) \tag{2,1}$$

for every v, $w \in E_x$. Fix a point x_0 and lets$_1,\ldots,s_k$: $U \to E$ be a local basis of sections of E defined on a neighbourhood U of x_0. If $g_{ij}(x) = g(s_i(x),s_j(x))$, $hj(x) = h(s_i(x),s_j(x))$, and $(h^{ij}(x)$ is the inverse matrix to $(h_{ij}(x)$ then one can easily make sure that

$$f_x(v) = \sum_{\beta}\sum_{i.j} h^{\beta j}(x) g_{ij}(x) v^i s_\beta(x) \tag{2.2}$$

where $v = \sum v^i s_i(x) \in E_x$. By (2.2) it is seen that fx depends analytically on x. Next, using (2.1) it is seen that

$$h_x(f_x(v), w) = h_x(v, f_x(w))$$

which means that there exists a basis $v_1(x), \ldots, v_k(x)$ of eigenvectors for fx which is orthonormal with respect to h. Clearly such a basis is orthogonal with respect to g.

Let $\lambda_1(x), \ldots, \lambda_k(x)$ denote the corresponding eigenvalues. Since $g(v_i(x), v_i(x)) = h(f_x(v_i(x)), v_i(x)) = \lambda_i(x)$ and ind$(g) = l$, it means that, possibly after reordering $\lambda_i(x)$'s, we have

$$\lambda_1(x), \ldots, \lambda_l(x) < 0, \quad \lambda_{l+1}(x), \ldots, \lambda_k(x) > 0$$

for $x \in U$. Finally, for $x \in U$ let us set

Notice that the spaces E_x^-, E_x^+ have an invariant meaning for f_x, i.e. they do not depend on the choice of the basis $v_1(x), \ldots, v_k(x)$, and moreover E_x^+ is the g_x - orthogonal complement of E_x^-. To end the proof we must show that the spaces E_x^-, E_x^+ depend analytically on x in a neighbourhood of x_0

Let

$$P(x;\lambda) = \det(f_x - \lambda Id) = (-1)^k(\lambda - \lambda_1(x))\ldots(\lambda - \lambda_k(x)). \tag{2.4}$$

The case $l = 1$ is easy, because then $\lambda_1(x)$ is a root of multiplicity 1, and hence, by implicit function theorem, is analytic (or smooth in a smooth category). Consequently, E_x^- = ker $(f_x - \lambda_1(x)Id)$ is an analytic subbundle.

Now suppose that $l > 1$. $P(x; \lambda)$ obviously is a pseudopolynomial, i.e. a polynomial with respect to the powers of λ with coefficients being analytic functions of x. Without changing the set of roots of P we may assume that the discriminant of P, which is an analytic function of x, does not vanish identically [21] in a neighbourhood of x_0 (we just get rid of multiple factors). If x_0 does not belong to the discriminant set of P, then clearly roots are of multiplicity 1, so they are real analytic functions around x_0.

Suppose then that x_0 belongs to the discriminant set of P. Consider a complexification of (2.4), i.e. a pseudopolynomial $P(z; \lambda)$ where $z \in \tilde{U}$, $\tilde{U}$ is an open subset of C^n such that $\tilde{U} \cap R^n = U$, $\lambda \in C$. From now on we suppose a neighbourhood U (resp. $\tilde{U}$) of x_0 to be sufficiently small. Using continuity of roots of pseudopolynomials (see [21]) and (2.3), we can find an $\varepsilon > 0$ such that

$$\lambda_1(z),\ldots,\lambda_l(z) \in \{\lambda \in \mathbb{C} : \mathrm{Re}\,\lambda < -\varepsilon\}, \quad \lambda_{l+1}(z),\ldots,\lambda_k(z) \in \{\lambda \in \mathbb{C} : \mathrm{Re}\,\lambda > \varepsilon\} \tag{2.5}$$

for $z \in \tilde{U}$. This proves that $P(z; \lambda)$ is reducible. Indeed, by [6], all roots of an irreducible pseudopolynomial $H(z, \lambda) = \lambda_k + A_1(z)\lambda^{k-1} + \cdots + A_k(z)$, $z \in \tilde{U}$, form a closed system, that is to say any root can be obtained from any other root by analytic continuation in $\tilde{U}$. In our case by (2.5) this cannot be true. Let

$$P(z;\lambda) = H_1(z;\lambda)\ldots H_m(z;\lambda) \tag{2.6}$$

be a decomposition into irreducible pseudopolynomials. Using what we have said above, after suitable reordering of factors in (2.6) we are led to the situation where for a certain number s, $1 \le s \le l$, the pseudopolynomial $H_1(z; \lambda) \ldots H_s(z; \lambda)$ has only negative roots when restricted to U while

$H_{s+1}(z;\lambda)\ldots H_m(z;\lambda)$ has only positive roots when restricted to U. In this way we have come to the conclusion that $\{\lambda_1(x),\ldots,\lambda_l(x)\}$ is the set of all roots of a real pseudopolynomial. In particular any function of the form $x\to G(\lambda_1(x),\ldots,\lambda_l(x))$, where $G(y_1,\ldots,y_l)$ is analytic and symmetric, is analytic in a neighbourhood of $x0$. Now, the observation

$$E_x^- = \ker\left((f_x-\lambda_1(x)Id)\circ\ldots\circ(f_x-\lambda_l(x)Id)\right)$$

$x\in U$, ends the proof, since $(f_x-\lambda_1(x)Id)\circ\ldots\circ(fx-\lambda_l(x)Id)$ is an analytic vector bundle morphism of constant rank. The decomposition in the hypothesis of Theorem 1.1 is of course not unique—it essentially depends on the choice of h.

2.2 Proof of Theorem 1.2

Having in mind the decomposition result, now it is enough to apply to $E^\pm$ the result by Sussmann [25]: Let $E\to M$ be a (in general singular) subbundle of the vector bundle $V\to M$ Let $r=\max_{x\in M}\dim E_x$. Then there exist global sections $s_1,\ldots,s_m : M\to E$, $m\le r(n+1)$, such that $E_x=\mathrm{Span}\{s_1(x),\ldots,s_m(x)\}$ for every $x\in M$.

3 APPLICATIONS TO THE GLOBAL SUB-LORENTZIAN GEOMETRY

For all details and facts concerning the sub-Lorentzian geometry the reader is referred to [12] and its reference section, and also to [17]. Let (M,H,g) be a fixed subLorentzian manifold. Recall that a time orientation of (H,g) is, by definition, a continuous timelike vector field on M (not every sub-Lorentzian metric admits a time orientation). Suppose our (M,H,g) to be time-oriented by a field X. A time orientation permits to divide all non-zero nonspacelike vectors into two disjoint classes, namely a nonspacelike $v\in Hq$ is called future directed if $g(v,X(q))<0$, and is called past directed if $g(v,X(q))>0$. Now, an absolutely continuous curve $\gamma:[a,b]\to M$ is called nonspacelike (future, past directed) if $\gamma\,(t)$ is a nonspacelike (future, past directed) a.e. on $[a,b]$. Analogously one defines timelike and null (future, past directed) curves. Such curves can be applied to the study of affine control systems—see [15]. For a fixed point $x_0\in M$ and its neighbourhood U we will denote by $J^+(x_0,U)$ the (future) nonspacelike reachable set from x_0 which, by definition, is equal to the set of all points $x\in U$ that can be reached from x_0 by a nonspacelike future directed curve contained in U. If we replace the word 'nonspacelike' with the word 'timelike' (resp. 'null') then we obtain the definition of the (future) timelike (resp. null) reachable set from x_0 which are denoted respectively by $I^+(x_0, U)$ and $N^+(x_0, U)$. In order to ensure better properties of reachable sets we assume that U is a normal neighbourhood of x_0, that is to say if we extend, around x_0, a metric g to a Lorentzian metric $\tilde{g}$, then U is a normal convex neighbourhood of x_0 with respect to $\tilde{g}$ and its closure $\bar{U}$ is contained in some other normal convex neighbourhood of 302 *M.* Grochowski x_0 with respect to $\tilde{g}$; see also [13] for a constructive definition of normal neighbourhoods. Now, if U is a normal neighbourhood of x_0, then $J^+(x_0, U)$ is closed with respect to U, $J^+(x_0, U) = cl_U(\text{int } I^+(x_0, U)) = \text{clU}(\text{int } N^+(x_0, U))$, and finally $\partial U J^+(x_0, U) = \partial U\, I^+(x_0, U) = \partial U\, N^+(x_0, U)$, where cl_U is the closure and ∂_U is the boundary, both taken with respect to U—see [12] for proofs. Replacing 'future directed' by 'past directed' we obtain dual notions, for instance in an obvious manner we define $J^-(x_0, U)$, the (past) nonspacelike reachable set from x_0. In case $U = M$ we use the notation $I^+(x) = I^+(x, M)$, $J^+(x) = J^+(x, M)$, $I^-(x) = I^-(x, M)$ etc. If $S \subset M$ is a set then we define $I^+(S) = \bigcup_{x \in S} I^+(x)$, and similarly for $J^+(S)$, $I^-(S)$, $J^-(S)$.

Using Theorem 1.1 we obtain the following

Proposition 3.1 Let (M, H, g) be a sub-Lorentzian manifold. Moreover let $H = H^- \oplus H^+$ be the decomposition from Theorem 1.1. Then (H, g) is time-orientable if and only if H^- is orientable as a vector bundle. Proof Since a rank 1 vector bundle is orientable if and only if it is trivial, orientability of H^- implies time-orientability of (H, g). Now suppose that (H, g) is time-orientable and let $X : M \to H$ be a time orientation. Denote by $P_x : H_x \to H_x^-$ the orthogonal (in sense of gx) projection. The formula $Y(x) = P_x(X(x))$ defines a smooth global non-vanishing vector field with values in H^-, hence H^- is an orientable vector bundle.

Since every bundle over a simply-connected base is orientable we have

Corollary 3.1 Every simply-connected sub-Lorentzian manifold is time-orientable. Remark 3.1 Note in this place that in the similar manner we can speak about timeorientability of sub-semi-Riemannian manifolds (cf. [19]; see [23] for the semiRiemannian case). To be more precise, if (M, H, g) is a sub-semi-Riemannian manifold, and $H = H^- \oplus H^+$ is the corresponding decomposition from Theorem 1.1, then we can say that (M, H, g) is time orientable if H^- is an orientable vector bundle. It is easy to see that such a notion does not depend on the particular choice of the decomposition. Indeed, if $H^- \oplus H^+ = \tilde{H}^- \oplus \tilde{H}^+$ are two decompositions, then the orthogonal projection $P : H \to H-$ defines a vector bundle isomorphism $P_{|\tilde{H}} : \tilde{H} \to H$. Now, as a corollary we have: every simply-connected sub-semi-Riemannian manifold is time-orientable.

Next immediate consequence of Theorem 1.1 is the following result

Proposition 3.2 Every simply-connected sub-Lorentzian manifold (M, H, g) with dim $M = 3$ is parallelizable. Proof Indeed, $TM = H \oplus L$, where L is a suitable rank 1 subbundle of TM. Now $TM = H^- \oplus H^+ \oplus L$ where, due to our assumptions, H^+, H^+, and L are rank 1 trivial bundles.

Thus, for instance, sub-Lorentzian metrics do not exist on the manifold R × S^2, although Lorentzian ones do exist. Let us state another fact that easily follows from Theorem 1.1.

Proposition 3.3 Let (M, H, g) be a simply-connected sub-Lorentzian manifold. If rank $H = 2$ then (H, g) admits a global basis.

Proof Pick a time orientation X. Then H = Span $\{X\} \oplus H^+$ where, by assumption, H^+ is trivial.

fact from control theory that can be found for instance in [16], Proposition 1.1 (or in any textbook devoted to the subject

Lemma 3.1 Suppose that we are given a control system (Σ) $\tilde{x} = f(x, u)$, $u \in U$, on a manifold M. Let $x \in M$ and denote by $A(x)$ the reachable set from x for (Σ). Finally, let $\gamma : [0, T] \to$ M be an arbitrary trajectory of (Σ) such that $\gamma(0) = x$. If $\gamma(t_0) \in$ int $A(x)$ for a $t_0 \in [0, T)$ then $\gamma(t) \in$ int $A(x)$ for all $t \in (t_0, T]$

Next we present some facts concerning reachable sets which are well-known in the Lorentzian geometry.

Proposition 3.4 Let $x \in M$. If $y \in$ int $I^+(x)$ then there exists a timelike future directed curve $\gamma : [0, 1] \to M$ such that $\gamma(0) = x$, $\gamma(1) = y$, $\gamma((0, 1]) \subset$ int $I^+(x)$. Proof The result is by no means obvious since the structure of the sets $I^+(x)$ from the global point of view is not known even in the Lorentzian case; in addition, in the sub-Lorentzian case the sets $I^+(x)$ need not be open. We come to the proof now. It is known by Krener's theorem (see e.g. [1]) that the sets int $I^+(x)$, int $I^-(x)$ are non-empty for every x. Moreover, it follows e.g. from [10] that if U is a normal neighbourhood of x then the set of points $z \in I^+(x, U)$ that can be reached from x by a timelike future directed curve that is not abnormal, is dense and open in $I^+(x, U)$. Let $\eta : [0, 1] \to M$ be a timelike future directed curve such that $\eta(0) = x$, $\eta(1) = y$. Take a normal neighbourhood U of x. Then $\eta(t_1) \in U$ for a $t_1 \in (0, 1)$. Next there exists a pointz $\in U \cap I^-(\eta(t_1))$that can be reached from x by a timelike future directed curve which is not abnormal. Denote this curve by σ. Since every timelike curve starting from x and contained in $\partial_U I^+(x, U)$ is abnormal (it is enough to apply Pontriagin maximum principle and the definition of abnormal curves—see [12]), it follows that $\sigma \setminus \{x\} \subset$ int $I^+(x, U)$. To sum up, there exists a timelike future directed curve joining x to y such that its initial segment is contained in int $I^+(x, U) \subset$ int $I^+(x)$. To terminate the proof we use Lemma 3.1.

Note that unlike the Lorentzian case a version of Proposition 3.4 for the sets N+(x) is not true. Consider as an example the Heisenberg sub-Lorentzian metric on R3 which is defined by supposing the fields $X = \frac{\partial}{\partial x} + \frac{1}{2}y\frac{\partial}{\partial z}$, $Y = \frac{\partial}{\partial y} - \frac{1}{2}x\frac{\partial}{\partial z}$ to be an orthonormal frame with the time orientation X. As it was computed in [11], $N^0(0) = J^+(0) = \{\eta_1 \leq 0, x \geq 0, z \geq 0\}$ ⍰ $\{\eta_2 \leq 0, x \geq 0, z \leq 0\}$ with $\eta_1 = z - \frac{1}{4}(x^2 - y^2)$ $\eta_2 = -z - \frac{1}{4}(x^2 - y^2)$, and as it is not difficult to check, int $N^+(0)$ contains no null future directed curves initiating at 0—an initial segment of any such curve must be contained in $\partial N^+(0)$

Proposition 3.5 On every compact time-oriented sub-Lorentzian manifold there exists a closed timelike future directed curve. 304 M. Grochowski Proof See e.g. [2], page 58 for the proof in the Lorentzian case. In our case, consider the family of open (non-empty) sets of the form int $I^+(x, U)$ where $x \in U$ and U is a normal neighbourhood of x. First let us notice that this family is indeed a covering of M. To this end take a point $x \in M$. Let us issue a timelike past directed curve $\sigma(t)$ from x, $\sigma(0) = x$, which is not abnormal. Denote by U a normal neighbourhood of x. Supposing that $t > 0$ is sufficiently small we can assume that U is a normal neighbourhood of $\sigma(t)$—this follows from the construction of normal neighbourhoods in the semi-Riemannian case. Now, the reasoning from the last part of the proof of Proposition 3.4 and Lemma 3.1 show that $x \in$ int $I^+(\sigma(t), U)$

Chose a finite subcovering int $I^+(x_1, U_1), \ldots,$ int $I^+(x_m, U_m)$. Then $x_1 \in$ int $I^+(x_{i1}, U_{i1})$ for a certain $i_1 \in \{1,\ldots m\}$. Next $x_{i1} \in$ int $I^+(x_{i2}, U_{i2})$, again $i_2 \in \{1,\ldots m\}$ etc. In this way we obtain a sequence of points $x_{i1}, x_{i2}, \ldots, x_{ik}, \ldots$ but since there is only a finite number of distinct points, repetitions must occur, i.e. we will have a situation like this: $x_{ik} \in$ int $I^+(x_{ik+1}, U_{ik+1})$, $x_{ik+1} \in$ int $I^+(x_{ik+2}, U_{ik+2}), \ldots, x_{ik+l-1} \in$ int $I^+(x_{ik+l}, U_{ik+l})$, $x_{ik+l} \in$ int $I^+(x_{ik}, U_{ik})$ for certain k and l. So x_{ik+l} can be joined from x_{ik} by a timelike future directed curve contained in U_{ik} ; x_{ik+l-1} can be joined from x_{ik+l} by a timelike future directed curve in $U_{ik+l}, \ldots$, and finally xik can be joined from x_{ik+1} by a timelike future directed curve in U_{ik+1} . The proof is over.

In the proof of Proposition 3.5 we could cover M with sets of the form int $I^+(x)$, $x \in M$. However, coverings by sets int $I^+(x, U)$, where U is a normal neighbourhood of x, are needed in the proofs below.

Corollary 3.2 On every compact time-oriented sub-Lorentzian manifold there exists a closed null future directed curve

Proof This follows from the previous proof and from the fact that int $I^+(x, U)$ = int $N^+(x, U) \subset N^+(x, U)$ (see [12], theorems 3.1, 3.2

A sub-Lorentzian manifold (M, H, g) is contact, if H is a contact distribution on M. It means that dim $M = 2n + 1$, rank $H = 2n$, and locally H can be described as the kernel of a 1-form ω such that $\omega \wedge (d\omega)^n$ is non-degenerate.

Corollary 3.3 Suppose that (M, H, g) is a compact and contact time-oriented subLorentzian manifold. Then there exists a closed piecewise smooth null future directed curve on M (with a finite number of non-smooth points).

Proof Since M may be covered by a finite number of normal neighbourhoods, it is enough, in view of Proposition 3.5, to make the following observation. Suppose that $y \in I^+(x, U)$ for a normal neighbourhood U of x. Then y can be reached from x by a piecewise smooth null future directed curve with a finite number of non-smooth points. Indeed, take any smooth null past directed vector field Z on U. Send a trajectory of Z from y. After some time it reaches $\partial U\, I^+(x, U)$ (cf. [10]). According to [12], $\partial U\, I^+(x, U) = \partial U\, J+(x, U)$ is made up of piecewise smooth null future directed curves starting from x and having a finite number of non-smooth points. Remark now important properties of closed causal curves.

Corollary 3.4 Let Γ be a closed timelike curve. Then, for any $x \in M$, $\Gamma \cap I^+(x) = \emptyset$ (resp. $\Gamma \cap I^-(x) = \emptyset$) if and only if $\subset I^+(x)$ (resp. $\Gamma \subset I^-(x)$). Also, for any x, $y \in$, $I^+(x) = I^+(y)$, $I^-(x) = I^-(y)$, and in fact $I^+(x) = I^+(\Gamma)$, $I^-(x) = I^-(\Gamma)$.

Corollary 3.5 Let Γ be a closed null curve. Then, for any $x \in M$, $\Gamma \cap N^+(x) = \emptyset$ (resp. $\Gamma \cap N^-(x) = \emptyset$) if and only if $\Gamma \subset N^+(x)$ (resp. $\Gamma \subset N^-(x)$). Also, for any x, $y \in$, $N^+(x) = N^+(y)$, $N^-(x) = N^-(y)$, and in fact $N^+(x) = N+(\Gamma)$, $N^-(x) = N^-(\Gamma)$.

Let us also note a result which is a straightforward corollary of known results in control theory (see e.g. [1], page 112).

Proposition 3.6 Let (M, H, g) be a time-oriented sub-Lorentzian manifold. If, for a point $x \in M$, $M = I^+(x)$, then $M = I^+(x) = J^+(x)$.

As the next application of Theorems 1.1, 1.2 we will construct a global affine control system describing nonspacelike curves.

Let (M, H, g) be a sub-Lorentzian manifold with dim $M = n$ and rank $H = k$. At the beginning of this section we defined nonspacelike curves. Now we will slightly generalize this definition. So a curve $\gamma : (a, b) \to M$, defined on a finite or infinite interval (a, b), is nonspacelike if it is absolutely continuous on every compact subinterval $[\alpha, \beta] \subset (a, b)$, and $\dot{\gamma}(t) \in H_{\gamma(t)}$, $g_{\gamma(t)}(\dot{\gamma}(t), \dot{\gamma}(t)) \leq 0$, both relations taking place a.e. on (a, b)

Let $H = H^- \oplus H^+$ be a decomposition from Theorem 1.1. By Theorem 1.2 there exist global sections $\sigma_1,\ldots,\sigma_d$ of H^-, $d \leq n^+ 1$, and global sections $\tau 1,\ldots,\tau_e$ of H^+, $e \leq (k - 1)(n + 1)$, which taken all together span H_x at any point $x \in M$. Let us consider smooth functions c_{ij}^-, c_{rs}^+ $M \to$ R defined by: $c_{ij}^-(x) = g_x(\sigma_i(x), \sigma_j(x))$, $i, j = 1,\ldots, d$, and $c_{rs}^+(x) = g_x(\tau_r(x), \tau_s(x))$, $r,s = 1,\ldots, e$. Let $\gamma : (a, b) \to$ M be a nonspacelike curve. Then we have a global representation

$$\dot{\gamma}(t) = \sum_{i=1}^{d} \eta_i(t)\sigma_i(\gamma(t)) + \sum_{r=1}^{e} \xi_r(t)\tau_r(\gamma(t)) \tag{3.1}$$

where η_i, $\xi_r : (a, b) \longrightarrow \mathbb{R}$ are locally integrable, and

$$\sum_{i,j=1}^{d} c_{ij}^{-}(\gamma(t))\eta_i(t)\eta_j(t) + \sum_{r,s=1}^{e} c_{rs}^{+}(\gamma(t))\xi_r(t)\xi_s(t) \leq 0$$

a.e. on (a, b). Suppose further that (M, H, g) is time-oriented by a vector field $\sigma : M^{-}\to H \subset TM$. After modifying σ as in the proof of Proposition 3.1 and normalizing it, we can suppose that σ generates H^{-} and $g(\sigma, \sigma) = -1$. In such a case, (3.1) becomes

$$\dot{\gamma}(t) = \eta(t)\sigma(\gamma(t)) + \sum_{r=1}^{e} \xi_r(t)\tau_r(\gamma(t)) \tag{3.2}$$

with η, $\xi_r : (a, b) \longrightarrow \mathbb{R}$ locally integrable, and such that

$$-\eta^2(t) + \sum_{r,s=1}^{e} c_{rs}^{+}(\gamma(t))\xi_r(t)\xi_s(t) \leq 0$$

a.e. on (a, b). Finally, if in addition is future directed, then (3.2) simplifies to

$$\dot{\gamma}(t) = \sigma(\gamma(t)) + \sum_{r=1}^{e} \xi_r(t)\tau_r(\gamma(t)) \tag{3.3}$$

with $\xi_r : (a, b) \to \mathbb{R}$ locally integrable, and such that

$$\sum_{r,s=1}^{e} c_{rs}^{+}(\gamma(t))\xi_r(t)\xi_s(t) \leq 1 \tag{3.4}$$

a.e. on (a, b).

Let us observe that in general the expression (3.3) is not a control system, since the set of control parameters

$$\mathcal{U}_x = \left\{ \xi \in \mathbb{R}^e : \sum_{r,s=1}^{e} c_{rs}^{+}(x)\xi_r\xi_s \leq 1 \right\} \tag{3.5}$$

depends on a state point, and not only on time (control systems with the set of control parameters depending on time are treated for instance in [7]). Supposing that M is compact we can help it in the following way. The matrix $(c_{rs}^{+}(x))_{r,s=1,\dots,e}$ defines a non-negative definite quadratic form, hence there exists a number $R > 0$ such that for each $\xi \in \mathbb{R}^e$

$$0 \leq \sum_{r,s=1}^{e} c_{rs}^{+}(x)\xi_r\xi_s \leq R\left(\xi_1^2 + \ldots + \xi_e^2\right).$$

In this way we are led to the affine control system

$$\dot{x} = \sigma + \sum_{r=1}^{e} u_r \tau_r, \quad u \in \mathcal{U} \tag{3.6}$$

with $U = \{u \in \mathbb{R}^e : \sum_{r=1}^{e} u_r^2 \leq \frac{1}{R}\}$ Obviously, any trajectory of the system (3.6) is a nonspacelike future directed curve on (M, H, g). In some cases, however, for instance when rank $H = 2$ and M is time-oriented (but not necessarily compact), the trajectories of (3.6) (with $R = 1$) coincide with the set of all nonspacelike future directed curves—this follows from Proposition 3.3

4 PROOFS OF THEOREMS 1.3, 1.4

We start from the proof of Theorem 1.4. Suppose (M, H, g) to be a compact timeoriented sub-Lorentzian manifold. The existence of closed timelike orbits was already established in Proposition 3.5. Next we have

Lemma 4.1 Let be a closed timelike curve which is not a Goh curve. Then for any $x \in$, $x \in \operatorname{int} I^+(x)$. In particular, $\subset \operatorname{int} I^+(\Gamma)$

Proof Since any subarc of a Goh curve is a Goh curve, it follows that there exists an $x \in \Gamma$ and a normal neighbourhood U of x such that $\Gamma \cap U$ is not a Goh curve in U.Then using [12] we know that $\Gamma \cap I^+(x, U) \subset \operatorname{int} I^+(x, U)$. Fix a point $z \in \Gamma \cap \operatorname{int} I^+(x, U)$ and take an open set V such that $z \in V \subset I^+(x)$. Now, for any $y \in V$ it is easy to join y from z by a timelike future directed curve: we first travel from z to x along a suitable piece of , and then from x to y.

Thus $z \in \text{int } I^+(z)$. Using Lemma 3.1 this last condition is equivalent to the statement that $I^+(z)$ is open. This proves the assertion since $I^+(x) = I^+(z)$ for any $x \in \Gamma$ by Corollary 3.4.

Replacing 'future directed' by 'past directed' we obtain

Lemma 4.2 Let Γ be a closed timelike curve which is not a Goh curve. Then for any $x \in \Gamma$, $x \in \text{int } I^-(x)$. In particular, $\Gamma \subset \text{int } I^-(\Gamma)$.

Corollary 4.1 If is a closed timelike curve which is not a Goh curve then the sets I +(Γ), $I^-(\Gamma)$ are open.

Now, we finish the proof of Theorem 1.4. Fix a closed timelike curve Γ We will show that $U = I^+(\Gamma) \cap I^-(\Gamma)$ is a neighbourhood of Γ having the desired properties. Indeed, take $x, y \in U$. Since $y \in I^+(\Gamma)$, there exists a timelike future directed curve joining a point of to y; also, since $x \in I^-(\Gamma)$, there exists a timelike future directed curve joining x to a point of Γ Finally, any two points of Γ can be connected by a timelike future directed curve, namely a suitable piece of Γ In this way we have constructed a timelike future directed curve joining x to y, and from the very definition of the sets $I^+(\Gamma)$, $I^-(\Gamma)$ this curve stays in U.

We proved that any two points from U can be joined by a timelike future directed curve contained in U. Let us comment now on how Theorem 1.4 can be strengthened. First of all, instead of assuming that our closed timelike curve Γ is not a Goh curve, we can introduce a weaker condition, namely we suppose that there exist $x_1, x_2 \in$ such that

$$\Gamma \cap I^+(x_1) \subset int\, I^+(x_1), \quad \Gamma \cap I^-(x_2) \subset int\, I^-(x_2) \tag{4.1}$$

This is exactly the condition used in the proof of lemmas 4.1, 4.2, and this is indeed a weaker assumption—see [14] for examples of Goh timelike curves satisfying condition (4.1)

Next we can replace a closed timelike future directed curve which is not a Goh curve with a closed null future directed curve (which exists by Corollary 3.2) which satisfies the modification of condition (4.1): there exist points $x_1, x_2 \in$ such that

$$\Gamma\setminus\{x_1\} \cap N^+(x_1) \subset int\, N^+(x_1), \quad \Gamma\setminus\{x_2\} \cap N^-(x_2) \subset int\, N^-(x_2) \tag{4.2}$$

Now we can achieve local controllability by null curves by repeating the above arguments, where instead of the sets int $I^{\pm}(y)$ we use the sets int $N^{\pm}(y)$. In this way we prove that $N^+(y) = N^+(\Gamma)$, $N^-(y) = N^-(\Gamma)$ are open, whenever y belongs to Γ, and that $U = N^+(\Gamma) \cap N^-(\Gamma)$ is such a neighbourhood of Γ that if $y,z \in U$ are arbitrary then there exists a null future directed curve contained in U connecting y to z. Note that condition (4.2) is never satisfied on sub-Lorentzian manifolds with rank $H = 2$

We come to the proof of Theorem 1.3 now. Consider the control system

$$\dot{x} = -\sigma + \sum_{r=1}^{e} u_r(-\tau_r), \quad u \in \mathcal{U} \quad , \tag{4.3}$$

and denote by $A^+(x)$ (resp. by $A^-(x)$) the reachable set from x for (1.1) (resp. for (4.3)), that is to say $A^+(x)$ (resp. $A^-(x)$) is the set of endpoints of all trajectories of (1.1) (resp. (4.3)) that start from x and are generated by measurable and bounded controls. Due to bracket generating hypothesis, Krener's theorem ensures that int$A^+(x)$, int$A^-(x)$ are non-empty for every $x \in M$. Now it is enough to repeat all above arguments replacing the setsint $I^\pm(y)$ with int$A\pm(x)$. As a result we obtain a controllable neighbourhood $U = A^+(\Gamma) \cap A-(\Gamma)$ for the system (1.1). Note that if is not a Goh curve then one can find points x1, $x_2 \in \Gamma$such that

$$\Gamma \setminus \{x_1\} \cap \mathcal{A}^+(x_1) \subset int\, \mathcal{A}^+(x_1), \quad \Gamma \setminus \{x_2\} \cap \mathcal{A}^-(x_2) \subset int\, \mathcal{A}^-(x_2)$$

Finally, remark that the compactness of M is needed only in proving the existence of closed timelike (null) curves. The proof of local controllability around a closed orbit does not require the compactness assumption

At the very end of this paper we will explain why the proof of local controllability in Theorem 1.3 slightly generalizes the results from [22]. So, at first notice that, unlike our considerations, in [22] only smooth closed orbits are studied. Moreover, using e.g. [4] (see the formula for the image under the differential of the endpoint map) it is known that the assumptions made in [22] imply that a closed orbit in question is not abnormal with respect to H. But first of all, it is not abnormality of a closed orbit that can spoil local controllability around it but the fact that the orbit is a Goh curve (any trajectory of (1.1) that lies on $\partial A^+(x)$ and is generated by controls from the set $\{u \in \mathbb{R}^e : \sum_{r=1}^{e} u_r^2 < \frac{1}{R}\}$ is a Goh curve for H), and secondly, these are only necessary conditions, so their satisfaction does not forjudge whether local controllability holds or does not

ACKNOWLEDGMENTS

This work was partially supported by the Polish Ministry of Research and Higher Education, grant NN201 607540.

REFERENCES

1. Agrachev, A., Sachkov, Y.: Control theory from geometric viewpoint. Encyclopedia of Mathematical Science, vol. 87. Springer, Berlin (2004)
2. Beem, J.K., Ehrlich, P.E., Easley, K.L.: Global Lorentzian Geometry. Marcel Dekker, NY (1996)
3. Baum, H.: Spin-Strukturen und Dirac-Operatoren über pseudoriemannschen Mannigfaltigkeiten. Teubner, Leipzig (1981)
4. Bonnard, B., Kupka, I.: Théorie des singularités de l'application entrée/sortie et optimalité des trajectoires singulières dans le problème du temps minimal. Forum Math. **5**, 111–159 (1993)
5. Choi, Y.S., Suh, Y.J.: Remarks on the topology of Lorentzian manifolds. Commun. Korean Math. Soc. **15**(4), 641–648 (2000)
6. Fuks, B.A.: Theory of analytic functions of several complex variables, vol. 1. American Mathematical Society (1963)
7. Hermès, H., Lasalle, J.P.: Functional analysis and time optimal control. Mathematics in Sciences and Engineering, vol. 56. Academic Press, New York (1969)
8. Huang, T., Yang, X.: Geodesics in the Heisenberg group HnHn with Lorentzian metric. J. Dyn. Control Syst. **18**(4) (2012)
9. Grauert, H.: On Levi's problem and the imbedding of real analytic manifolds. Ann. Math. **68**(2), 460–472 (1958)
10. Grochowski, M.: Geodesics in the sub-Lorentzian Geometry. Bull. Polish Acad. Sci. **50**(2) (2002)
11. Grochowski, M.: Reachable sets for the Heisenberg sub-Lorentian metric on R3.R3. An estimate for the distance function. J. Dyn. Control Syst. **12**(2) (2006)
12. Grochowski, M.: Properties of reachable sets in the sub-Lorentzian geometry. J. Geom. Phys. **59**, 885–900 (2009)
13. Grochowski, M.: Normal forms and reachable sets for analytic martinet sub-Lorentzian structures of Hamiltonian type. J. Dyn. Control Syst. **17**(1) (2011)
14. Grochowski, M.: The structure of reachable sets for affine control systems induced by generalized Martinet sub-Lorentzian metrics. ESAIM Control Optim. Calculus Var. doi:
15. Grochowski, M.: Some properties of reachable sets for control affine systems. Anal. Math. Phys. **1**(1) (2011)
16. Hermes, H.: Local controllability and sufficient conditions in singular problems. J. Differ. Equ. **20**(1), 213–232 (1976)
17. Korolko, A., Markina, I.: Nonholonomic Lorentzian geometry on some HH-type groups. J. Geom. Anal. 19(4) (2009)
18. Korolko, A., Markina, I.: Geodesics on HH-type Quaternion Groups with sub-Lorentzian metric and their physical interpretation. J. Complex Anal. Oper. Theory **4**(3), 589–618 (2010)
19. Korolko, A., Markina, I.: Semi-Riemannian geometry with nonholonomic constraints. Taiwan. J. Math. **15**(4) (2011)
20. Liu, W., Sussmann, H.: Shortest paths for sub-Riemannian metrics on rank-two distributions. In: Memoires of the American Mathematical Society, vol. 118, No. 564 (1995)
21. Łojasiewicz, S.: Ensembles semi-analytiques. Inst. Hautes Études Sci, Bures-sur-Yvette, France (1964)
22. Nam, K., Arapostathis, A.: A sufficient condition for local controllability of nonlinear systems along closed orbits. IEEE Trans. Autom. Control **37**, 378–380 (1992)
23. O'Neill, B.: Semi-Riemannain Geometry with Applications to Relativity, Pure and Applied Ser. vol. 103. Academic Press, New York (1983)

24. Steenrod, N.E.: The Topology of Fiber Bundles. In: Princeton Mathematical Series, vol. 14. Princeton Univ Press, Princeton (1951)
25. Sussmann, H.: Smooth Distributions are Globally Finitelly Spanned, Analysis and Design of Nonlinear Control Systems, vol. 3–8. Springer, Berlin (2008)

CHAPTER 5

Convective wave breaking in the KdV equation

Mats K. Brun[1] and Henrik Kalisch[1]

1 Department of Mathematics, University of Bergen, P.O. Box 7800, 5020 Bergen, Norway

ABSTRACT

The KdV equation is a model equation for waves at the surface of an inviscid incompressible fluid, and it is well known that the equation describes the evolution of unidirectional waves of small amplitude and long wavelength fairly accurately if the waves fall into the Boussinesq regime. The KdV equation allows a balance of nonlinear steepening effects and dispersive spreading which leads to the formation of steady wave profiles in the form of solitary waves and cnoidal waves. While these wave profiles are solutions of the KdV equation for any amplitude, it is shown here that there for both the solitary and the cnoidal waves, there are critical amplitudes for which the horizontal component of the particle velocity matches the phase velocity of the wave. Solitary or cnoidal solutions of the KdV equation which surpass these amplitudes feature incipient wave breaking as the particle velocity exceeds the phase velocity near the crest of the wave, and the model breaks down due to violation of the kinematic surface boundary condition. The condition for breaking can be conveniently formulated as a convective breaking criterion based on the local Froude number at the wave crest. This breaking criterion can also be applied to time-dependent situations, and one case of interest is the development of an undular bore created by an influx at a lateral boundary. It is shown that this boundary forcing leads to wave breaking in the leading wave behind the bore if a certain threshold is surpassed.

1 INTRODUCTION

The Korteweg–de Vries (KdV) equation is a model equation describing the evolution of long-crested waves at the surface of a body of fluid. The equation is derived as a physical model equation under the assumption that there is an approximate balance between nonlinear steepening effects and dispersive spreading. In mathematical terms, this balance is expressed by introducing two small parameters α and β measuring the wave amplitude and the wave length, respect tively. Suppose that the undisturbed depth of the fluid is given by h_0. If a represents a typical wave amplitude, and ℓ represents a typical wavelength, the two parameters $\alpha = a/h_0$ and $\beta = h_0^2/\ell^2$ should be small and of the same order. This is the Boussinesq scaling. If in addition the wave motion is predominantly in a single direction, then the KdV equation is an approximate model describing the surface wave motion [9, 30]. If the undisturbed depth h0h0is taken as a unit of distance and $\sqrt{h_0/g}$ is taken as a unit of time, then the KdV equation appears in the form

$$\eta_t + \eta_x + \tfrac{3}{2}\eta\eta_x + \tfrac{1}{6}\eta_{xxx} = 0 \quad (1.1)$$

In the present article, the focus is on whether—or rather how—the KdV equation is able to describe incipient wave breaking. While the KdV equation does not admit the distinctive steepening and development of infinite gradients known from nonlinear hyperbolic equations, it will be shown here that the KdV equation features a different kind of wave breaking which is more closely related to spilling at the wave crest.

Let us recall that the KdV equation can be thought of as a combination of the simple nonlinear balance law

$$\eta_t + \eta_x + \tfrac{3}{2}\eta\eta_x = 0 \quad (1.2)$$

and the linear dispersive equation

$$\eta_t + \eta_x + \tfrac{1}{6}\eta_{xxx} = 0 \quad (1.3)$$

As indicated in Fig. 1, all waves featuring negative slope break at some point in the model (1.2), but no waves break in the model (1.3) (however, (1.3) features dispersive blow-up for some data [10, 27, 34]). The KdV Eq. (1.1) allows a balance of nonlinear steepening effects and dispersive spreading which arrests the typical hyperbolic wave breaking exhibited by

(1.2), and leads to the formation of steady traveling waves, such as solitary and cnoidal waves [1, 29] which propagate without a change in the wave profile.

In the context of the full Euler equations, the water-wave problem also admits various steady traveling-wave solutions. In this case, wave breaking can be induced by a number of instabilities, such as the superharmonic instability [43, 44] and transverse instabilities [22, 25]. In the KdV equation, both the solitary and cnoidal wave solutions are known to be stable [6, 14], and therefore wave breaking triggered by various instabilities is not reproduced by the KdV model.

Solitary-wave solutions of (1.1) are of the form

$$\eta(x,t) = H\operatorname{sech}^2\left(\tfrac{\sqrt{3H}}{2}(x - x_0 - ct)\right) \tag{1.4}$$

where the phase velocity is given by c=1+H2.c=1+H2. While these formulas define solutions of the KdV equation for all waveheights *H*, it turns out that solutions of large waveheight are inconsistent with the model in which the KdV equation is valid. In particular, as will be shown in Sect. 2, for waveheights exceeding the critical waveheight Hmax solitary=0.6879Hmax solitary=0.6879, the phase velocity of the wave is smaller than the particle velocity at the crest, a fact which may be interpreted as incipient wave breaking.

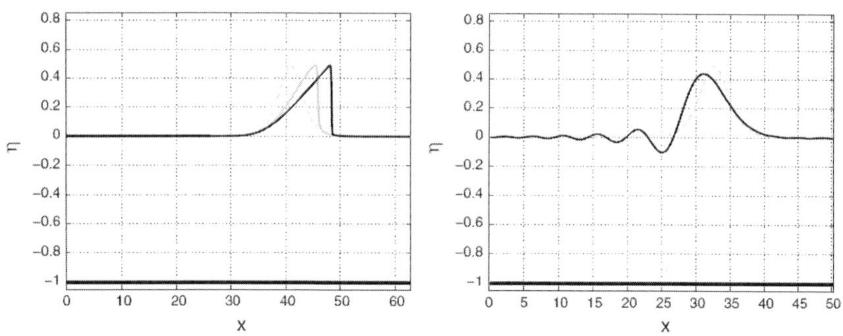

Figure. 1 *Left panel*: Solution of Burgers equation $\eta_t + \frac{3}{2}\eta\eta_x = 0$ which features hyperbolic wave breaking. *Right panel*: Solution of the Airy equation $\eta_t + \frac{1}{6}\eta_{xxx}$ which features dispersive spreading. Initial data are shown as *dashed curves*. If the *dashed curve* were used as initial data for the KdV Eq. (1.1), then this wave would propagate without change in shape

To elaborate on this argument, note that it is shown in [46] that the derivation of the equation as a surface water-wave model enables the reconstruction of an approximation of the fluid velocity field underneath the surface. In particular, it is possible to derive relations expressing the horizontal and vertical velocity components in terms of the principal unknown variable ηηwhich describes the shape of the free surface. The horizontal velocity component in the fluid is given by

$$u(x, y, t) = \eta - \tfrac{1}{4}\eta^2 + \left(\tfrac{1}{3} - \tfrac{y^2}{2}\right)\eta_{xx} \tag{1.5}$$

and this relation holds to the same order in the asymptotic parameters α and β as to which the KdV equation is valid.

Since the horizontal component of the particle velocity can be found approximately, it may be compared to the local phase velocity of the wave. This leads to one of the most fundamental breaking criteria used in the literature, the convective breaking criterion which predicts wave breaking if the particle velocity at the wavecrest exceeds the phase velocity of the wave (see for example [32, 45]). More specifically, this breaking criterion (often termed kinematic criterion) is based on the local Froude number, and predicts breaking if

$$\frac{U}{c} \geq 1 \tag{1.6}$$

where $U = u(x, \eta(x, t), t)$ is the horizontal component of the velocity field evaluated at the free surface, c is the local phase velocity of the wave. To illustrate this point of view, Fig. 2 shows two waves with corresponding fluid particles near the crest of the wave. In the left panel, a wave with sufficiently small amplitude is depicted. A fluid particle located near the wavecrest remains in the free surface and recedes from the crest in unison with the wave motion. However in the right panel, a wave of higher amplitude is shown, and particles located near the crest have higher horizontal velocity than the wave itself.

The utility of the convective criterion has been the subject of some discussions in the literature. For wave breaking in shallow water, such as in focus in the present article, some studies such as [47] conclude that the convective criterion does a fair job of predicting wave breaking even for three-dimensional waves. On the other hand, studies focusing on breaking of deep-water waves such as [40] observe that the kinematic criterion has been useful in some practical situations, while there is also evidence that it may not be a reliable indicator to pinpoint the onset of breaking in the case of deep water. In some of the cases where the convective breaking criterion performs badly, the difficulty of obtaining accurate readings for the phase velocity of the wave in experimental situations may be responsible [39].

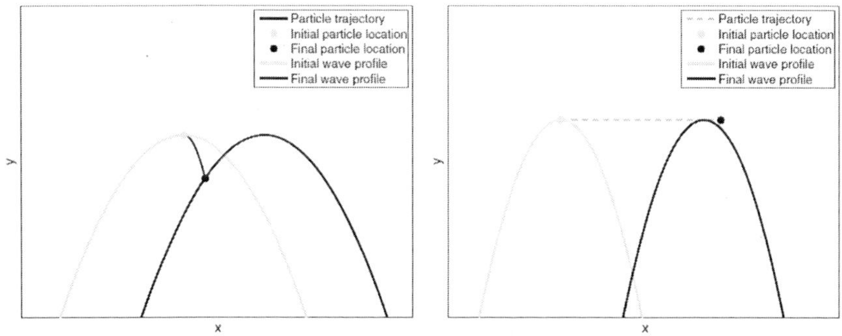

Figure. 2 *Left panel*: Surface profile of a regular wave, and trajectory of a particle in the free surface. *Right panel*: Surface profile of a breaking wave. A particle contained in the free surface is leaving the fluid domain, indicating incipient wave breaking

Note that the criterion (1.6) is stated in sufficient generality to be applicable to both experimental and numerical work. In the case of numerical modeling, and in particular in the case of phase-resolving models, there is no difficulty in evaluating the local wave and particle velocities. The local phase velocity can usually be approximated using a variety of methods. For example, Fourier techniques have been used in [36]. What is more, in the case of solitary and cnoidal waves, the phase and particle velocities are known in closed form, and it is straightforward to test the breaking criterion (1.6). These computations will be carried out in Sects. 2 and 3.

While it is relatively easy to evaluate the breaking criterion in several situations for the Eq. (1.1), it should also be noted that neither the hyperbolic steepening nor the convective wave breaking is likely to happen for free surface waves unless some focusing effect is present, such as an underlying current [40, 48], strong three-dimensional effects [47], or specially tailored initial data [10]. On the other hand wave breaking is likely to occur in the presence of forcing in the form of an uneven bottom topography or a discharge. These two possibilities are sketched in Fig. 3. In particular the left panel of Fig. 3 indicates one of the most widely known type of wave breaking, the development of breakers on a sloping beach. In this case, depending on a number of parameters such as waveheight, wavelength and bottom slope, a variety of breaking phenomena takes place, ranging from spilling at the crest, to overturning and surging breaking [19].

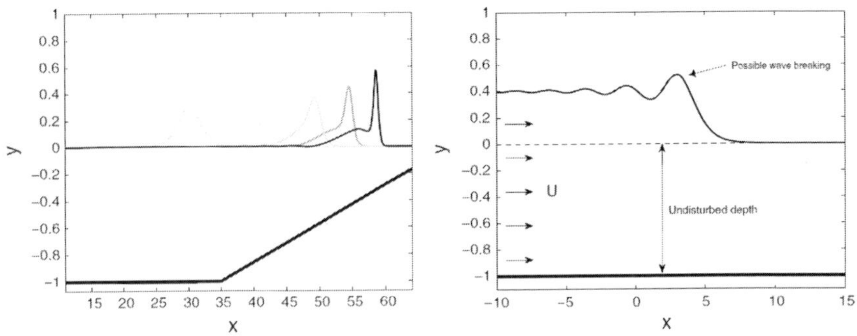

Figure. 3 *Left panel*: Breaking wave in Boussinesq model with sloping bottom. *Right panel*: Wave breaking due to a large discharge at left boundary

In the right panel of Fig. 3, the case of a forced inflow is indicated. In this case, a so-called bore generally emerges. Sufficiently small discharges lead to the appearance of an undular bore which is a characterized by a moderately steep front followed by a leading wave and trailing smaller waves. For larger inflows, the leading wave behind the bore may break, and for large enough inflows, the bore can be entirely turbulent [15, 28]. In the final section of the present paper, the case of a forced inflow and undular bore is discussed in some detail, and the question of the transition from purely undular to undular with spilling breaking is investigated. The numerical results are compared to experimental findings of Favre [21], who determined a critical inflow condition dividing between laminar flows and flows featuring breaking waves. Our numerical simulations show incipient breaking near the critical case of Favre, but some discrepancy remains. Further details on this comparison are given in Sect. 4.

Before we enter the main development of this paper, some remarks are in order. First, it should be noted that convective wave breaking of the type discussed here is unlikely to happen in hyperbolic equations or systems. While the simple conservation law $\eta_t + \frac{3}{2}\eta\eta_x = 0$ does not contain expressions for the local particle velocity, the nonlinear shallow-water equations

$$h_t + (uh)x = 0,$$

$$u_t + h_x + uu_x = 0,$$

include both the surface deflection $\eta(x,t) = h(x, t) - 1$, and an average horizontal particle velocity $u(x, t)$ in the description. In this case, a simple wave propagating to the right is given by the Riemann invariant $u+2\sqrt{h}$, and has phase velocity $u+\sqrt{h}$. As shown in [46], for a wave profile $h = H(x)$ with undisturbed depth 1, the fluid velocity is $u = 2\sqrt{h} - 2$, and the phase velocity of the wave is $C = 3\sqrt{H} - 2$. Thus we see that the phase velocity is always larger than the fluid velocity in this case. On the other hand, the typical hyperbolic steepening and eventual breaking will happen for appropriate initial conditions.

Secondly, it is important to note that wave breaking can be induced by instability even in cases without any forcing. This mechanism is well known in the case of waves on deeper water. Traditionally the wave steepness has been used as a predictor for the occurrence of crest instabilities and subsequent breaking of deep water waves [40], but recent work has indicated that criteria based on the energy flux in wave groups may be successful [38]. Breaking induced by crest instabilities may also occur in shallow water, such as in the situation at hand. First of all, solitary waves exist in the fully nonlinear Euler equations up to a waveheight of H = 0.8332. However, in [43], it was shown that the solitary wave is unstable in the context of the full Euler potential theory for waveheights exceeding 0.78, and the authors of [44] showed how this instability can be related to wave breaking. In [25], a three-dimensional situation was considered, and transverse instabilities were found at a nondimensional waveheight H = 0.713. For periodic wave in shallow water, similar results were provided in [33] and [22].

Finally, it should be mentioned that the relation (1.5) and similar expressions for the vertical velocity may also be used advantageously for a number of purposes, such as the definition of a family of Boussinesq models [8], describing particle paths beneath the surface [12], and the study of mechanical balance laws associated to the evolution equation [2, 3]. On the other hand, there is a variety of different strategies to construct the velocity field in the fluid, such as the analytic method used in [23] to find the velocities associated to a periodic cnoidal solution of the KdV equation. Particle paths under periodic traveling waves can also be found to rather high accuracy using the Lagrangian approach [18]. Concerning the notion of mechanical balance laws, we mention in particular that an analysis along these lines casts doubt on the conservation of mechanical energy in the KdV approximation [3]. The conservation of energy in both the KdV and higher-order KdV equations has also been questioned recently in a study which utilizes a Lagrangian framework [24].

The plan of the paper is as follows. In Sect. 2, the convective breaking criterion is applied to the solitary wave, and it is found that the limiting waveheight is *Hmax solitary* = 0.6879. In Sect. 3, the breaking criterion is applied to the cnoidal wave solutions. It is found that cnoidal wave solutions may feature breaking at a much smaller waveheight than the solitary wave. Finally, Sect. 4 is devoted to the study of wave breaking in a dynamic situation with a forced inflow. The resulting undular bore features a leading wave which may break, and the focus is on finding the critical case dividing the field of purely undular bores from partially turbulent breaking bores.

2 CONVECTIVE WAVE BREAKING FOR SOLITARY WAVES

The surface water-wave problem for an inviscid fluid is generally described by the Euler equations with an impenetrable bottom, and kinematic and dynamic boundary conditions at the free surface. Assuming weak transverse effects, the unknowns are the surface elevation $\eta(x, t)$, the horizontal and vertical fluid velocities $u(x, y, t)$ and $v(x, y, t)$, respectively, and the pressure $P(x, y, t)$. If the assumption of irrotational flow is made, then a velocity potential $\phi(x,y,t)$ can be used. In the non-dimensional variables mentioned in the introduction, the problem may be posed on the domain $\{(x, y) \in \mathbb{R}^2 \mid 0 < y < 1 + \eta(x, t)\}$. The derivation of the KdV equation is based on approximating the velocity potential by an asymptotic series. In suitably scaled variables, the horizontal velocity component takes the form

$$u = \phi_x = w - \beta \frac{y^2}{2} w_{xx} + O(\beta^2), \tag{2.1}$$

where $w(x, t)$ represents the horizontal velocity at the bottom. On the other hand, the one-way assumption inherent in the KdV equation necessitates that the velocity at the bed is given by

$$w = \eta - \tfrac{1}{4}\alpha\eta^2 + \tfrac{1}{3}\beta\eta_{xx} + O(\alpha^2, \alpha\beta, \beta^2)$$

By combining these two relations and neglecting terms of $O(\alpha^2, \alpha\beta, \beta^2)$, we obtain an expression for the horizontal component of the velocity field in terms of η and y:

$$u = \phi_x = \eta - \tfrac{1}{4}\alpha\eta^2 + \beta\left(\tfrac{1}{3} - \tfrac{y^2}{2}\right)\eta_{xx}.$$

In the original variables, the horizontal velocity is then expressed as

$$u(x, y, t) = \eta - \tfrac{1}{4}\eta^2 + \left(\tfrac{1}{3} - \tfrac{y^2}{2}\right)\eta_{xx} \tag{2.2}$$

and the breaking criterion (1.6) can be written as

$$\eta - \tfrac{1}{4}\eta^2 + \left(\tfrac{1}{3} - \tfrac{(1+\eta)^2}{2}\right)\eta_{xx} \geq c. \tag{2.3}$$

We now apply this breaking criterion to the solitary wave given by (1.4), where H is the waveheight, c is the wavespeed, and x_0 is the initial location of the wave crest. Differentiating twice with respect to x yields

$$\begin{aligned}\eta_{xx} &= 3H^2 \operatorname{sech}^2\left(\tfrac{\sqrt{3H}}{2}(x - ct - x_0)\right)\tanh^2\left(\tfrac{\sqrt{3H}}{2}(x - ct - x_0)\right) \\ &\quad - \tfrac{3}{2}H^2 \operatorname{sech}^4\left(\tfrac{\sqrt{3H}}{2}(x - ct - x_0)\right).\end{aligned}$$

Since the solitary wave retains its shape for all time we may evaluate at $x - ct - x_0 =$, in which case we obtain $\eta = H$ and $\eta_{xx} = -\frac{3}{2}H^2$. Substituting these values into Eq. (2.2) and evaluating at the surface yields the breaking criterion (1.6)

$$H - \tfrac{1}{4}H^2 - \tfrac{3}{2}\left(\tfrac{1}{3} - \tfrac{(1+H)^2}{2}\right)H^2 \geq 1 + \tfrac{1}{2}H. \tag{2.4}$$

Setting the left hand side equal to the right hand side and rearranging terms yields

$$\mathcal{P}(H) = \tfrac{3}{4}H^4 + \tfrac{3}{2}H^3 + \tfrac{1}{2}H - 1 = 0 \tag{2.5}$$

for the critical value of the waveheight. We see that P is a fourth order polynomial in H, and since $\mathcal{P}'(H) = 3H^3 + \frac{9}{2}H^2 + \frac{1}{2} > 0$ for $H \geq 0$ and $P(0) < 0$ while $P(1) > 0$, it can have only one positive root which lies in [0, 1]. This root can be found numerically to obtain the following value of the maximum allowable waveheight for the solitary wave:

$$H\text{max solitary} = 0.6879. \tag{2.6}$$

Even though the KdV equation admits solutions with any waveheight, waves with a waveheight larger than $H_{\text{max solitary}}$ do not describe actual surface waves since these waves already feature incipient wave breaking. Of course the waveshape of a solitary-wave solution of the KdV equation may not be close to a surface water wave already for smaller waveheights than $H_{\text{max solitary}}$ (cf. [13]).

3 MAXIMUM WAVEHEIGHT FOR PERIODIC WAVES

The cnoidal waves are defined with the help of the incomplete elliptic integral of the first kind. For a given elliptic parameter $0 < m < 1$, this integral is

$$u = \int_0^{\phi} \frac{d\theta}{\sqrt{1-m\sin^2\theta}}.$$

The elliptic functions sn$(\cdot|m)$, cn$(\cdot|m)$ and dn$(\cdot|m)$ are defined as

$$\begin{aligned} \mathrm{sn}(u|m) &= \sin\phi, \\ \mathrm{cn}(u|m) &= \cos\phi, \\ \mathrm{dn}(u|m) &= \sqrt{1-m\sin^2\phi} \end{aligned}$$

On the other hand, the complete elliptic integrals of the first and second kind, respectively, are defined by

$$K(m) = \int_0^{\pi/2} \frac{d\theta}{\sqrt{1-m\sin^2\theta}} \quad \text{and} \quad E(m) = \int_0^{\pi/2} \sqrt{1-m\sin^2\theta}\, d\theta.$$

As shown in [20], the cnoidal travelling wave solutions of (1.1) are given in terms of three parameters f_1, f_2 and f_3 by

$$\eta(x,t) = f_2 + (f_1 - f_2)\,\mathrm{cn}^2\left(\tfrac{\sqrt{3(f_1-f_3)}}{2}(x - ct - x_0)\middle|m\right), \tag{3.1}$$

where $H = f_1 - f_2$ is the wave height, $c = 1 + \frac{1}{2}(f_1 + f_2 + f_3)$ is the wavespeed, and $m = (f_1 - f_2)/(f_1 - f_3)$ is the elliptic parameter. Any cnoidal wave is completely determined as long as these three constants are fixed. If σ is defined by $\sigma^2 = \frac{4}{3(f_1 - f_3)}$, then the wavelength can be written as $\lambda = 2\sigma K(m)$.

Since the wave oscillates about the equilibrium level of the surface, the mean value of the surface displacement over one wavelength should be zero. This requirement can be expressed by integrating over one wavelength, viz.

$$\int_0^{\lambda} \eta\, dx = 0$$

and it can be shown that this integral is

$$\int_0^{\lambda} \eta\, dx = 2\sigma\Big((f_1 - f_3)E(m) + f_3 K(m)\Big) = 2\sigma\left(\frac{f_1 - f_2}{m} E(m) + f_3 K(m)\right). \quad (3.2)$$

Since $H = f_1 - f_2$, we therefore have $f_3 = -\frac{H}{m}\frac{E(m)}{K(m)}$. The parameters f_1 and f_2, can then also be expressed in terms of wave height H and the elliptic parameter m as follows:

$$f_1 = \frac{H}{m}\left(1 - \frac{E(m)}{K(m)}\right),$$

$$f_2 = \frac{H}{m}\left(1 - m - \frac{E(m)}{K(m)}\right).$$

Thus, fixing H and m is sufficient to specify any cnoidal wave, as long as the undisturbed water level is set to zero.

To determine whether a given cnoidal wave is a reasonable description of a water wave in the KdV approximation, we test the wave breaking criterion (2.3). By differentiating (3.1) twice with respect to x we obtain

$$\begin{aligned}\eta_{xx} =& \tfrac{3}{2}(f_1 - f_2)(f_1 - f_3)\Big\{ \mathrm{sn}^2\left(\tfrac{\zeta}{\sigma}|m\right)\mathrm{dn}^2\left(\tfrac{\zeta}{\sigma}|m\right) \\ &+ m^2\,\mathrm{sn}^2\left(\tfrac{\zeta}{\sigma}|m\right)\mathrm{cn}^2\left(\tfrac{\zeta}{\sigma}|m\right) - \mathrm{cn}^2\left(\tfrac{\zeta}{\sigma}|m\right)\mathrm{dn}^2\left(\tfrac{\zeta}{\sigma}|m\right)\Big\},\end{aligned}$$

where the argument is given by $\zeta = x - ct - x_0$. Evaluating at $\zeta = 0$ yields

$$\eta = f_1,$$

$$\eta_{xx} = -\tfrac{3}{2}(f_1 - f_2)(f_1 - f_3).$$

Substituting these expressions into Eq. (2.3) transforms the breaking criterion (2.3) into

$$f_1 - \tfrac{1}{4}f_1^2 - \tfrac{3}{2}\left(\tfrac{1}{3} - \tfrac{(1+f_1)^2}{2}\right)(f_1 - f_2)(f_1 - f_3) \geq 1 + \tfrac{1}{2}(f_1 + f_2 + f_3). \quad (3.3)$$

Defining the constants

$$a = \tfrac{1}{m}\left(1 - \tfrac{E(m)}{K(m)}\right),\ b = \tfrac{1}{m}\left(1 - m - \tfrac{E(m)}{K(m)}\right) \text{ and } c = \tfrac{1}{m}\tfrac{E(m)}{K(m)},$$

we can write the constants f_1, f_2 and f_3 as

$$f_1 = Ha, f_2 = Hb \text{ and } f_3 = -Hc.$$

Substituting these definitions into Eq. (3.3) and setting the left and right hand side equal yields

$$\begin{aligned} Q_m(H) = {} & \frac{3}{4}H^4(a^4 + a^3c - a^3b - a^2bc) \\ & + \frac{3}{2}H^3(a^2c + a^3 - a^2b - abc) \\ & + \frac{1}{4}H^2(ac - bc - ab) \\ & + \frac{1}{2}H(a - b + c) - 1 = 0. \end{aligned} \quad (3.4)$$

This is a fourth-order polynomial in H, and by fixing a value for m it can be solved numerically to obtain the maximum allowable wave height for the cnoidal wave, $H_{\text{max cnoidal}}(m)$. Since the periodic wave reduces to the solitary wave in the nonlinear limit ($m \to 1$), and $m \to 0^+$ is the linear limit, we search for real roots of (3.4) in the interval [0, 1].

Different values of m and corresponding values of $H_{\text{max cnoidal}}(m)$ are listed in Table 1. The table also shows corresponding values of wave length, λ

and the dimensionless parameters α and β. In addition to Stokes' number SS, defined by the ratio α/β is tabulated.

Figure 4 shows a plot of $H_{max\ cnoidal}$(m) obtained using a finer resolution in m than displayed in the table above. It can be seen that the maximum wave height seems to approach zero as m approaches zero, which fact is related to a going 'outside' of the domain of validity of the Boussinesq approximation, where α and β are assumed to be of the same order of magnitude. It is also worth noting that the limiting values of a, b and c as $m \to 1$ are:

$$a = 1, b = 0, \text{ and } c = 0$$

Applying this to Eq. (3.4) gives the breaking criterion for the solitary wave, Eq. (2.5). This is also easy to see if we plot the maximum wave height of the cnoidal wave as a function of wavelength λ.

Table 1 Critical waveheight for the cnoidal solution of the KdV-equation, calculated for various values of the elliptic parameter m. Corresponding values of wavelength and the dimensionless parameters α and β, in addition to the Stokes number SS are also listed

m	Hmax cnoidalHmax cnoidal	λλ	αα	ββ	SS
0.01	0.0196	2.591	0.0098	0.1489	0.0661
0.1	0.1698	2.857	0.0849	0.1224	0.6933
0.2	0.2909	3.178	0.1454	0.0990	1.4689
0.3	0.3820	3.507	0.1910	0.0812	2.3499
0.4	0.4548	3.849	0.2274	0.0674	3.3702
0.5	0.5152	4.218	0.2575	0.0562	4.5834
0.6	0.5667	4.632	0.2833	0.0465	6.0813
0.7	0.6114	5.128	0.3056	0.0380	8.0399
0.8	0.6504	5.781	0.3252	0.0299	10.869
0.9	0.6841	6.829	0.3420	0.0214	15.952

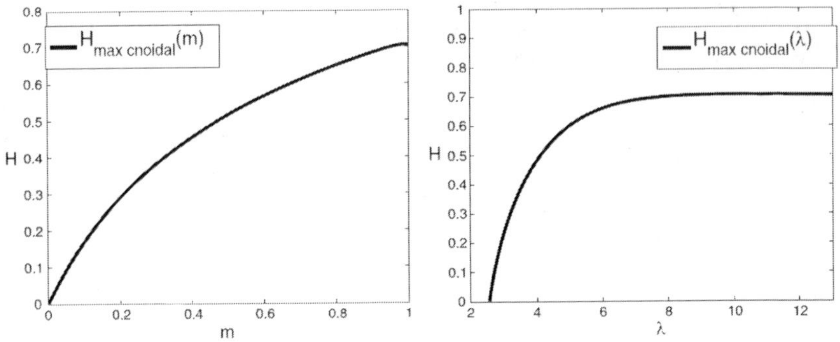

Figure. 4 *Left*: Maximum allowable wave height for the cnoidal solution as a function of *m*. *Right*: Maximum allowable wave height for the cnoidal solution as a function of wavelength

4 NUMERICAL STUDY OF WAVE BREAKING IN UNDULAR BORES

A bore or traveling hydraulic jump is the transition between two uniform flows with different depths. Assuming that one of these flow depths is the undisturbed water level h_0, and the incident water level is given by $h_0 + a_0$, we denote by $\alpha = a_0/h_0$ the strength of the bore. With the nondimensionalization explained in the introduction, the undisturbed depth is 1, and the incident water level is α. In this section we aim to find the threshold for which the bore transitions from being purely undular to one which exhibits breaking in the leading wave. This threshold is defined in terms of the ratio α, and was found experimentally to be $\alpha = 0.28$ [21].

As already mentioned, if the kinematic breaking criterion is used in connection with phase-resolving numerical models, there is no difficulty in evaluating the local wave and particle velocities. When coupled with the Boussinesq ansatz for the horizontal velocity, the convective criterion allowed for a somewhat accurate description of the breaking of the leading wave behind an undular bore [7]. In this study, numerical integration of a Boussinesq system for undular bores indicated that breaking should occur if the bore strength is greater than 0.379. While this result is qualitatively in the right direction, the critical bore strength is overestimated by about 30%30%. In the following, it will be shown how this result can be improved.

In order to find the critical bore strength in the numerical KdV model, we will therefore use the value of $\alpha = 0.25$ as a starting point, and then proceed with small increments until the breaking criterion is reached, i.e.

when the horizontal particle speed of the fluid exceeds the phase speed. The initial position of the bore front will be set to be the origin, and the bore will then travel downstream in the positive x-direction. Initial data are given by

$$\eta_0(x) = \tfrac{1}{2}a_0\big[1 - \tanh(kx)\big] \tag{4.1}$$

where k is a model parameter denoting the steepness of the initial bore slope. The numerical approach to obtaining approximate solutions of (1.1) with the appropriate boundary conditions is detailed in the appendix.

In order to test the convective breaking criterion (1.6), at each time step the horizontal particle velocity and the phase speed of the bore front have to be calculated numerically. First, given the numerical solution $\eta(x_j, t_n) = \eta_j^n$, we approximate the second spatial derivative of the surface, by the second-order central-difference formula

$$\eta_{xx}(x_j, t) \approx \frac{\eta_{j-1}^n - 2\eta_j^n + \eta_{j+1}^n}{\delta x^2}$$

where δx is the spatial grid size. We then define the approximate horizontal particle velocity at a grid point x_j and time tn, and evaluated at the free surface as $u_j^n = u(x_j, \eta_j^n, t_n)$, such that Eq. (2.2) yields

$$u_j^n = \eta_j^n - \tfrac{1}{4}(\eta_j^n)^2 + \left(\tfrac{1}{3} + \frac{(1+\eta_j^n)^2}{2}\right)\frac{\eta_{j-1}^n - 2\eta_j^n + \eta_{j+1}^n}{\delta x^2}. \tag{4.2}$$

The phase speed of the first wave behind the bore front can be approximated numerically. If the location of the maximum of the wave, say X_n is found numerically and recorded at each time t_n, the velocity c_n of the leading wave can be computed approximated using a finite-difference approximation of the first derivative: $c_n \sim (X_n - X_{n-1})/\delta t$. The left panel of Fig. 5 shows the initial profile with $a_0 = 0.25$ and $k = 1$ as a dashed curve, and the resulting bore profile at $t = 6$ as a solid curve. The figure also shows the horizontal component of the velocity field underneath the first wavecrest, and it is apparent that this bore does not break as the particle velocity remains smaller than the propagation velocity of the leading wave.

After increasing α by increments of 0.001 breaking is obtained for α_{crit} = 0.353. A computation for α = 0.5 (which is already past breaking) is shown in the right panel of Fig. 5. It can be seen that the horizontal particle velocity is greater than the front velocity after the solution has developed for some time. As expected, the critical bore strength is larger than the one found experimentally, but is fairly close to the one found in [7].

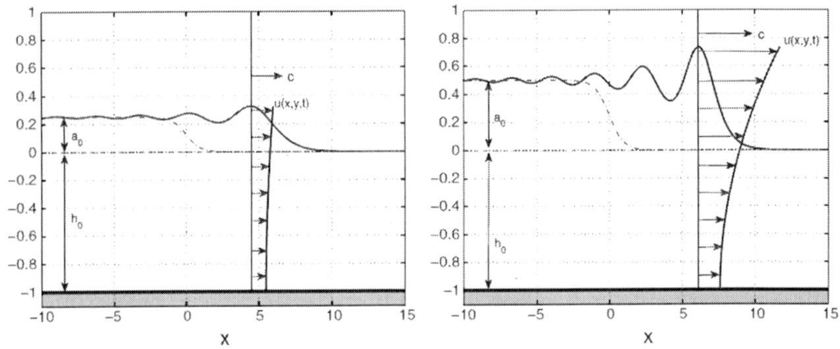

Figure. 5 The surface profiles and the horizontal velocities below the wave-crests at nondimensional time t = 6 for two sets of initial data indicated with dashed-dotted curves. *Left panel*: Phase velocity c = 1. is greater than u at the surface; no breaking. *Right panel*: c = 1.35 is smaller than u at the surface: breaking

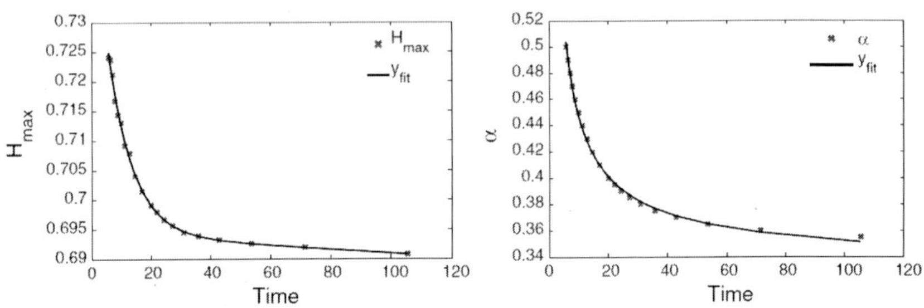

Figure. 6 *Left panel*: Maximum wave height as a function of time. *Right panel*: Bore strength as a function of corresponding breaking times

The maximum waveheight $H_{\max}$ of the solution may be extracted at the particular time step where breaking first occurs. If this waveheight is compared to the critical waveheight of the solitary wave from Sect. 2, it is found that the leading wave behind the bore breaks at a waveheight slightly larger than the solitary wave. However, the values obtained for maximum wave height of the bore seem to approach the maximum height of the solitary wave as the bore strength decreases. Since smaller bore strengths will have larger lead times before the largest wave reaches the point of breaking, in these cases, the leading wave has already separated as an individual solitary wave, and the waveheight at breaking matches closely the value obtained for $H_{max\ solitary}$.

The figure below shows a plot of the calculated values of H_{max} as a function of time. A curve of the form $y_{fit} = At^B + C$ has also been fitted to the data.

The parameters A, B and C in the fitted curve are given by

$$A = 0.1764,\ B = -0.8582 \text{ and } C = 0.6863,$$

and the residual sum of squares is = $2.54 \cdot 10^{-5}$. The curve fits the data fairly well, and it is evident from the form of the equation that it approaches the value $C = 0.6863$ for large t. This value is very close to the maximum possible height of the solitary wave $H_{max\ solitary} = 0.6879$. We can therefore confirm that the bore eventually disintegrates into several solitary waves, but as the bore strength, α, increases, the bore will break before this happens and thus cannot be described in terms of solitary waves. This observation also points to the possibility of obtaining a lower value of the critical bore strength α_{crit} if a similar curve fit is used for this parameter. To this end, we plot the bore strength as a function of corresponding breaking times. If a curve of the same form as above is used, the parameters A, B and C in the fitted curve are given by

$$A = 0.5751,\ B = -0.6908 \text{ and } C = 0.3283.$$

The residual is = $7.74 \cdot 10^{-5}$, and visual inspection of Fig. 6 indicates that the fitted curve fits the data rather well also in this case. The numerical value of C indicates an asymptotic critical bore strength of = 0.3283 for large values of t, and this value is somewhat lower than the value of α_{crit} obtained previously.

In order to assess whether the values found are reasonable, the results should be compared to experimental findings. One set of controlled experiments was run by H. Favre, and is described in his monograph [21]. In these experiments, it was found that the critical bore strength was close to $\alpha = a_0/h_0 = 0.28$. Thus given a certain undisturbed depth h_0, if the bore strength was chosen in such a way that the ratio a_0 to h_0, we less than 0.28, the result was a purely undular bore. If the bore strength was chosen larger than that, the leading wave behind the bore front would break, leading to a much lower waveheight behind the bore front. It was also found that much larger bore strengths led to turbulent bores, but since the KdV equation is only valid for laminar flows, this regime is beyond the scope of the present investigation.

5 CONCLUSION

It has been shown that the derivation of the KdV equation as a model for surface water waves admits the definition of a local particle velocity field. The horizontal component of this velocity field may be evaluated near the crest of a wave. For solitary and cnoidal waves, evaluating the local Froude number leads to a convective wave breaking criterion which shows that there is a limiting waveheight beyond which the waves feature incipient wave breaking. Thus even though the solitary wave and the cnoidal waves are solutions of the KdV equation for any waveheight, they do not represent valid approximate surface waves for waveheights beyond the critical waveheight.

The critical waveheight for solitary waves is $H_{\text{max solitary}} = 0.6879$ is quite large, and it is not surprising that these waves do not represent reasonable approximations to real surface waves. For the cnoidal waves, the situation is more complex. For waves near Stokes number 1, the limiting waveheight according to the convective breaking criterion is in the range 0.15 to 0.3 (see Table 1), and these waves are in the range where it is generally thought that solutions of the KdV equation give a faithful approximation of real surface waves. As the elliptic parameter m approaches 1, the critical waveheight of the cnoidal waves approaches $H_{\text{max solitary}} = 0.6879$. On the other hand, the limit in the case when $m \to 0$ is the linear case in which solutions have negligible amplitude, and the critical waveheight in this case is indeed 0.

The convective breaking criterion has also been used to determine the critical bore strength α_{crit} for which an undular bore first exhibits breaking. As can be seen in Fig. 7, the result $\alpha = 0.323$ is still somewhat higher than the experimentally determined value $\alpha = 0.28$, which was obtained using wave tank experiments [21], but lower than the ratio obtained using a similar numerical model in [7]. The discrepancy with the experimental findings could be ascribed to the assumptions made on the physical system, i.e. the absence of transverse motion of the fluid and an irrotational and inviscid flow. While recent studies have highlighted the importance of these effects [15, 35], it is not clear that the inclusion of any of these will improve the quantitative agreement between the current analysis and the experimental findings of [21]. For example frictional damping due to molecular viscosity and boundary effects are more likely to delay the onset of breaking than to facilitate it. It therefore seems likely that the limiting procedure used in obtaining the model equations is responsible for some of the discrepancy, and a possible avenue for improving the quantitative description of wave breaking in an undular bore would be to include some of the higher-order terms which were neglected in the KdV model. Such higher-order models including both nonlinear and dispersive effects are well known (see [24] and the references therein).

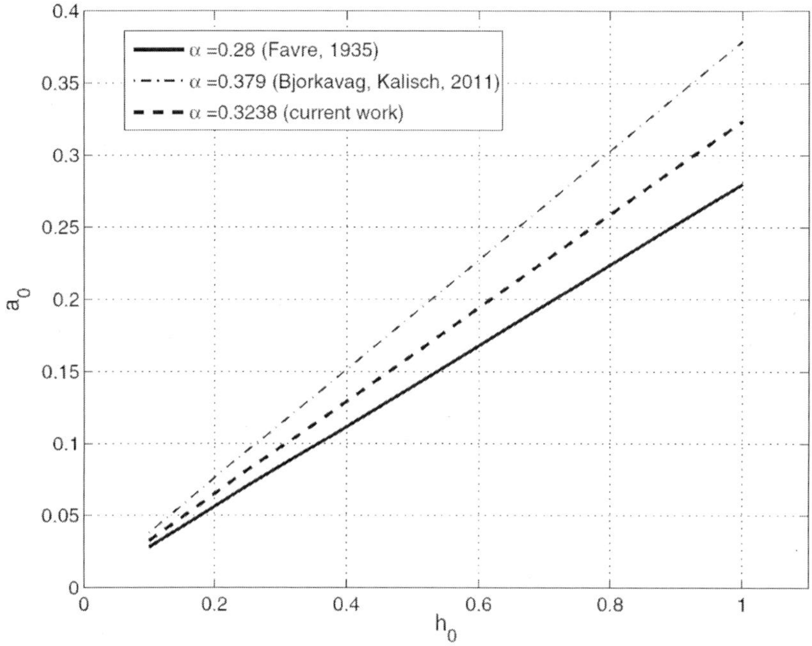

Figure. 7 The critical amplitude a0a0 dividing the purely undular case from the case where the leading wave features wave breaking graphed against the undisturbed depth h0h0. The slope of the curves is the non-dimensional critical bore strength. The value $\alpha = 0.28$ was found in Favre's experimental work [21]. The value $\alpha = 0.379$ was found in previous work [7]. The value $\alpha = 0.3283$ is found in the current work

Another possible improvement is the relaxation of the breaking criterion recently advocated in [4, 5]. In these works, several breaking criteria were used in connection with numerical Boussinesq models, and with the aim of quantifying the onset of breaking in shoaling waves. It was found that the convective criterion (1.6) did not perform as well as other criteria based on local waveheight [26, 42], and a relaxation constant was introduced in order to improve the analysis. However, since this constant has to be determined empirically from a given set of experiments, we have chosen not to pursue such a development here.

ACKNOWLEDGEMENTS

This work was supported in part by the Research Council of Norway through Grant No. 213474/F20.

APPENDIX: THE NUMERICAL SCHEME FOR THE KDV EQUATION

The numerical approximation of solutions of the Eq. (1.1) is based on a finite-difference method for the spatial derivatives with appropriate boundary conditions, and a hybrid Adams–Bashforth/ Crank–Nicolson time integration scheme. Since the KdV equation is of third order in spatial derivatives, three boundary conditions are needed for the numerical scheme. In the far field downstream of the bore front the surface elevation approaches zero, while in far field upstream it approaches αα. These far field conditions are chosen to be the boundary conditions which are exact up to machine precision as long as the spatial domain is large enough. Thus care has to be taken to ensure this is the case. For the third boundary condition a homogeneous Neumann condition is chosen on the right. To summarize, the problem to be solved is given by

$$\begin{aligned} \eta_t + \eta_x + \tfrac{3}{4}\left(\eta^2\right)_x + \tfrac{1}{6}\eta_{xxx} &= 0, \; x \in [-l, l], \; t \geq 0 \\ \eta(x, 0) &= \eta_0(x), \\ \eta(-l, t) &= \alpha, \\ \eta(l, t) &= 0, \\ \eta_x(l, t) &= 0. \end{aligned} \qquad . \qquad (5.1)$$

In order to incorporate non-homogeneous boundary conditions, we define the auxiliary function

$$\xi\,(x,t) \equiv \eta(x,t) - \eta_0(x). \qquad (5.2)$$

Upon substituting for $\eta(x,t)$ in (5.1) we have the problem in terms of ξ and η_0 which now has homogeneous Dirichlet conditions along with the same homogeneous Neumann condition on the right. The problem (5.1) now reads

$$\xi_t + \xi_x + \tfrac{3}{4}(\xi^2)_x + \tfrac{3}{2}(\eta_0\xi)_x + \tfrac{1}{6}\xi_{xxx} = -F, \; x \in [-l, l], \; t \geq 0,$$

where $F \equiv \eta_0' + \frac{3}{2}\eta_0\eta_0' + \frac{1}{6}\eta_0'''$, and homogeneous boundary and initial conditions are imposed.

To approximate the spatial derivatives a finite difference scheme is applied, while to advance in time, an explicit Adams-Bashforth method is used on the nonlinear terms and an implicit Crank-Nicolson method is used on the linear terms. Both methods are of second order, and while this does not automatically yield a second-order scheme for the full equation, numerical experiments in [37] suggest that this is in fact the case.

We continue by discretizing the spatial domain, $[-l, l]$ uniformly using a finite set of points, $\{x_j\}_{j=0}^{N} \subset [-l, l]$, where $x_0 = -l$ and $xN = l$, and $\delta x = 2l/N$ is the distance between two neighboring grid points. The time domain is also discretized uniformly using $t_n = n\delta t$, where $t_0 = 0$. We define the approximate function value at time tntn and grid point x_j to be $v_j^n = \xi(x_j, t_n)$. The solutions $\eta(x_j, t_n) = \eta_j^n$ is then given by (5.2).

The first and third derivatives at a point xjxj are approximated as follows using the central difference formulas:

$$\xi_x(x_j, t) \approx \frac{v_{j+1} - v_{j-1}}{2\delta x}$$

$$\xi_{xxx}(x_j, t) \approx \frac{v_{j+2} - 2v_{j+1} + 2v_{j-1} - v_{j-2}}{2\delta x^3}$$

From the Dirichlet conditions we have that $v_0 = 0$ and $vN = 0$ so we need only solve the equation for the grid points $\{x_j\}_{j=1}^{N-1}$. This leaves us with only two points for which the third derivative approximation is not valid. From the Neumann condition we have that $\xi x(l,t) = 0$, so by writing this as a central difference approximation we get that

$$\frac{v_{N+1} - v_{N-1}}{2\delta x} = 0 \implies v_{N+1} = v_{N-1}$$

This enables us to use the third derivative approximation at grid point xN−1xN−1 as follows

$$\xi_{xxx}(x_{N-1}, t) \approx \frac{v_{N+1} - 2v_N + 2v_{N-2} - v_{N-3}}{2\delta x^3} = \frac{v_{N-1} + 2v_{N-2} - v_{N-1}}{2\delta x^3}$$

As there is no Neumann condition on the left we use a forward difference formula to approximate the third derivative at grid point x1x1:

$$\xi_{xxx}(x_1, t) \approx \frac{-v_4 + 6v_3 - 12v_2 + 10v_1 - 3v_0}{2\delta x}$$

Defining difference matrices in this way, we may set up the following difference equation:

$$\frac{\mathbf{v}^{n+1} - \mathbf{v}^n}{\delta t} = -\frac{3}{4} D_1 (\mathbf{v}^n)^2 - \frac{3}{2} D_1 \mathbf{v}^n \eta_0 - \frac{1}{6} D_3 \mathbf{v}^n - D_1 \mathbf{v}^n - \mathbf{F},$$

where $\mathbf{v}^n = (v_1^n, v_2^n, \ldots, v_{N-1}^n)$, $\eta_0 = (\eta_0(x_1), \eta_0(x_2), \ldots, \eta_0(x_{N-1}))$ $\mathbf{F} = (F(x_1), F(x_2), \ldots, F(x_{N-1}))$

By applying the Adams–Bashforth method to the two first terms on the right, and the Crank–Nicolson method to the next two terms we get

$$\begin{aligned}\frac{\mathbf{v}^{n+1} - \mathbf{v}^n}{\delta t} = &- \frac{3}{4} D_1 \Big[3\big(\tfrac{1}{2}(\mathbf{v}^n)^2 + \mathbf{v}^n \eta_0\big) - \big(\tfrac{1}{2}(\mathbf{v}^{n-1})^2 + \mathbf{v}^{n-1}\eta_0\big) \Big] \\ &- \frac{1}{2}\Big(D_1\big[\mathbf{v}^{n+1} + \mathbf{v}^n\big] + \tfrac{1}{6} D_3 \big[\mathbf{v}^{n+1} + \mathbf{v}^n\big] \Big) - \mathbf{F}.\end{aligned}$$

At each time step this equation has to be solved for v^{n+1} to advance in time. By defining the matrix $E = (I + \frac{\delta t}{2} D_1 + \frac{\delta t}{12} D_3)$ for convenience, we do this as follows (I denotes the $(N-1) \times (N-1)$ identity matrix):

$$\begin{aligned}\mathbf{v}^{n+1} = &E^{-1}\Big[I - \tfrac{\delta t}{2} D_1 - \tfrac{\delta t}{12} D_3\Big]\mathbf{v}^n \\ &- \frac{3\delta t}{4} E^{-1} D_1 \Big[3\big(\tfrac{1}{2}(\mathbf{v}^n)^2 + \mathbf{v}^n \eta_0\big) - \big(\tfrac{1}{2}(\mathbf{v}^{n-1})^2 + \mathbf{v}^{n-1}\eta_0\big) \Big] - \delta t E^{-1}\mathbf{F}.\end{aligned}$$

This method requires function values at two previous time steps to calculate the next one, so at the very first iteration a different approach is needed. One way of fixing the problem is to use a forward Euler method on the nonlinear terms at the first time step. This method has a local truncation error of order two and any instability issues will not be a problem for one single time step. Thus we get the following difference equation for the first time step:

$$\frac{\mathbf{v}^2 - \mathbf{v}^1}{\delta t} = -\frac{3}{2} D_1 \left[\tfrac{1}{2} (\mathbf{v}^1)^2 + \mathbf{v}^1 \eta_{\mathbf{0}} \right] - \frac{1}{2} \left(D_1 \left[\mathbf{v}^2 + \mathbf{v}^1 \right] + \tfrac{1}{6} D_3 \left[\mathbf{v}^2 + \mathbf{v}^1 \right] \right) - \mathbf{F}$$

.

REFERENCES

1. Ablowitz, M., Segur, H.: Solitons and the inverse scattering transform. SIAM Studies in Applied Mathematics 4. SIAM, Philadelphia (1981)
2. Ali, A., Kalisch, H.: Mechanical balance laws for Boussinesq models of surface water waves. J. Nonlinear Sci. **22**, 371–398 (2012)
3. Ali, A., Kalisch, H.: On the formulation of mass, momentum and energy conservation in the KdV equation. Acta Appl. Math. **133**, 113–131 (2014)
4. Bacigaluppi, P., Ricchiuto, M., Bonneton, P.: A 1D stabilized finite element model for non-hydrostatic wave breaking and run-up. In: Finite Volumes for Complex Applications VII-Elliptic. Parabolic and Hyperbolic Problems Springer Proceedings in Mathematics & Statistics vol. 78, pp. 779–790 (2014)
5. Bacigaluppi, P., Ricchiuto, M., Bonneton, P.: *Upwind stabilized finite element modelling of non-hydrostatic wave breaking and run-up*, hal-00990002 (2014)
6. Benjamin, T.B.: The stability of solitary waves. Proc. R. Soc. Lond. A **328**, 153–183 (1972)
7. Bjørkavåg, M., Kalisch, H.: Wave breaking in Boussinesq models for undular bores. Phys. Lett. A **375**, 1570–1578 (2011)
8. Bona, J.L., Chen, M., Saut, J.-C.: Boussinesq equations and other systems for small-amplitude long waves in nonlinear dispersive media. I: Derivation and linear theory. J. Nonlinear Sci. **12**, 283–318 (2002)
9. Bona, J.L., Colin, T., Lannes, D.: Long wave approximations for water waves. Arch. Ration. Mech. Anal. **178**, 373–410 (2005)
10. Bona, J.L., Saut, J.-C.: Dispersive blowup of solutions of generalized Korteweg–de Vries equations. J. Differ. Equ. **103**, 3–57 (1993)
11. Bona, J.L., Sun, S.M., Zhang, B.-Y.: A nonhomogeneous boundary-value problem for the Korteweg–de Vries equation posed on a finite domain. Commun. Partial Differ. Equ. **28**, 1391–1436 (2003)
12. Borluk, H., Kalisch, H.: Particle dynamics in the KdV approximation. Wave Motion **49**, 691–709 (2012)
13. Borluk, H., Kalisch, H., Nicholls, D.P.: A numerical study of the Whitham equation as a model for steady surface water waves. J. Comput. Appl. Math. **296**, 293–302 (2016)
14. Bottman, N., Deconinck, B.: KdV cnoidal waves are spectrally stable. Discrete Contin. Dyn. Syst. **25**, 1163–1180 (2009)
15. Chanson, H.: Current knowledge in hydraulic jumps and related phenomena: a survey of experimental results. Eur. J. Mech. B/Fluids **28**, 191–210 (2009)
16. Constantin, A., Escher, J.: On the blow-up rate and the blow-up set of breaking waves for a shallow water equation. Math. Z. **233**, 75–91 (2000)
17. Chawla, A., Özkan-Haller, H.T., Kirby, J.T.: Spectral model for wave transformation and breaking over irregular bathymetry. J. Waterw. Port Coast. Ocean Eng. **124**, 189–198 (1998)
18. Chen, Y.-Y., Hsu, H.-C., Chen, G.-Y.: Lagrangian experiment and solution for irrotational finite-amplitude progressive gravity waves at uniform depth. Fluid Dyn. Res. **42**, 045511 (2010)
19. Dean, R.G., Dalrymple, R.A.: Water Wave Mechanics for Engineers and Scientists. World Scientific, Singapore (1991)
20. Dingemans, M.W.: Water Wave Propagation Over Uneven Bottoms: Linear Wave Propagation, vol. 13. World Scientific, Singapore (1997)
21.

22. Favre, H.: Ondes de Translation. Dunod, Paris (1935)
23. Francius, M., Kharif, C.: Three-dimensional instabilities of periodic gravity waves in shallow water. J. Fluid Mech. **561**, 417–437 (2006)
24. Henry, D.: Steady periodic flow induced by the Korteweg-de Vries equation. Wave Motion **46**, 403–411 (2009)
25. Karczewska, A., Rozmej, P., Infeld, E.: Energy invariant for shallow-water waves and the Korteweg–de Vries equation: doubts about the invariance of energy. Phys. Rev. E **92**, 053202 (2015)
26. Kataoka, T., Tsutahara, M.: Transverse instability of surface solitary waves. J. Fluid Mech. **512**, 211–221 (2004)
27. Kazolea, M., Delis, A., Synolakis, C.: Numerical treatment of wave breaking on unstructured finite volume approximations for extended Boussinesq-type equations. J. Comput. Phys. **271**, 281–305 (2014)
28. Kharif, C., Pelinovsky, E.: Physical mechanisms of the rogue wave phenomenon. Eur. J. Mech. B/Fluids **22**, 603–634 (2003)
29. Koch, C., Chanson, H.: Unsteady Turbulence Characteristics in an Undular Bore, in River Flow 2006. Taylor & Francis Group, London (2006)
30. Korteweg, D.J., de Vries, G.: On the change of form of long waves advancing in a rectangular channel and on a new type of long stationary wave. Philos. Mag. **39**, 422–443 (1895)
31. Lannes, D.: The Water Waves Problem. AMS, Providence (2013)
32. Lawden, D.F.: Elliptic Functions and Applications. Springer, New York (1989)
33. Longuet-Higgins, M.S.: On wave breaking and equilibrium spectrum of wind. Proc. R. Soc. Lond. Ser. A **310**, 151–159 (1969)
34. McLean, J.W.: Instabilities of finite amplitude gravity waves on water of finite depth. J. Fluid Mech. **114**, 331–341 (1982)
35. Rauch, J.: Partial Differential Equations. Springer, New York (1991)
36. Richard, G., Gavrilyuk, S.: The classical hydraulic jump in a model of shear shallow-water flows. J. Fluid Mech. **725**, 492–521 (2013)
37. Sato, S., Kabiling, M.B.: A numerical simulation of beach evolution based on a nonlinear dispersive wave-current model. In: Proceedings of the Twenty-Fourth International Conference held in Kobe, Japan, Coastal Engineering 1994, New York (1995), pp. 2557–2570
38. Skogestad, J.O., Kalisch, H.: A boundary value problem for the KdV equation: Comparison of finite-difference and Chebyshev methods. Math. Comput. Simul. **80**, 151–163 (2009)
39. Song, J.B., Banner, M.L.: On determining the onset and strength of breaking for deep water waves. Part I: unforced irrotational wave groups. J. Phys. Oceanogr. **32**, 2541–2558 (2002)
40. Stansell, P., MacFarlane, C.: Experimental investigation of wave breaking criteria based on wave phase speeds. J. Phys. Oceanogr. **32**, 1269–1283 (2002)
41. Tian, Z., Perlin, M., Choi, W.: Evaluation of a deep-water wave breaking criterion. Phys. Fluids **20**, 066604 (2008)
42. Tissier, M., Bonneton, P., Marche, F., Chazel, F., Lannes, D.: A new approach to handle wave breaking in fully non-linear Boussinesq models. Coast. Eng. **67**, 54–66 (2012)
43. Tonelli, M., Petti, M.: Simulation of wave breaking over complex bathymetries by a Boussinesq model. J. Hydraul. Res. **49**, 473–486 (2011)
44. Tanaka, M.: The stability of solitary waves. Phys. Fluids **29**, 650–655 (1986)
45. Tanaka, M., Dold, J.W., Lewy, M., Peregrine, D.H.: Instability and breaking of a solitary wave. J. Fluid Mech. **185**, 235–248 (1987)
46. Tulin, M.P., Li, J.J.: On the breaking of energetic waves. Int. J. Offshore Polar Eng. **2**, 8 (1992)
47. Whitham, G.B.: Linear and Nonlinear Waves. Wiley, New York (1974)MATHGoogle Scholar

48. Wu, C.H., Nepf, H.M.: Breaking criteria and energy losses for three-dimensional wave breaking. J. Geophys. Res. Oceans **107**, C10 (2002)
49. Yao, A., Wu, C.H.: Incipient breaking of unsteady waves on sheared currents. Phys. Fluids **17**, 082104 (2005)

CHAPTER 6

Modeling for Collapsing Cavitation Bubble near Rough Solid Wall by Mulit-Relaxation-Time Pseudopotential Lattice Boltzmann Model

Minglei Shan[1,2], Yipeng Zhu1, Cheng Yao[1,2], Qingbang Han[1,2], Changping Zhu[1,2]

[1] *College of Internet of Things Engineering, Hohai University, Changzhou, China*

[2] *Jiangsu Key Laboratory of Power Transmission and Distribution Equipment Technology, Hohai University, Changzhou, China*

ABSTRACT

Cavitation bubble collapse near rough solid wall is modeled by the multi-relaxation-time (MRT) pseudopotential lattice Boltzmann (LB) model. The modified forcing scheme, which can achieve LB model's thermodynamic consistency by tuning a parameter related with the particle interaction range, is adopted to achieve desired stability and density ratio. The bubble collapse near rough solid wall was simulated by the improved MRT pseudopotential LB model. The mechanism of bubble collapse is studied by investigating the bubble profiles, pressure field and velocity field evolution. The eroding effects of collapsing bubble are analyzed in details. It is found that the process and the effect of the interaction between bubble collapse and rough solid wall are affected seriously by the geometry of solid boundary. At the same time, it demonstrates that the MRT pseudopotential LB model is a potential tool for the investigation of the interaction mechanism between the collapsing bubble and complex geometry boundary.

Keywords: Cavitation Bubble, Bubble Collapse, Lattice Boltzmann Method, Pseudopotential Model, Rough Solid Wall

1. INTRODUCTION

Cavitation is ubiquitous in liquid, and happens when the local pressure is below the saturated vapor pressure. As cavitation bubble collapse near a solid wall, the associated phenomena include instant high pressure, high velocity jets and high temperature, which closely relate with the cavitation erosion of the solid material surface. On the other side, the collapse of the cavitation bubble has been applied in environmental protection, ultrasonic therapy, lab on chip and material surface cleaning [1] [2].

The mechanism of the cavitation bubble collapse near solid wall is a fundamental issue for the above applications. However, as too many phenomena are involved, theoretical model of cavitation bubble collapse is difficult to be established, and for complex geometry boundary, the analytical solution is even impossible. Many publications are devoted to the investigation of the bubble collapse near the planar wall starting from the experimental work [3]. In recent years, many works have been developed to investigate the interaction between collapsing bubble and non-planar solid wall [1] [4] , and numerical methods are becoming more and more important tools to study the bubble collapse near solid wall [5] [6]. Commonly used numerical methods include the finite volume method (FVM) [4] , the finite element method (FEM) [7] , and the boundary element method (BEM) [8]. These macroscopic numerical modeling methods based on solving partial differential equations are limited in processing the multiphase flows and complex geometry boundary conditions. As describing the multiphase flow, macroscopic methods need the assistance of the schemes of the interface tracking or capturing, which will reduce the computational efficiency. For the complex geometry boundary, it is difficult and inefficient to implement by macroscopic methods.

Owing to the flexibility for complex geometry boundary and the simplicity of the algorithm, the lattice Boltzmann method (LBM) has been developed into a powerful tool for the flow, heat and mass transfer simulations relating with com- plex geometry boundary [9] [10]. In LBM community, many multiphase models have been presented, which can be generally classified as the color-gradient model [11] , the pseudopotential model (or Shan-Chen model) [12] [13] , the free-energy model [14] and the phase-field model. The pseudopotential model is widely and successfully used in the LBM multiphase community due to its conceptual simplicity and computation efficiency [15] [16]. In the pseudopotential model, the fluid interactions are mimicked by an interparticle potential, from which a non-monotomic equation of state (EOS) can be obtained. As a result, the separation of fluid phases or components can be achieved automatically in this method, and the methods to track or capture the interfaces are not required.

In recent years, the pseudopotential model, as the top choice of the multiphase LB model, was introduced into the issue of cavitation by Sukop and Or [17]. Subsequently, several research efforts emerged to investigate the mechanism of cavitation. Chen [18] simulated the cavitation bubble growth using the modified pseudopotential LB model with the exact difference method (EDM) force scheme. The results in quiescent flows agree fairly well with the solution of Rayleigh- Plesset equation. Mishra [19] introduced a model of cavitation based on the pseudopotential LB model that allows for coupling between the hydrodynamics of a collapsing cavity and supported solute chemical species. Unfortunately, the above researches do not involve the interaction between bubble and solid wall. Until, the pseudopotential model was introduced into the researches on the mechanism investigations of cavitation bubble collapse near solid wall by Shan et al. [20] [21]. However, these previous works involve just planar wall. In engineering cases, the rough walls are ubiquitous. The mechanism investigation of cavitation bubble collapse near rough wall is the basic issue for the efficient applications of cavitation bubble collapse.

One of the key challenges of the simulation of cavitation bubble collapse near rough wall is the stability of the LB model. For pseudopotential multiphase model, the stability is closely related to the thermodynamic consistency [22] , and great efforts have been made for this issue [23] [24] [25]. Recently, Li et al. [24] [25] [26] founded that there exists a suitable forcing scheme, which can meet the thermodynamic consistency requirement in an efficient way. In order to investigate the interaction mechanism between the collapsing bubble and complex geometry boundary, in the present work, the improved forcing scheme for the pseudopotential MRT LB model developed by Li et al. [25] is adopted to achieve the sufficient density ratio and stability of the numerical model.

2. NUMERICAL METHOD

The pseudopotential LB model, also well known as Shan-Chen model, was developed by Shan and Chen in 1993 [12]. In pseudopotential model, the fluid interactions are mimicked by an interparticle potential, which is now widely called pseudopotential. In original pseudopotential LB model, the Single-Relaxation- Time (SRT) collision operator was employed. In recent years, the MRT collision operator has been verified that it is superior to the SRT operator in terms of numerical stability. The MRT LB evolution equation can be given as follows:

$$f_\alpha\left(\boldsymbol{x}+\boldsymbol{e}_\alpha\delta_t,t+\delta_t\right)=f_\alpha\left(\boldsymbol{x},t\right)-\left(\boldsymbol{M}^{-1}\boldsymbol{\Lambda}\boldsymbol{M}\right)_{\alpha\beta}\left(f_\beta-f_\beta^{eq}\right)+\delta_t\boldsymbol{F}'_\alpha, \tag{1}$$

where f_a is the density distribution function, f_α^{eq} is its equilibrium distribution, t is the time, x is the spatial position, e is the discrete velocity along the ath direction, δ_t is the time step, $\boldsymbol{F}'_\alpha$ is the forcing term in the velocity space, and $\boldsymbol{M}$ is an orthogonal transformation matrix. $\boldsymbol{\Lambda}$ in Equation (1) is a diagonal matrix, and for D2Q9 lattice, $\boldsymbol{\Lambda}$ is given by

$$\boldsymbol{\Lambda}=diag\left(\tau_\rho^{-1},\tau_e^{-1},\tau_\zeta^{-1},\tau_j^{-1},\tau_q^{-1},\tau_j^{-1},\tau_q^{-1},\tau_v^{-1},\tau_v^{-1}\right) \tag{2}$$

Through the transformation matrix $\boldsymbol{M}$, f and f_α^{eq} can be projected onto the moment space via $m = mf$ and $m^{eq} = mf^{eq}$, and the collision step of MRT LB equation (Equation (1)) can be rewritten as [25]

$$\boldsymbol{m}^*=\boldsymbol{m}-\Lambda\left(\boldsymbol{m}-\boldsymbol{m}^{eq}\right)+\delta_t\left(\boldsymbol{I}-\frac{\boldsymbol{\Lambda}}{2}\right)\boldsymbol{S} \tag{3}$$

where I is the unit tensor, and S is the forcing term in the moment space with $(I - 0.5\,\boldsymbol{\Lambda})S = MF'$. For D2Q9 lattice, m^{eq}can be given by

$$\boldsymbol{m}^{eq}=\rho\left(1,-2+3|\boldsymbol{v}|^2,1-3|\boldsymbol{v}|^2,v_x,-v_x,v_y,-v_y,v_x^2-v_y^2,v_xv_y\right)^{\mathrm{T}}. \tag{4}$$

Here $\rho=\sum_\alpha f_\alpha$ is the macroscopic density, v is the macroscopic velocity which satisfies $|\boldsymbol{v}|^2=v_x^2+v_y^2$ and is calculated by

$$v = \frac{\sum_{\alpha} \boldsymbol{e}_{\alpha} f_{\alpha} + 0.5\delta_t \boldsymbol{F}}{\rho}, \quad (5)$$

where $F = (F_x, F_y)$ for two dimensional space is the force action on the fluid system. Then the streaming step of the MRT-LB equation can be formulated as

$$f_{\alpha}\left(\boldsymbol{x} + \boldsymbol{e}_{\alpha}\delta_t, t + \delta_t\right) = f_{\alpha}^{*}\left(\boldsymbol{x}, t\right), \quad (6)$$

Where $f^{*} = M^{-1}m^{*}$

For the pseudopotential LB model in D2Q9 lattice case, the F in Equation (5) is given by

$$\boldsymbol{F} = -G\psi\left(\boldsymbol{x}\right)\sum_{\alpha=1}^{8} w_{\alpha}\psi\left(\boldsymbol{x} + \boldsymbol{e}_{\alpha}\right)\boldsymbol{e}_{\alpha} \quad (7)$$

where $\psi(x)$ is the interparticle potential, G is the interaction strength, and $w_{1,2,3,4,} = 1/3$ and $w_{5,6,7,8,} = 1/12$ are the weights. In the present work, the form of ψ proposed by Yuan et al. [27] is adopted which can be formulated as

$$\psi = \sqrt{\frac{2\left(p_{EOS} - \rho c_S^2\right)}{Gc^2}}, \quad (8)$$

where p_{EOS} is the pressure calculated by equation of state (EOS). The G in Equation (8) loses the meaning of the interaction strength and is used to ensure that the whole term inside the square root is positive [27]. The Carnahan-Starl- ing (CS) EOS is adopted in the present work, which can be given by [27]

$$p_{EOS} = \rho RT \frac{1 + b\rho/4 + \left(b\rho/4\right)^2 - \left(b\rho/4\right)^3}{\left(1 - b\rho/4\right)^3} - a\rho^2, \quad (9)$$

where $a = 0.4963R^2T_c^2/p_c$ and.$b = 0.1873RT_c/P_c$ Here T_c and p_c are the critical temperature and pressure, respectively.

F in Equation (7) can be incorporated in evolution equation via S with specific forcing scheme. Li et al. [25] proposed a MRT version forcing scheme to achieve thermodynamic consistency. For the D2Q9 lattice, Li's forcing scheme can be given by

$$S = \begin{pmatrix} 0 \\ 6\left(v_x F_x + v_y F_y\right) + \dfrac{0.75\varepsilon |F|^2}{\psi^2 \delta_t (\tau_e - 0.5)} \\ -6\left(v_x F_x + v_y F_y\right) - \dfrac{0.75\varepsilon |F|^2}{\psi^2 \delta_t (\tau_e - 0.5)} \\ F_x \\ -F_x \\ F_y \\ -F_y \\ 2\left(v_x F_x - v_y F_y\right) \\ v_x F_y + v_y F_x \end{pmatrix}, \tag{10}$$

where ε is a parameter related with the particle interaction range [28] and is proved by Li et al. to have the function of adjusting the thermodynamic consistency [25]. Li's forcing scheme has the following advantages: a) maintaining a uniform layout with the a general form of the LB forcing scheme; b) achieving thermodynamic consistency only by tuning one constant parameter; and c) fully retaining the LBM's advantages of simple and efficient. In [21] it is found that the thermodynamic consistency is independent of kinematic viscosity for Li's improved forcing scheme, and the surface tension is independent of the relaxation time.τ_v These features make it more convenient to investigate the physical mechanism of the multiphase flows.

Unless otherwise specified, the unit adopted in this paper is the lattice unit of LBM, i.e. the units of the length, time, mass, velocity, density and pressure are lu(lattice unit), ts(time step), mu(mass unit), mu lu^{-2}, lu ts^{-1} and mu ts^{-2}.

3. CAVITATION BUBBLE COLLAPSE NEAR ROUGH SOLID WALL

It is confirmed that the geometry is a crucial role on the process of the cavitation bubble collapse [29]. The interaction between collapse bubble and rough solid wall is a common issue abstracted from the cavitation applications involving the interactions between bubble and complex geometry boundary. In the present work, the rough solid wall is described by the wall with the geometries of periodic grooves with equal widths as shown in Figure 1. In this diagram, D_g and W_g are the depth and the width of groove, respectively. W_b is the width of bulge, and $W_P = W_g + W_b$ is the period width of the periodically arranged geometry.

The computational domain is shown in Figure 2. R_0, $P_v$and P_∞ are the initial radius, the pressure inside bubble and the ambient pressure, respectively. d_g and d_b are the distance from bubble center to groove bottom and bulge top, respectively. Based on the concept of the stand-off parameter [30] , two non- dimensional parameters are introduced as follows to describe the positional relations between bubble and rough solid wall

$$\lambda_b = \frac{d_b}{R_0}, \tag{11}$$

$$\lambda_g = \frac{d_g}{R_0}, \tag{12}$$

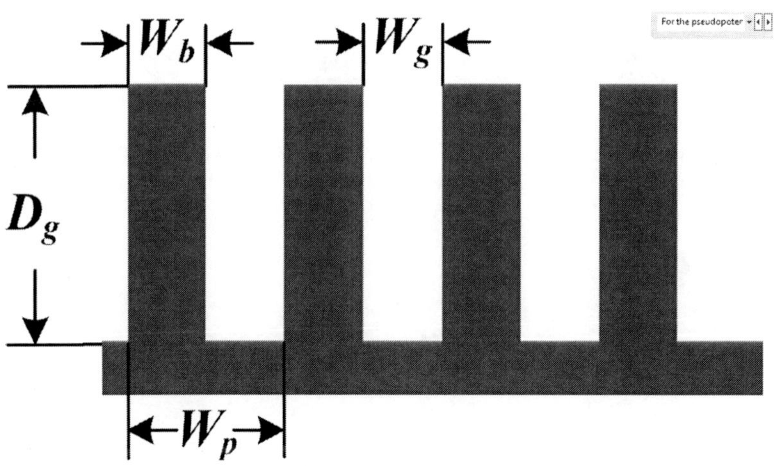

Figure 1. Schematic diagram of rough wall geometry.

Here, , γ_h, $\gamma_g$express the stand-off distances between bubble and bulge top, groove bottom, respectively.

The pressure boundary conditions, implemented by the nonequilibrium extra- polation scheme [31] , are adopted in the directions of left and right.

And, a constant pressure boundary condition is adopted in the top of computational domain by Zou-He scheme [32].

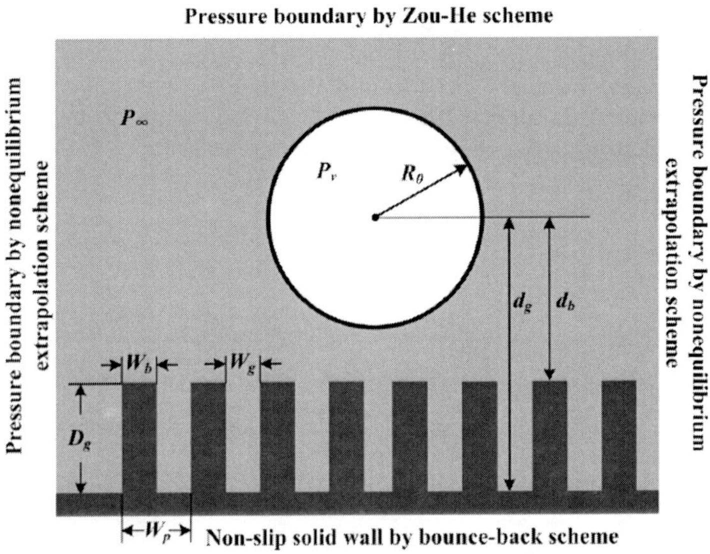

Figure 2. Computational domain for single solid wall case.

A 401×401 lattice is adopted in the simulations of this section. A vapor bubble with radius of R = 80 is initially placed at the center of the domain. The density field is initialized as [33]

$$\rho(x,y)=\frac{\rho_l+\rho_g}{2}+\frac{\rho_l-\rho_g}{2}\tanh\left(\frac{2\left(\sqrt{(x-x_0)^2+(y-y_0)^2}-R_0\right)}{W}\right) \quad (13)$$

where (x_0,y_0) is the center of the bubble, W is the prescribed width of the phase interface and is set as 5 in present work, tanh is the hyperbolic tangent function. The parameters of solid wall geometry are set as D_g = 144 $W_b = W_g$ =10, W_p = 20, dg = 160 and d_b=15, respectively.

The parameters for the present MRT pseudopotential LB model are chosen as follows: $\tau_\rho^{-1}=\tau_j^{-1}=1.0$, $\tau_e^{-1}=\tau_\zeta^{-1}=1.0$, $\tau_q^{-1}=1.1$, $\tau_\nu=0.6$ and ε =1.86 .For CS EOS, a = 0.5, = 4 and R = 1. The temperature T= 0.7T_c. In order to simulate the bubble collapse process, a positive pressure difference

$\Delta P = P_\infty - P_v$ is achieved by artificially tuning the initial liquid density based on the equilibrium state. In this section, the pressure difference is $\Delta P = 0.0116$.

4. RESULTS

4.1. Evolution of Density Field

The evolution of density field is shown in Figure 3. The deformation of bubble profile can be investigated from density field. The initial spherical bubble begins to collapse motivated by the pressure difference from $t = 0$. The velocity of collapse is slow at the starting stage. Until $t = 202$, the bubble profile appears obvious deformation, i.e., the radius of curvature reduce towards the rough solid wall.

Then, the bubble becomes an ellipsoid and the density just over the bubble increases. Along with the diffusion of the higher density area, the top of the bubble is flatten, and then concaved directing the solid wall. At the same time, the more dense and concentrated density area appears at the concave and accelerate the shift of the bubble profile. As a result, the jet is formed. The first collapse happens when the upper bubble wall clashes the bottom wall. The jet perforates the bottom wall and produces a crucial bubble. The shock wave generated by the first collapse propagates rapidly towards the rough solid wall and bounce back.

The bubble experiences the complex and volatile deformation at the last stage under the jet which includes shock wave and bounce wave.

Figure 3 describes the whole process of bubble collapse with time t. Comparing with the flat solid wall case, the bubble collapse near rough solid wall appears the similar dynamics process [20]. The differences are mainly reflected in two aspects.

The first one is the bounce of the shock wave in rough solid wall case is weaker than the one in the planar wall case. The strength and the reflection path of the shock wave are affected by the geometry of the rough solid wall. The second one is the geometry of the rough solid wall affects the bubble deformation, especially at the last stage.

In order to be more intuitive, pressure field evolution of the collapsing bubble is investigated in the next section.

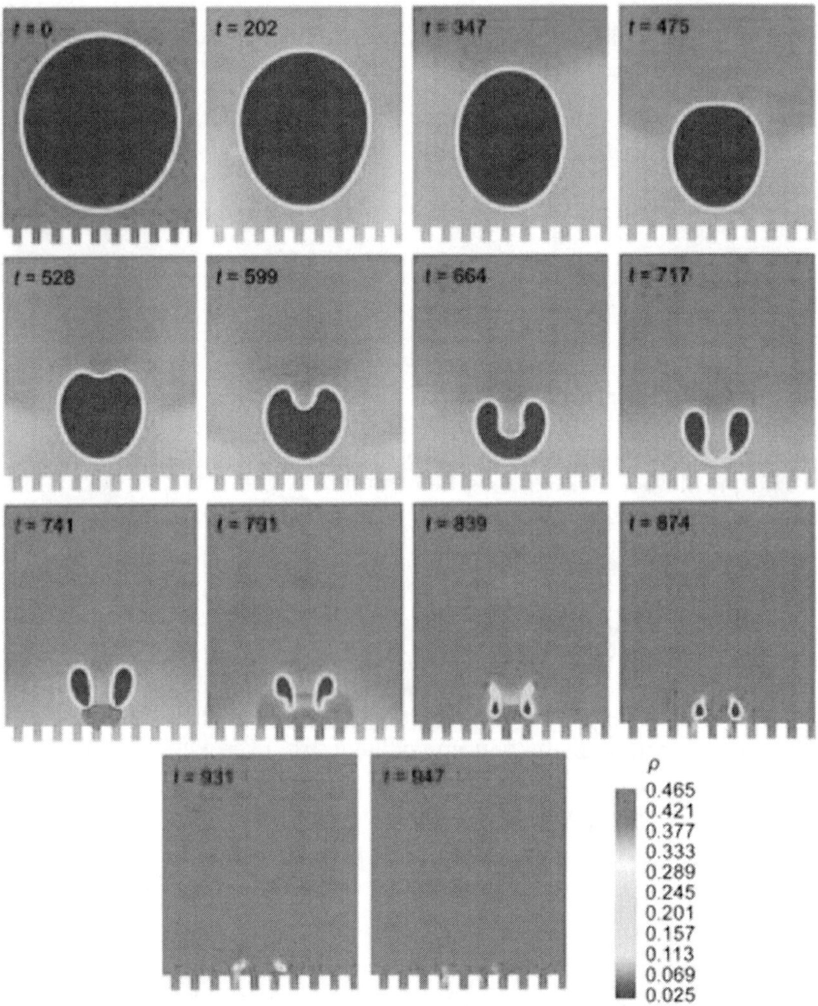

Figure 3. Density field evolution of the collapsing bubble near rough solid wall.

4.2. Evolution of Pressure and Velocity Field

Several representative moments of the pressure and velocity evolution are shown in Figure 4. At t = 528, one high pressure area emerges at the concave of bubble top. It is consistent with the analysis of the density distribution at the same moment. At the same time, the fluid velocity is increased towards the solid wall which formats the inceptive jet. The jet velocity at the concave is significantly higher than that of other areas in the moment. Along with the sagging of the upper bubble wall, the jet velocity is higher and higher. At t = 717, the first collapse occurs. The clash between the upper and the bottom walls induces shock wave as shown in t = 741. And at the same moment, the bubble is broken by the high velocity jet. Affected by the vortexes, which are induced by the jet and the bouncing back shock wave, the crucial bubble deforms and then collapses close to the rough solid wall. From Figure 4, we can find that the shock waves bounce intensively at the solid surfaces of the bulges top, and propagate unceasingly into the grooves with decreasing strength. The fluid jet follows the similar mode. It is inevitable that the shock wave and the jet will erode and impact the rough solid wall.

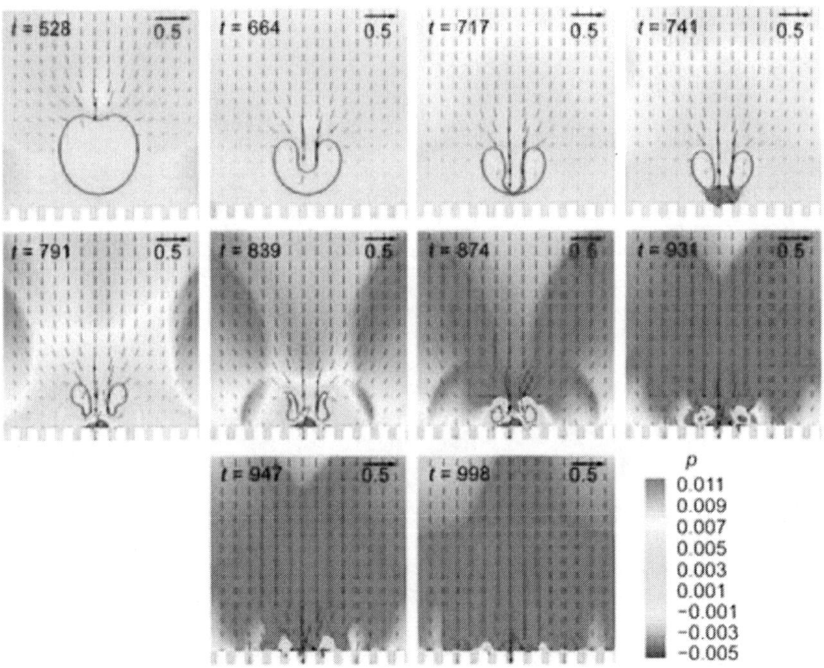

Figure 4. Pressure and velocity field evolution of the collapsing bubble near rough solid wall.

Figure 5 gives the finer detail display of the interaction between the pressure, velocity and the rough solid wall at t = 998. It is found that in the groove and near the bulge, which are off-center the collapse position, the vortexes are formed. Conversely, the laminar flows are formed in the groove just under the collapse position. The vortexes and the laminar flows induce the remarkable eroding effect on the solid surface of the bulge and the groove, respectively.

4.3. Analysis of Eroding Effect

In order to investigate the eroding process and effect induced by bubble collapse close to the rough solid wall, four test sections are set at the bulges top and the grooves side walls as shown in Figure 5. The coordinates of these four test sections are $(a \mid x = 191, y \in [257, 270])$, $(b \mid x = 211, y \in [257, 270])$, $(c \mid y = 256, x \in [200, 210])$ and $(d \mid y = 256, x \in [220, 230])$, respectively. The pressure evolution and the velocity evolution of these four test sections, from 730 ts to 1000 ts, are displayed in Figure 6 and Figure 7.

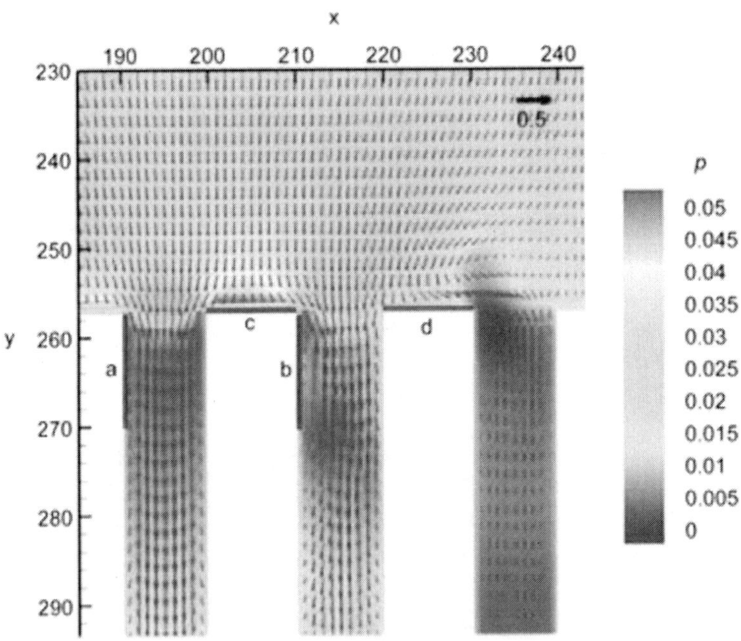

Figure 5. Pressure and velocity distribution at rough solid wall after the bubble collapse.

Form Figure 6, we can find that the test section a experiences three pressure peaks during the time interval of observation. The closer to the groove entrance, the greater the pressure amplitude is. The first pressure peak with the greatest amplitude is the result of shock wave. The second is the water hummer effect of jet. However, as the test section is parallel with jet direction, the pressure amplitude is weaker than the first pressure peak. The third pressure peak is induced by the second collapse. The pressure evolution at the test section b displays similar pattern. Due to the impeding effect of the bulge, the lower pressure area is induced near the test section b. The test section c is just under the collapse position, so the three pressure peaks are greater than the test section a and b. The pressure pattern of the test section d is similar as that of b on account of the larger deviation from the collapse station. But the pressure amplitudes are higher than the test section b because of the closer distance from the solid wall. The pressure evolution at the four test sections demonstrates that no matter the bulge solid wall or the groove entrance experience the oscillation of pressure.

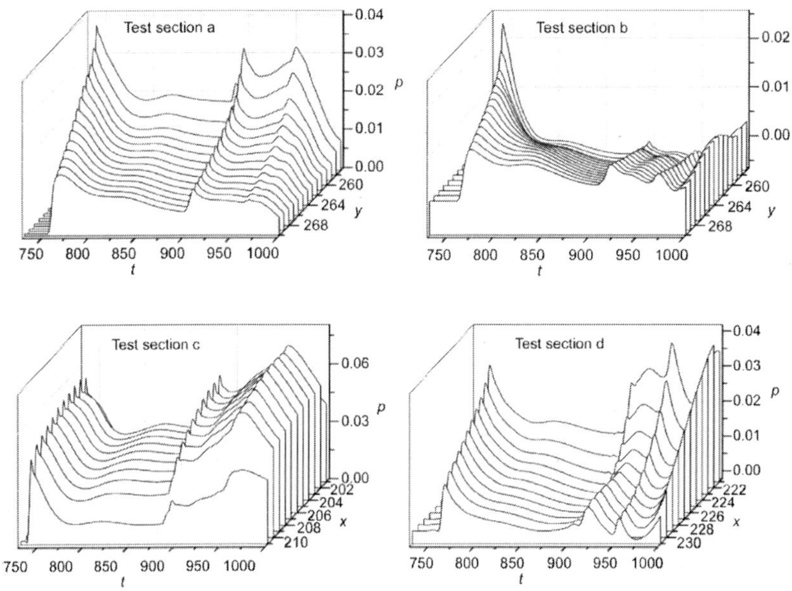

Figure 6. Pressure evolution at the test sections.

The velocity evolution of the four test sections are shown in Figure 7. For the test section a, every test point experiences the similar process of the fluid velocity, which increases sharply and then gradually rises to the peak. The pattern of the velocity is same as the pressure, i.e., the closer to the groove entrance, the greater the velocity is. It demonstrated that the groove geometry impedes both of the pressure and the velocity at the groove side walls. Although the same near the entrance, the test section b shows a different pattern. Under the combined action of shock wave and jet, the vortexes are induced at b. The velocity evolution pattern of the test section c is symmetric in x direction. Due to the non-slid effect of solid wall, the fluid velocity at center of the bulge is lower than the sides, which are closer to the groove entrances. For the test section d, the velocity pattern is more diversified with lower fluid velocity than other test sections. The oscillation of velocity is more violent for farther distance from collapse position. The last velocity peaks are related with the collapse of the crucial bubble.

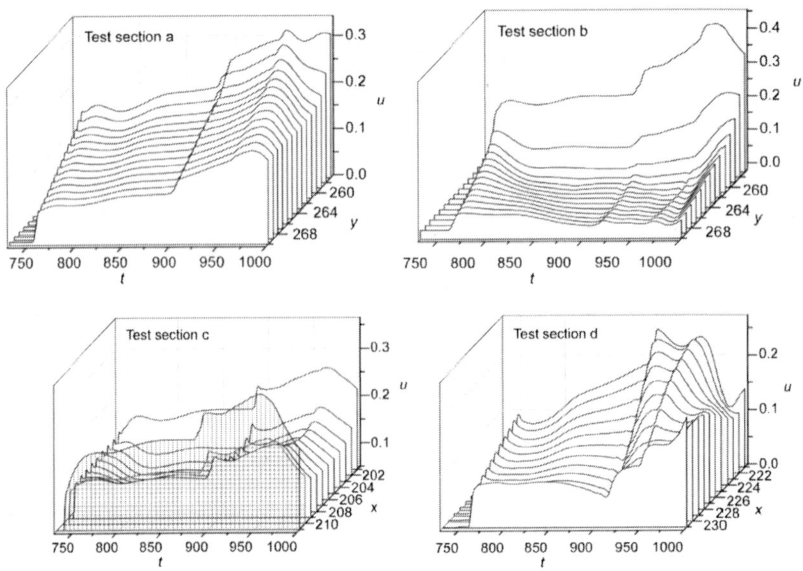

Figure 7. Velocity evolution at the test sections.

In summary, the eroding effect at the solid wall of the bulge part is weaker than that at the side walls, and is more obvious at the side walls of the grooves than other sections.

From the pressure and velocity evolution patterns, it demonstrates that the fluid velocity accounts for the eroding effect at the side walls of the grooves, and the pressure accounts for the eroding effect at the bulge part of the rough solid wall.

5. CONCLUSIONS

For the modeling of the collapsing bubble near rough solid wall, an improved MRT pseudopotential LB model was adopted with the modified forcing scheme by Li et al. Then the bubble collapse near rough solid wall was simulated. The bubble collapse mechanism was investigated from the dynamic process including bubble profiles evolution, pressure and velocity field evolution. The eroding effects of shock wave and jet were analyzed in details. It is found that the process and the effect of bubble collapse are affect seriously by the geometry of solid boundary. In the present work, the interaction between the collapsing bubble and the geometries boundary structured with periodic grooves was investigated. We found that the fluid velocity is the major cause for the eroding effect at the side walls of the grooves, and the pressure is the major cause for the eroding effect at the bulge part of the rough solid wall.

The improved forcing scheme proposed by Li et al. provides a convenient and efficient approach to achieve thermodynamic consistency. By the modelling of the bubble collapse near the rough solid wall, it is demonstrated that the MRT pseudopotential LB model improved by the modified forcing scheme has enough stability to describe the whole process of bubble collapse and the interaction between the collapsing bubble and complex geometry boundary.

ACKNOWLEDGEMENTS

The present work is supported by the National Key Research and Development Program of China (Grant No. 2016YFC0401600), the Primary Research & Development Plan of Jiangsu Province, China (Grant No. BE2016056) and the Natural Science Foundation of Jiangsu Province, China (Grant No. SBK2014043338).

REFERENCES

1. Abboud, J.E. and Oweis, G.F. (2013) Themicrojetting Behavior from Single Laser-Induced Bubbles Generated above a Solid Boundary with a Through Hole. Experiments in Fluids, 54, 1430. https://doi.org/10.1007/s00348-012-1438-6
2. Hashmi, A., Yu, G., Reilly-Collette, M., Heiman, G. and Xu, J. (2012) Oscillating Bubbles: A Versatile Tool for Lab on a Chip Applications. Lab on a Chip, 12, 4216-4227. https://doi.org/10.1039/c2lc40424a

3. Naudé, C.F. and Ellis, A.T. (1961) On the Mechanism of Cavitation Damage by Nonhemispherical Cavities Collapsing in Contact with a Solid Boundary. Journal of Basic Engineering, 83, 648-656. https://doi.org/10.1115/1.3662286
4. Li, B.B., Jia, W., Zhang, H.C. and Lu, J. (2014) Investigation on the Collapse Behavior of a Cavitation Bubble near a Conical Rigid Boundary. Shock Waves, 24, 317-324. https://doi.org/10.1007/s00193-013-0482-3
5. Aganin, A.A., Ilgamov, M.A., Kosolapova, L.A. and Malakhov, V.G. (2016) Dynamics of a Cavitation Bubble near a Solid Wall. Thermophysics and Aeromechanics, 23, 211-220. https://doi.org/10.1134/S0869864316020074
6. 6. Magaletti, F., Gallo, M., Marino, L. and Casciola, C.M. (2016) Shock-Induced Collapse of a Vapor Nanobubble near Solid Boundaries. International Journal of Multiphase Flow, 84, 34-45. https://doi.org/10.1016/j.ijmultiphaseflow.2016.02.012
7. Hsu, C.Y., Liang, C.C., Nguyen, A.T. and Teng, T.L. (2014) A Numerical Study on the Underwater Explosion Bubble Pulsation and the Collapse Process. Ocean Engineering, 81, 29-38. https://doi.org/10.1016/j.oceaneng.2014.02.018
8. Zhang, A.M. and Liu, Y.L. (2015) Improved Three-Dimensional Bubble Dynamics Model Based on Boundary Element Method. Journal of Computational Physics, 294, 208-223. https://doi.org/10.1016/j.jcp.2015.03.049
9. Chen, S. and Doolen, G.D. (1998) Lattice Boltzmann Method for Fluid Flows. Annual Review of Fluid Mechanics, 30, 329-364. https://doi.org/10.1146/annurev.fluid.30.1.329
10. Succi, S. (2001) The Lattice Boltzmann Equation: For Fluid Dynamics and Beyond. Oxford University Press, Oxford.
11. Gunstensen, A.K., Rothman, D.H., Zaleski, S. and Zanetti, G. (1991) Lattice Boltzmann Model of Immiscible Fluids. Physical Review A, 43, 4320. https://doi.org/10. 1103/ physreva.43.4320
12. Shan, X. and Chen, H. (1993) Lattice Boltzmann Model for Simulating Flows with Multiple Phases and Components. Physical Review E, 47, 1815. https://doi.org/ 10.1103/ physreve.47.1815
13. Shan, X. and Chen, H. (1994) Simulation of Non Ideal Gases and Liquid-Gas Phase Transitions by the Lattice Boltzmann Equation. Physical Review E, 49, 2941. https://doi.org/10.1103/physreve.49.2941
14. Swift, M.R., Osborn, W.R. and Yeomans, J.M. (1995) Lattice Boltzmann Simulation of Non-Ideal Fluids. Physical Review Letters, 75, 830. https://doi.org/10.1103/ physrevlett.75.830
15. Chen, L., Kang, Q., Mu, Y., He, Y.L. and Tao, W.Q. (2014) A Critical Review of the Pseudo Potential Multiphase Lattice Boltzmann Model: Methods and Applications. International Journal of Heat and Mass Transfer, 76, 210-236.
16. Huang, H., Sukop, M. and Lu, X. (2015) Multiphase Lattice Boltzmann Methods: Theory and Application. John Wiley & Sons Ltd., Chichester. https://doi.org/10. 1002/9781118971451
17. Sukop, M.C. and Or, D. (2005) Lattice Boltzmann Method for Homogeneous and Heterogeneous Cavitation. Physical Review E, 71, Article ID: 046703. https://doi.org/ 10.1103/physreve.71.046703
18. Chen, X.P., Zhong, C.W. and Yuan, X.L. (2011) Lattice Boltzmann Simulation of Cavitating Bubble Growth with Large Density Ratio. Computers & Mathematics with Applications, 61, 3577-3584.
19. Mishra, S.K., Deymier, P.A., Muralidharan, K., Frantziskonis, G., Pannala, S. and Simunovic, S. (2010) Modeling the Coupling of Reaction Kinetics and Hydrodynamics in a Collapsing Cavity. Ultrasonics Sonochemistry, 17, 258-265.
20. Shan, M.L., Zhu, C.P., Xi, Z.H.O.U., Cheng, Y.I.N. and Han, Q.B. (2016) Investigation of Cavitation Bubble Collapse near Rigid Boundary by Lattice Boltzmann Method. Journal of Hydrodynamics, 28, 442-450.
21. Shan, M.L., Zhu, C.P., Yao, C., Yin, C. and Jiang, X.Y. (2016) Pseudopotential Multi-Relaxation-Time Lattice Boltzmann Model for Cavitation Bubble Collapse with High Density Ratio. Chinese Physics B, 25, Article ID: 104701. https://doi.org/10.1088/1674-1056/25/10/104701

22. Li, Q. and Luo, K.H. (2014) Thermodynamic Consistency of the Pseudopotential Lattice Boltzmann Model for Simulating Liquid-Vapor Flows. Applied Thermal Engineering, 72, 56-61.
23. Kupershtokh, A.L., Medvedev, D.A. andKarpov, D.I. (2009) On Equations of State in a Lattice Boltzmann Method. Computers & Mathematics with Applications, 58, 965-974.
24. Li, Q., Luo, K.H. and Li, X.J. (2012) Forcing Scheme in Pseudopotential Lattice Boltzmann Model for Multiphase Flows. Physical Review E, 86, Article ID: 016709. https://doi.org/10.1103/physreve.86.016709
25. Li, Q., Luo, K.H. and Li, X.J. (2013) Lattice Boltzmann Modeling of Multiphase Flows at Large Density Ratio with an Improved Pseudopotential Model. Physical Review E, 87, Article ID: 053301. https://doi.org/10.1103/physreve.87.053301
26. Li, Q., Luo, K.H., Kang, Q.J., He, Y.L., Chen, Q. and Liu, Q. (2016) Lattice Boltzmann Methods for Multiphase Flow and Phase-Change Heat Transfer. Progress in Energy and Combustion Science, 52, 62-105.
27. Yuan, P. and Schaefer, L. (2006) Equations of State in a Lattice Boltzmann Model. Physics of Fluids, 18, Article ID: 042101. https://doi.org/10.1063/1.2187070
28. Shan, X. (2008) Pressure Tensor Calculation in a Class of Nonideal Gas Lattice Boltzmann Models. Physical Review E, 77, Article ID: 066702. https://doi.org/10.1103/physreve.77.066702
29. Shi, D.Y., Wang, Z.K. and Zhang, A.M. (2014) Study on Coupling Characteristics between Bubble and Complex Walls at the Same Scale. Acta Physica Sinica, 63, 533-538.
30. Plesset, M.S. and Chapman, R.B. (1971) Collapse of an Initially Spherical Vapour Cavity in the Neighbourhood of a Solid Boundary. Journal of Fluid Mechanics, 47, 283-290. https://doi.org/10.1017/S0022112071001058
31. Guo, Z.-L., Zheng, C.-G. and Shi, B.-C. (2002) Non-Equilibrium Extrapolation Method for Velocity and Pressure Boundary Conditions in the Lattice Boltzmann Method. Chinese Physics, 11, 366. https://doi.org/10.1088/1009-1963/11/4/310
32. Zou, Q. and He, X. (1997) On Pressure and Velocity Boundary Conditions for the Lattice Boltzmann BGK Model. Physics of Fluids, 9, 1591-1598. https://doi.org/10.1063/1.869307
33. Huang, H., Krafczyk, M. and Lu, X. (2011) Forcing Term in Single-Phase and Shan-Chen-Type Multiphase Lattice Boltzmann Models. Physical Review E, 84, Article ID: 046710. https://doi.org/10.1103/physreve.84.046710

CHAPTER 7

Physics on the Adiabatically Changed Finslerian Manifold and Cosmology

Anton A. Lipovka

Centro de Investigacion en Fisica, Universidad de Sonora, Sonora, Mexico

ABSTRACT

In present paper, we confirm our previous result [4] that the Planck constant is adiabatic invariant of the electromagnetic field propagating on the adiabatically changed Finslerian manifold. Direct calculation of the Planck constant from the first principles with using of the actually measured cosmological parameters, gives value $h = 6\times10^{-27}$.(erg.s). We also confirm that Planck constant (and hence other fundamental constants which depend on h) is varied on time due to changing of geometry. As an example the variation of the fine structure constant is calculated. Its relative variation $(d\alpha/dt)/)$ consist 1.0×1.0^{-18} (1/s). We show that on the Finsler manifold characterized by adiabatically changed geometry, the classical free electromagnetic field is quantized geometrically, from the properties of the manifold in such manner that the adiabatic invariant of field is $ET = 6\times10^{-27} = h$. Equations of electrodynamics on the Finslerian manifold are obtained. It is stressed that quantization naturally appears from these equations and is provoked by adiabatically changed geometry of the manifold. We consider in details two direct consequences of the equations: i) cosmological redshift of photons and ii) effects of Aharonov-Bohm, that immediately follow from obtained equations. It is shown that quantization of systems consists of electromagnetic field and baryonic component (like atoms) is obvious and has clear explanation.

Keywords: Aharonov-Bohm Effect, Electrodynamic, Cosmology, Foundation of Quantum Theory

1. INTRODUCTION

The problem of formulation of general theory which could naturally unify General Relativity (GR) and Quantum Theory (QT) is of most fundamental and actual problem of modern physics. But to resolve this problem, we should first to know what the Planck constant is. This is the question which opens the door and allows us to find unified theory for all branches of physics. To obtain this key we have first to have in mind some important facts.

1) In quantum mechanics, Planck constant always appears together with derivatives in the same power. This fact clearly points out on the possible relation between Planck constant and geometry.

2) Einstein [1] and later Debye [2] have shown from thermodynamical approach that electromagnetic field is quantized alone, without any assumption on the nature of oscillators. So this is another hint that it should be quantized from geometry, without axiomatic introduction of the wave function.

3) Recently the cornerstone result was announced [3] in respect to the quantization at very small scales. It was found that at the small scales a quantum system behaves as a classical one (see paper [3] for details). This result also argues that QT is pure geometrical phenomenon and it disappears at small scales when geometry becomes to be approximately Euclidean.

4) In 2011, Zhotikov showed [4] that quantum postulates of Bohr and quantization rules of Bohr-Sommerfeld, follow directly from the geometric structure of space-time.

5) Recently, we have published the paper in which we clearly points out the geometrical origin of the Planck constant [5]. In this paper, we have shown that the Planck constant is adiabatic invariant of electromagnetic field on the adiabatically changed Finslerian manifold. From this fact, the variation of the Planck constant on the time directly follows (and hence variations of fundamental constants, for example variation of the fine structure constant, due to adiabatically changed geometry [5] [6]).

On the one hand, we have these serious arguments for the geometric origin of quantization, but on the other hand, we have also serious problems with Riemannian geometry. General relativity was created as theory on (pseudo-) Riemannian manifold (we will use farther in this paper "Riemannian" instead of "pseudo-Remannian"). Such choice was not favored by some serious arguments but only by the fact the real geometry is curved. Metric tensor $g^{\mu\nu}(x)$ in this case depends only on the coordinates and this fact applies some restrictions on the theory, and leads to the serious problem with singularity. However, as it follows from experiments on measurement of the cosmological constant, our Uni- verse expands with acceleration and for this reason complete geometry of Finsler should be favoured. On the Finslerian manifold, the metric tensor $g^{\mu\nu}(x, \dot{x})$depends not only on coordinates, but also on velocities and (as it will be shown in this paper) this fact allows us to introduce the cosmological constant in natural way and to calculate from geometry correct value of the Planck constant, unify gravity, electrodynamics and QT. The Hubble constant, cosmological constant, quantization and much more follow from the Finsler geometry in a natural way and we can conclude that Finsler geometry naturally complies with all observational data.

In this paper, we also obtain classical equations of motion for a system on the manyfold with adiabatically changed geometry and supplied by Finslerian metrical function $L(x, \dot{x}) = g_{\mu\nu}(x, \dot{x})dx^{\mu} dx^{\nu}$. We show how the Planck's constant naturally appears from geometry and, within the (3 + 1) formalism, we write an exact relation between the Planck constant on the one hand and scalar curvature and cosmological constant on the other hand.

We show how the classical electromagnetic field is naturally quantized due to existence of adiabatic invariant of the field on adiabatically changed Finslerian manifold.

Finally, we write equations of electrodynamics, which classically (geometrically) describe quantization of electromagnetic field (and hence of any electromagnetically interacting system). To illustrate how it works, we clearly considered two important particular cases which describe these equations: i) case of free electromagnetic field, when quantization appears from geometry and leads to losses of energy as photon propagates (so called cosmological redshift), and ii) the Aharonov-Bohm effects, which immediately follow from obtained electrodynamic equations.

All-around in this paper we suppose that Latin indexes i,j,k,l,m = 1,2,3 and greek $\alpha,\beta,\ldots\mu,\nu\ldots$= 0,1,2,3. Signature of metric is (1,-1,-1,-1).

2. ADIABATIC INVARIANT AND GENERAL FORMALISM

In classical physics the equations of motion follow immediately from the relation $S_m(x) = S_0$, where $S_m(x)$ is action of matter and S_0 is a constant or zero. By varying this equation we obtain usual classical Hamilton (or Lagrange-Euler) equations [7].

In the case of General Relativity we put into the right hand part of Equation (1) the only invariant we have in pseudo-Riemannian geometry—scalar curvature R of the manifold and this way obtain the Einstein equations.

More general case appears when we consider a manifold supplied by Finslerian metric. The case of static Finslerian manifold is well studied and can be found for example in textbooks [8] [9] and papers [10] [11]. We should also mention here very interesting discussion regarding unusual properties of equations of motion on Finsler manifold suggested in [12]. Let us consider what happens in general case, when the right-hand term of Equation (1) contains some adiabatically changed parameters which characterize the Finslerian manifold, i.e. when the geometry of the manifold is changed adiabatically on time.

Let M be an 3-dimensional, class C^3 manifold characterized by scalar curvature R in the point x^k, where x^k is a local coordinate on an open subset $U \subset M$. Let also suppose M be supplied by Finsler metric $L(x, \dot{x})$ and write a 1-parameter family of hypersurfaces on the M defined by equation

$$S_m(x) = S_M(x) \tag{1}$$

Here S_m stay for matter action and S_M corresponds to the 1-parameter family of the adiabatically changed hypersurfaces on the$M$. Our more general case differs from that examined in [8] by the fact that now hypersurface is changed adiabatically.

Our aim is to write the geodesic equations (Hamilton or Lagrange-Euler equations) for this general case. By varying (1)

$$\delta S_m = \int \delta \mathcal{L}_m \left(x, \dot{x}^k \right) \mathrm{d}t = \int \delta \left(p_k \mathrm{d}x^k - H \mathrm{d}t \right) = \delta S_M = \int \delta \mathcal{L}_M \left(x^k, \dot{x}^k \right) \mathrm{d}t \tag{2}$$

we immediately obtain Hamilton-like equations

$$\frac{\mathrm{d}p_k}{\mathrm{d}t} = -\frac{\partial H}{\partial x^k} - \frac{\partial \mathcal{L}_M}{\partial x^k} \tag{3}$$

and

$$\frac{dx^k}{dt} = \frac{\partial H}{\partial p_k} + \frac{\partial \mathcal{L}_M}{\partial p_k} \tag{4}$$

One can see that due to expansion of the Universe (adiabatic changes of right hand term in (1)) there appears an additional force in the Equation (3) and an additional velocity in (4), which we naturally can attribute to cosmological constant (acceleration) and to the Hubble constant (v Hx). It actually corresponds to the fact that absolutely closed systems do not exist and those additional terms appear due to adiabatic changes of geometry (changes of metric tensor) which take place because of the Universe expansion.

Absolutely the same way we can write equations of Lagrange-Euler by varying (1):

$$\frac{\partial \mathcal{L}_m}{\partial x^k} - \frac{d}{dt}\frac{\partial \mathcal{L}_m}{\partial \dot{x}^k} = \frac{\partial \mathcal{L}_M}{\partial x^k} - \frac{d}{dt}\frac{\partial \mathcal{L}_M}{\partial \dot{x}^k} \tag{5}$$

As one can see in right part of this equation again appear two additional terms due to expansion of the Universe (due to changing of geometry of the manifold, as the system under consideration is moving).

3. EXACT PLANCK'S CONSTANT VALUE AND QUANTIZATION OF ELECTROMAGNETIC FIELD

Let us calculate value of the Planck constant from the parameters which characterize the Finslerian manifold. Consider a generalized system distributed over volume. Let Tp(M) and $T_p^*(M)$ be respectively tangent and cotangent bundles on M, where and $p_a \in T_p(M)$ amd $p^\alpha \in T_p^*(M)$ are covariant and contravariant components of corresponding momentum.

We are interested here in the variation of the photon momentum due to adiabatic change of geometry of M. This variation can be obtained directly from the geometry [6] , but also from Einstein's equations. Variation of the momentum density of our system due to changes of M, in unit volume summed over all directions is given by expression

$$\delta p = \frac{c^3}{8\pi G}\delta R = \frac{c^3}{4\pi G}\delta\frac{1}{R^2} \tag{6}$$

Here, the scalar curvature $R = 2/R^2$ and R is the radius of curvature (we note here that we are Rinterested in the evolution of the electromagnetic field, which is described by a tensor of rank = 2. Such tensors are known to correspond to a 2D surface). In this section, and later in this article we are interested in the electromagnetic field, or more precisely—its variation due to the change of the metric. Therefore, we identify the coordinate x with the size of the field resonator (a "box"). Accordingly, the value $\dot{x}$ will characterize the speed of change for this "box" due to adiabatic changes in geometry of the manifold M.

It is well known that on the Finsler manifold the metric tensor depends on x and $\dot{x}$ for this reason the momentum $p^2 = g^{\mu\nu}(x,\dot{x})p_\mu p_\nu$ also depends on x and $\dot{x}$. Let us assume that the geometry changes are adiabatic and due to this fact we can represent x as $x = x_0 + \theta t$, where x_0 and θ are constant. In this case we have S and, hence (see Equation (6)) $\delta R(t,\dot{x})$ too. So we can write

$$\delta p = \frac{-c^3}{2\pi G}\frac{1}{R^3}\delta R(t,\dot{x}) = \frac{-c^3}{2\pi G}\frac{1}{R^3}\left(\delta R(t) + \frac{\partial R(H)}{\partial H}\delta H\right) \tag{7}$$

here we note that $R(t, \dot{x}) = R(t, Hx_0)$ where H is the Hubble constant.

But by taking into account that (see [6])

$$\frac{\partial R}{\partial t} = \frac{\partial}{\partial t}\frac{c}{2H} = -\frac{c}{2H^2}\frac{\partial H}{\partial t} \tag{8}$$

one can write

$$\frac{\partial R}{\partial H} = \frac{\partial R}{\partial t}\frac{\partial t}{\partial H} = -\frac{c}{2H^2} \tag{9}$$

Let us consider variation of Hubble constant.

From the relation $\dot{x} = Hx$ we can find

$$\delta \dot{x} = x\delta H + H\delta x \tag{10}$$

or

$$x\delta H = \ddot{x}\delta t - H\delta x \tag{11}$$

But the only cosmological acceleration we have experimentally measured is that, associated with the cosmological constant Λ, so we can write for this variation

$$\delta H = \left(c^2\Lambda - H^2\right)\delta t \tag{12}$$

Substituting these expressions into (7) we find

$$\delta p = \frac{2c^3 H}{\pi G}\left(\frac{2H^2}{c^2} - \Lambda\right)\delta t \tag{13}$$

or taking into account that for electromagnetic field $R = 2/R^2$ and by changing volume from the spherical coordinates to the euclidean ones, we obtain

$$\delta p = \frac{3c^3 H}{8\pi^2 G}\left(\mathcal{R} - 4\Lambda\right)\delta t \tag{14}$$

This is a variation of momentum (in unit volume in 3 directions) due to adiabatically changed geometry, written for our generalized system localized on the Finslerian manifold.

Now we are ready to write a complete adiabatic invariant for a free propagating electromagnetic field. The components of the 4-momentum p of free electromagnetic field propagating on the Finslerian manifold with adiabatically changed geometry are varied on time. This variation proceeds adiabatically and can be considered as linear function i.e. for energy ε of the field, for example, we have

$$\frac{\delta\varepsilon}{\varepsilon} = -\frac{\delta t}{t} \tag{15}$$

so, the adiabatic invariant we are interested in is given by the expression

$$\varepsilon t = -\frac{\delta \varepsilon}{\delta t} t^2 \tag{16}$$

But

$$\delta \varepsilon = c \delta p_k \tag{17}$$

By substituting (14) into (16) we can write finally (we divide δ_P by factor 3 because in the case of the electromagnetic plane wave, we are interesting only in one direction of the momentum)

$$\varepsilon t = -ct^2 \frac{\partial p}{\partial t} = -\frac{c^4 H}{8\pi^2 G}(\mathcal{R} - 4\Lambda)t^2 = \eta_0 \tag{18}$$

so, by taking into account $R = 2/R^2$,$R = c/2H$ (see comments made before and also [6]), measured values of H = 73 kms^{-1} = 2.4×10^{-18} s^{-1}and Λ = 1.7×10^{-56} [13] we have for this adiabatic invariant $n_0 = h = 6 \times 10^{-27}$ (erg.s.) for one second and in cubic centimeter, as it should be. In the same way we can obtain similar relations for other components of 4-momentum:

$$p_\gamma x^\gamma = \eta_\gamma \tag{19}$$

(there is no summation over γ in this relation and for the photon propagating in direction x^3 the components $p_1 = p_2 = 0$). In the case of plane electromag- netic wave, introduced here 4-vector n_γ has components in unit volume:

$$n_\gamma = (h, 0, 0 , h).$$

And for general case we can write

$$n_\gamma = (h, h, h ,h) \tag{20}$$

Here the adiabatic invariant for electromagnetic field (Planck constant), which depends clearly on the parameters of the manifold $\mathcal{R}$ and Λ (and consequently depends on time) is:

$$h = -\frac{c^4 H}{8\pi^2 G}(\mathcal{R} - 4\Lambda)t^2 \tag{21}$$

that gives for unit time and unit volume at present epoch h = 6×10-27 (erg.S) as it was mentioned above. From this relation it is easy to see that the Planck constant depends on time as $h \sim 1/t$.

3.1. Hilbert Integral

Now let's consider integral of Hilbert for particular case of the free electromagnetic field propagating along geodesic on the adiabatically changed Finslerian manifold:

$$\int p_\alpha \mathrm{d}q^\alpha = \Delta S_M \tag{22}$$

where P_a is 4-momentum of the field, $q^a$is generalized coordinate and right hand termδS_M, as before, corresponds to the changing of the 1-parameter family of hypersurfaces due to adiabatic variation of geometry as system under consideration is moving on *M*. For electromagnetic field described by Lagrangian $\mathcal{L}_m = F_{\mu\nu}F^{\mu\nu}/16\pi$ we have from (22)

$$\frac{1}{c}\int \frac{\partial \mathcal{L}_m}{\partial A_{\mu,\nu}} A_{\mu,\sigma} \mathrm{d}x^\sigma = \Delta S_M \tag{23}$$

By taking into account that tensor of energy-momentum is

$$T_\sigma^\nu = \frac{\partial \mathcal{L}_m}{\partial A_{\mu,\nu}} A_{\mu,\sigma} - \delta_\sigma^\nu \mathcal{L}_m \tag{24}$$

we can write for propagating classical electromagnetic field

$$\frac{1}{c}\int \left(T_\sigma^\nu + \delta_\sigma^\nu \mathcal{L}_m\right) \mathrm{d}x^\sigma = \Delta S_M \tag{25}$$

If the field propagates in the directionx_3, than electric field $E = E_1$ and magnetic field $H = H_2$. So the only non-zero components of the field tensor $F_{\mu\nu}$ are $F_{01}= -F_{10} = E_1$; $F_{13}= -F_{31} = H_2$ and hence $T^{00}= T^{33} = T^{30}= (E^2 + H^2)/8\pi$, where for the plane wave we have $E = E_0 \cos(k_\alpha x^\alpha)$ and $H = H_0 \sin(k_\alpha x^\alpha)$.

In this case for the 00-component, for example, by taking into account (18) and after elementary integration we obtain

$$\frac{E_0^2 + H_0^2}{8\pi} = \frac{h}{T} = h\nu \tag{26}$$

Similar relations one can write for other components.

So, as one can see the classical electromagnetic field is quantized due to adiabatic variation of the Finslerian manifold and we do not need some artificial methods to quantize it.

3.2. Variation of the Fine Structure Constant

As it was shown before [6] even on the adiabatically changed Riemannian manifold the value of the fine structure constant is changed adiabatically on time. This variation appears due to a change in the metric of space, in which the atom is localized. In the case of Finslerian manifold this variation is smaller by factor 2/3 due to the presence of the cosmological constant. To show this let's start from (14). For one direction (divided by factor 3) we have from (14):

$$\delta p = \frac{c^3 H}{8\pi^2 G}(\mathcal{R} - 4\Lambda)\delta t \tag{27}$$

But the fine structure constant is $a = V/c$ where V is electron velocity at the first Bohr orbit. Momentum in this case we can write as

$$P = \frac{m\alpha c}{\sqrt{1-\alpha^2}} \tag{28}$$

so

$$\delta P = \frac{mc}{\left(1-\alpha^2\right)^{3/2}}\delta\alpha \tag{29}$$

and

$$\delta\alpha = \frac{\left(1-\alpha^2\right)^{3/2} Hc^3}{mc8\pi^2 G}\left(\mathcal{R}-4\Lambda\right)\delta t \tag{30}$$

that give us value $\dot{a}/a$ = -1.03×10^{-18} (for 1 second), $R = 2/R^2$,$R = c/2H$, (see [6], H = 73 kms^{-1} = Mpc^{-1} = 2.4×10^{-18} s^{-1}and Λ = 1.7×10^{-56} cm^{-2})

4. ELECTRODYNAMICS ON THE FINSLERIAN MANIFOLD

In Riemannian geometry, the first pair of equations of electrodynamics follows directly from the properties of the field tensor.

$$F_{\mu\nu} = A_{\nu;\mu} - A_{\mu;\nu} = A_{\nu,\mu} - A_{\mu,\nu} \tag{31}$$

where

$$A_{\mu;\nu} = A_{\mu,\nu} - \Gamma^{\sigma}_{\mu\nu} A_{\sigma} \tag{32}$$

And for this reason $(\Gamma^{\sigma}_{\mu\nu} = \Gamma^{\sigma}_{\nu\mu})$ the first pair of equations on Riemannian manifold with constant scalar curvature is

$$\partial_{\sigma} F_{\mu\nu} + \partial_{\mu} F_{\nu\sigma} + \partial_{\nu} F_{\sigma\mu} = 0 \tag{33}$$

On the Finsler manifold, we can obtain the first pair in the same way, but in this case the field tensor is

$$\tilde{F}_{\mu\nu} = A_{\nu;\mu} - A_{\mu;\nu} \tag{34}$$

where covariant differentials DA_μ include now terms with the Cartan connections $C_{\mu\nu\sigma} = \frac{1}{2}\frac{\partial g_{\mu\nu}}{\partial \dot{x}^\sigma}$ like this $C^\sigma_{\mu\nu} A_\sigma \mathrm{d}\dot{x}^\nu$ and also $\Gamma^\sigma_{\mu\nu} A_\sigma \mathrm{d}x^\nu$. In the most important case, we are interested in, when the scalar curvature is small and the metric tensor has spatial structure described by the Robertson-Walker metric, the additional terms $C^\sigma_{\mu\nu} A_\sigma \mathrm{d}\dot{x}^\nu$ and $\Gamma^\sigma_{\mu\nu} A_\sigma \mathrm{d}x^\nu$ in covariant derivative can be evaluated easily as $\approx A_\sigma \mathrm{d}x^\nu / R$ (here $1/R = H/c$ 5×10^{-29} is the inverse radius of scalar curvature at the point of observation) so we have

$$\tilde{F}_{\mu\nu} = A_{\nu,\mu} - A_{\mu,\nu} - t_{\mu\nu} \tag{35}$$

where our estimation for the small components consist $t_{\mu\nu} \approx A_\mu \mathrm{d}x^\nu / R \approx A_\mu \mathrm{d}x^\nu \times 10^{-29}$.

As one can see these components probably will be significant only in the vicinity of black holes and can be omitted in our present consideration.

For this reason the first pair of the equations can be written as

$$\tilde{F}_{\mu\nu;\sigma} + \tilde{F}_{\nu\sigma;\mu} + \tilde{F}_{\sigma\mu;\nu} = 0 \tag{36}$$

or, by taking into account our estimations discussed above, we can write

$$\partial_\sigma F_{\mu\nu} + \partial_\mu F_{\nu\sigma} + \partial_\nu F_{\sigma\mu} = O\left(t_{\mu\nu,\sigma}\right) \tag{37}$$

So one can see that the first pair of electrodynamic equations remains to be the first pair of the Maxwell equations with high precision.

The second pair of equations of electrodynamics follows directly from variation of functional (1) if we consider a charge characterized by 4-current j^a, and the electromagnetic field on the Finslerian manifold:

$$S_m = S_M \tag{38}$$

Here S_M as before, corresponds to the family of the hypersurfaces on the expanded manifold. By varying S_m we have

$$\delta S_m = -\frac{1}{c}\int_\Omega \left[\frac{1}{c} j^\alpha \delta A_\alpha + \frac{1}{16\pi}\delta\left(F_{\mu\nu}F^{\mu\nu}\right)\right] \mathrm{d}\Omega \tag{39}$$

here we put $F_{\mu\nu}$ instead of $\tilde{F}_{\mu\nu}$ because, as we have seen, small additional terms corresponding to small components of $t_{\mu\nu}$ are insignificant in the case of Robertson-Walker metric. Integrating the second term by parts, we obtain

$$\delta S_m = -\frac{1}{c}\int_\Omega \left[\frac{1}{c} j^\mu + \frac{1}{4\pi}\frac{\partial F^{\mu\nu}}{\partial x^\nu}\right] A_{\mu,\sigma}\delta x^\sigma \mathrm{d}\Omega \tag{40}$$

By varying S_M we have (see Equations (16)-(20))

$$\delta S_M = \int_\Omega \frac{\eta_\sigma}{\left(x^\sigma\right)^2}\delta x^\sigma \mathrm{d}\Omega \tag{41}$$

where, as it was shown before, $n\sigma$ = (h, h, h,h)in unit volume (here h is the Planck constant) and Ω is a 4-volume. The equations under discussion one can write as follows

$$\frac{1}{c}\int_\Omega \left[\frac{1}{c} j^\mu + \frac{1}{4\pi}\frac{\partial F^{\mu\nu}}{\partial x^\nu}\right] A_{\mu,\sigma}\delta x^\sigma \mathrm{d}\Omega = -\int_\Omega \frac{\eta_\sigma}{\left(x^\sigma\right)^2}\delta x^\sigma \mathrm{d}\Omega + O\left(\eta_\sigma^2\right) \tag{42}$$

or finally

$$\frac{1}{c}\left[\frac{1}{c} j^\mu + \frac{1}{4\pi}\frac{\partial F^{\mu\nu}}{\partial x^\nu}\right] A_{\mu,\sigma} = -\frac{\eta_\sigma}{\left(x^\sigma\right)^2} + O\left(\eta_\sigma^2\right) \tag{43}$$

(there is no summation over σ here).

This is the second pair of equations of electrodynamics on the adiabatically changed Finslerian manifold. The bounded electromagnetic field (second term) in this case is explicitly included into consideration, as it takes place in the case of Bohmian formalism when this field appears in QT as quantum potential (see [14] for details and also results of paper [15]). It is easy to see that when the Planck constant tends to zero, the expression (43) (as it should be) is converted into the Maxwell equations. (Especially emphasize here that in this extreme case, the topology becomes simple, space is isotropic, flat and described by the Minkowski geometry). Let us consider two important cases which immediately follow from these equations.

4.1. Cosmological Redshift

It is well known that as the photon propagates through expanding universe its frequency (or wave length) is changed. This loss of energy by free electromagnetic field, named as cosmological redshift, appears in our equations by natural way as losses of the energy by photon due to adiabatically changed geometry of manifold.

$$\frac{1}{4\pi c}\frac{\partial F^{\mu\nu}}{\partial x^{\nu}}A_{\mu,\sigma}=\frac{1}{8\pi c}\frac{\partial\left(F_{\mu\nu}F^{\mu\nu}\right)}{\partial x^{\sigma}}=-\frac{\eta_{\sigma}}{\left(x^{\sigma}\right)^{2}}+O\left(\eta_{\sigma}^{2}\right) \tag{44}$$

(there is no summation over σ).

The meaning of these equations is most obvious if we put $\sigma = 0$. In this case, the left side will be the electromagnetic field energy loss with time, and the right side became h/t^2—small value associated with the geometry of the universe (see expression (18)), i.e.

$$\frac{1}{8\pi}\frac{\partial\left(F_{\mu\nu}F^{\mu\nu}\right)}{\partial t}\approx-\frac{h}{t^{2}}=\frac{c^{4}H}{8\pi^{2}G}\left(\mathcal{R}-4\Lambda\right)$$

4.2. The Aharonov-Bohm Effects

Another important case that follows directly from the second pair of equations is the Aharonov-Bohm effects. As it is known, a necessary condition for the existence of the Aharonov-Bohm (AB) effects is the presence, in the overall structure of the equations, of the "zero field" potentials which cannot be removed by gauge transformations and they do not create electromagnetic fields [16] [17]. These "zero-potentials" are the result of "non-trivial topology" of the area on which the particle moves [16] [18] [19]. Such a situation arises in electrodynamics of anisotropic media where the structure of Maxwell's equations eliminates the possibility of satisfying the boundary conditions. To satisfy regularly the boundary conditions in anisotropic media, usually the zero-potential is introduced, which do not create electromagnetic fields (see [16] and references there-in). In the case of adiabatically expanding Finslerian manifold, the anisotropy of space occurs for any moving body automatically as right part of Equaiton (43). Therefore, it is safe to say that in the case of AB effects we are dealing directly with the anisotropy of space, which appears due to adiabatically changed Finslerian manifold as the particle moves along its trajectory. In this case, the role of the zero potentials (which do not generate electromagnetic fields) performs variable geometry of space as it follows directly from (43).

In absence of electric and magnetic fields en route of propagation of the particle under consideration, the second term disappears (but it still take place inside of the solenoid and affects our particle: "In spite of the fact that the magnetic field vanishes out of the solenoid, the phase shift in the wave functions is proportional to the corresponding magnetic flux inside of the solenoid" [20]) and we obtain (we neglect here by the small term $O(\eta_\sigma^2)$)

$$\frac{1}{c^2} j^\mu A_{\mu,\sigma} = -\frac{\eta_\sigma}{\left(x^\sigma\right)^2} \tag{45}$$

but $j^\mu = (pc, pV^k)$ (here p is charge density and V^k is 3-velocity) so if we put $p = e$ and remember that $Sj^\mu = 0$ (for this reason $A_\mu \partial_\sigma j^\mu = 0$ and $A_{\mu,\sigma} j^\mu = \partial_\sigma \left(A_\mu j^\mu\right)$) we obtain by using the Gauss theorem

$$\frac{e}{c} A_0 = -\frac{\eta_0}{\left(x^0\right)}, \quad \frac{e}{c} A_k V^k = -\frac{\eta_k}{\left(x^k\right)} \tag{48}$$

These equations describe the electric and magnetic effects of Aharonov-Bohm (here x^0 and $x^k$are fixed). Namely for $\mu = 0$ we have for the phase variation $\Delta\Phi$

$$e \int_t \varphi \mathrm{d}t = -h \frac{\Delta t}{1\,\text{sec}.} = -h\Delta\Phi \tag{49}$$

electric effect of Aharonov-Bohm, and when $\mu = k$ (here $k = 1,2,3$) we have relation

$$\frac{e}{c} \int_l A_k \mathrm{d}x^k = -h \frac{\Delta x}{1cm.} = -h\Delta\Phi \tag{50}$$

describes magnetic effect of Aharonov-Bohm.

To conclude this part we would like to stress again that whereas the bounded field (second term in (43)) do not appears in these relations, it actually affects the moving particle through potentials A_μ [20] and this bounded field corresponds to quantum potential in the Bohmian formalism [14] [15].

5. COMPLETE THEORY

In previous part we have obtained electrodynamic equations. They are applied in the case when the movement of charge (or 4-current j^μ) is defined. To construct self-consistent theory we should treat j^μ as variable from the beginning.

We consider now a charge characterized by 4-current j^a, and the electromagnetic field on the Finslerian manifold:

$$S_m = S_M \tag{51}$$

Here S_M as before corresponds to family of the hypersurfaces on the expanded manifold. In complete form, when $\delta j^\mu \neq 0$, we have the action

$$-\sum\int mc\mathrm{d}s - \frac{1}{c}\int_\Omega\left[\frac{1}{c}j^\alpha A_\alpha + \frac{1}{16\pi}\left(F_{\mu\nu}F^{\mu\nu}\right)\right]\mathrm{d}\Omega = S_M \tag{52}$$

which describes quantum properties of our system. It is clear there are a lot of different systems, particular cases and applications which can not be considered here because of their huge amount. For this reason let us consider here the hydrogen atom as an example. In order to coincide with the Schrödinger formulation of quantum mechanics, we should neglect the third term in (52) which corresponds to quantum potential [14] [15] in Bohmian formulation, and gives the zero energy correction (for example in the case of harmonic oscillator it gives term 1/2 in expression for energy [5] [14]). In this case by varying (52) we have

$$-\delta\int\left(mc\mathrm{d}s + \frac{e}{c}A_\alpha \mathrm{d}x^\alpha\right) = \delta S_M \tag{53}$$

and we can write

$$\int\left[mc\frac{du_{\mu}}{ds}-\frac{e}{c}\left(\frac{\partial A_{\nu}}{\partial x^{\mu}}-\frac{\partial A_{\mu}}{\partial x^{\nu}}\right)u^{\nu}\right]\delta x^{\mu}ds=-\int\frac{\eta_{\mu}}{\left(x^{\mu}\right)^{2}}\delta x^{\mu}ds \tag{54}$$

So, finally we obtain equations of motion

$$mc\frac{du_{\mu}}{ds}-\frac{e}{c}F^{\mu\nu}u_{\nu}=-\frac{\eta_{\mu}}{\left(x^{\mu}\right)^{2}} \tag{55}$$

By taking into account that classical period for orbital movement of electron is

$$T=\pi e^{2}\sqrt{\frac{m}{2\left|E\right|^{3}}} \tag{56}$$

in classical limit $v \ll c$ the straightforward calculations give the energy for first Bohr orbit, obtained from classical electrodynamic on Finslerian manifold:

$$E_{1}=\frac{me^{4}}{2h^{2}} \tag{57}$$

that coincides with quantum calculations. Relativistic corrections are obvious and follow directly from (55).

6. CONCLUSIONS

In this paper, we confirm our previous result [5] that Planck constant is adiabatic invariant of electromagnetic field propagating on the adiabatically changed Finslerian manifold. Direct calculation of the Planck constant value made from cosmological parameters gives $h = 6\times10^{-27}$ (erg.s) that is in excellent agreement with the measured value. We also confirm that Planck constant (and hence other fundamental constants which depend on h) is varied on time due to changing of geometry of the manifold.

As an example we suggest a calculation of the fine structure constant variation. The obtained value consists $\dot{\alpha}/\alpha$ = -1.03×10^{-18} (for 1 second) and

this variation is expected to be measured in nearest future.

We show that on the Finslerian manifold characterized by adiabatically changed geometry, classical free electromagnetic field is quantized geometrically, from the properties of the manifold.

Equations for electrodynamics on the Finslerian manifold are suggested. It is shown that quantization naturally appears from these equations and is resulted from adiabatically changed geometry of manifold. We consider in details two direct consequences of the equations—cosmological redshift of photons and effects of Aharonov-Bohm.

Finally we show appearance of quantization for the systems consisting of electromagnetic field and charged baryonic components (like atoms and molecules).

ACKNOWLEDGEMENTS

I would like to express my deepest appreciation and gratitude to my teachers: Choban E.A., Dubov V.V., Dubrovich V.K., Finkelstain B.A., Gosachinsky I.V., Ivanov V.K., Khersonsky V.K., Okunev A.A., Tichomirov S.R., Toptygin I.N., Rukolaine A.V., Varshalovich D.A., Zlatin A.N., and many others. I also have to thank my wife Olga. Thanks to her efforts, I was able to complete this work for the summer vacation 2016.

REFERENCES

1. Einstein, A. (1905) Uber einen die Erzeugung und Verwandlung des Lichtes betreffenden heuristischen Gesichtspunkt. Annalen der Physik, 17, 132-148. https://doi.org/10.1002/andp.19053220607
2. Debye, P. (1910) Der Wahrscheinfichkeitsbegriff in der Theorie der Strahlung. Annalen der Physik, 33, 1427-1434. https://doi.org/10.1002/andp.19103381617
3. De Gosson, M. and Hiley, B. (2013) Short-Time Quantum Propagator and Bohmian Trajectories. Physics Letters A, 377, 3005-3008. https://doi.org/10.1016/ j. physleta.2013.08.031
4. Zhotikov, V.G. (2011) N. Bohr's Postulates and Bohr-Sommerfeld's Quantization Rules as Consequences from Finsler Geometry. Proceedings of International Meeting "Physical Interpretation of Relativity Theory", Moscow, 4-7 July 2011.
5. Lipovka, A.A. (2014) Planck's Constant as Adiabatic Invariant Characterized by Hubble's and Cosmological Constants. Journal of Applied Mathematics and Physics, 2, 61-71. https://doi.org/10.4236/jamp.2014.25009

6. Lipovka, A.A. and Cardenas, I.A. (2016) Variation of the Fine Structure Constant. arXiv: 1608.04593.
7. Rund, H. (1966) The Hamilton-Jacobi Theory in the Calculus of Variations: Its Role in Mathematics and Physics. D. Van Nostrand Co., New York.
8. Rund, H. (1959) The Differential Geometry of Finsler Spaces. (Grundlehren der mathematischen Wissenschaften (Book 101)). Springer, Berlin.
9. Zhotikov, V.G. (2014) Introduction to the Finsler Geometry and Its Generalizations (for Physicists). MPTI, Moscow.
10. Zhotikov, V.G. (2014) About the Modern Point of View on the A. Einstein's Principle of General Relativity. Hypercomplex Numbers in Geometry and Physics, 1, 21-36.
11. Zhotikov, V.G. (2009) Finsler Geometry (According to Wagner) and the Equations of the Movement in the Relativistic Dynamics. Proceedings of 15th International Meeting: Physical Interpretation of Relativity Theory, Moscow, 6-9 July 2009, 133-144.
12. Zhotikov, V.G. (2016) On Influence of Derivatives Higher Order of Lagrange Function Dynamical System on Its Center Mass Movement. Inzhenernaya Fisica, 12, 11-14.
13. Marchc, M.C., Trotta, R., Berkes, P., Starkman, G.D. and Vaudrevange, P.M. (2011) Improved Constraints on Cosmological Parameters from Type Ia Supernova Data. Monthly Notices of the Royal Astronomical Society, 418, 2308-2329. https://doi.org/10.1111/j.1365-2966.2011.19584.x
14. Lipovka, A.A. (2016) Nature of the Quantum Potential. Journal of Applied Mathematics and Physics, 4, 897-902. https://doi.org/10.4236/jamp.2016.45098
15. Philippidis, C., Bohm, D. and Kaye, R.D. (1982) The Aharonov-Bohm Effect and the Quantum Potential. Il Nuovo Cimento B, 71, 75-88. https://doi.org/10.1007/BF02721695
16. Chirkov, A.G. and Ageev, A.N. (2001) On the Nature of the Aharonov-Bohm Effect. Technical Physics, 46, 147-153. https://doi.org/10.1134/1.1349267
17. Kobe, D.H. (1979) Aharonov-Bohm Effect Revisited. Annals of Physics, 123, 381-410. https://doi.org/10.1016/0003-4916(79)90344-0
18. Bagrov, V.G., Gitman, D.M. and Skarzhinsky, V.D. (1986) Aharonov-Bohm Effect for Stationary and Coherent States of an Electron in Homogeneous Magnetic Field. In: Komar, A.A., Ed., Classical and Quantum Effects in Electrodynamics, Izdatel'stvo Nauka, Moscow, 151-165.
19. Bagrov, V.G., Gavrilov, S.P., Gitman, D.M. and Meira Filho, D.P. (2011) Coherent and Semiclassical States in a Magnetic Field in the Presence of the Aharonov-Bohm Solenoid. Journal of Physics A: Mathematical and Theoretical, 44, Article ID: 055301. https://doi.org/10.1088/1751-8113/44/5/055301
20. Wu, T.T. and Yang, C.N. (1975) Concept of Nonintegrable Phase Factors and Global Formulation of Gauge Fields. Physical Review D, 12, 3845-3857.

CHAPTER 8

Fluxes, bundle gerbes and 2-Hilbert spaces

Severin Bunk[1] Richard J. Szabo[2]

[1]*Department of ematics, Maxwell Institute for ematical Sciences, Heriot-Watt University, Edinburgh EH14 4AS, UK*

[2]*Department of ematics, Maxwell Institute for ematical Sciences, The Higgs Centre for Theoretical Physics, Heriot-Watt University, Edinburgh EH14 4AS, UK*

ABSTRACT

We elaborate on the construction of a prequantum 2-Hilbert space from a bundle gerbe over a 2-plectic manifold, providing the first steps in a programme of higher geometric quantisation of closed strings in flux compactifications and of M5-branes in *C*-fields. We review in detail the construction of the 2-category of bundle gerbes and introduce the higher geometrical structures necessary to turn their categories of sections into 2-Hilbert spaces. We work out several explicit examples of 2-Hilbert spaces in the context of closed strings and M5-branes on flat space. We also work out the prequantum 2-Hilbert space associated with an M-theory lift of closed strings described by an asymmetric cyclic orbifold of the SU(2)SU(2) WZW model, providing an example of sections of a torsion gerbe on a curved background. We describe the dimensional reduction of M-theory to string theory in these settings as a map from 2-isomorphism classes of sections of bundle gerbes to sections of corresponding line bundles, which is compatible with the respective monoidal structures and module actions.

Keywords: Bundle gerbes, Higher geometric quantisation, Fluxes in string and M-theory, Higher geometric structures

1 INTRODUCTION AND SUMMARY

The geometry of fluxes in string theory and M-theory has over the years spawned many new directions in classical geometry and quantisation, see, for example, [2, 8, 9, 14, 30, 35]. In many of these developments the need for higher geometrical structures together with higher notions of quantisation is becoming ever more prominent (see, for instance, [45] for a survey of the literature and further references).

Physical theories are usually formulated at the classical level using geometry and variational principles. Quantisation can then be understood as a ematical construction applied to the classical data that describes a quantum version of the physical theory in question. Seeking new formalisms for quantisation is crucial in understanding how to extract physical properties from a ematical quantum theory and to properly explore the passage from classical mechanics to quantum mechanics. In particular, beyond its intrinsic interest in string theory and M-theory, seeking the appropriate receptacles for higher versions of quantisation can offer interesting new perspectives on quantum theory itself and a move towards a more systematic theory of quantisation.

Evidence that such higher quantisations are possible has been provided in a number of works, including [10, 12, 19, 20, 39]. In this series of works, higher prequantisation has been investigated not just for the 2-plectic case which is central to this paper, but for any degree $n \geq 2$ of the higher symplectic form. In the present paper, we make some of the abstract objects in these references very explicit, using the more differential geometric language of bundle gerbes rather than the powerful but abstract language of simplicial sheaves. At the same time, using the 2-categorical theory developed in [51], we are able to use a more global framework as compared to [42] and develop further the ideas about using twisted vector bundles as quantum states outlined there. To our knowledge, the notion of *sections of a bundle grebe L* in terms of morphisms from a trivial bundle gerbe to L has not yet appeared in the literature, though it has been around as folklore, see [50]. It has also been pointed out there that the resulting categories are Kapranov–Voevodsky 2-vector spaces, and we extend this towards higher prequantisation by showing that these are actually 2-Hilbert spaces (cf. Sect. 8).

The basic problem of quantisation begins with a symplectic manifold M, with symplectic form $\omega\omega$, which is the classical phase space of a given physical system. Quantisation can then be loosely regarded as a procedure for constructing a "quantum version" of symplectic geometry. In most approaches to quantisation, the first step (either explicitly or implicitly) is to pick the additional structure of a prequantum line bundle $L \rightarrow M$, which is a hermitean line bundle with connection whose magic flux is proportional to the 2-form ω. In this paper we deal with geometric quantisation [28, 29], which is one of the best established quantisation procedures. We review its basic features in Sect. 2 and then confront it with string theory and M-theory in Sect. 3: we explain why geometric quantisation is not appropriate for the quantisation of closed strings in flux compactifications or of M5-branes in C-field backgrounds of M-theory, focusing on the particular cases that involve replacing the symplectic form $\omega\omega$ with a suitable 3-form. As we discuss in Sect. 4, the relevant geometric structure is that of a gerbe on M, and in subsequent sections we attempt to find a suitable modification of geometric quantisation appropriate to this setting. In Sects. 5 and 6 we explain in detail the sense in which gerbes may be regarded as categorifications of line bundles, which le to a better understanding of gerbes and of higher geometry in general. In particular, the discussion of 2-bundle metrics in Sect. 6.2 refines the current understanding of bundle gerbes as higher line bundles to show that hermitean bundle gerbes can naturally be considered as higher *hermitean* line bundles. The higher analogue of Hilbert spaces of physical states in quantum theory should now be a suitable notion of 2-Hilbert spaces, which we describe in some generality in Sect. 7 in the form that we need in this paper. In Sect. 8 we then construct the higher version of the prequantum Hilbert space of geometric quantisation and illustrate our approach by working out a number of explicit examples of these 2-Hilbert spaces in Sect. 9 which are of relevance in closed string theory and M-theory.

In Sect. 10 we highlight some of the open questions which are not yet addressed by our formalism. One of these concerns is an already notoriously difficult problem in ordinary geometric quantisation: the second step in that procedure involves the extra choice of a polarisation, which corresponds to locally representing the symplectic manifold M as a cotangent bundle $T^{*}U$. Global polarisations are not guaranteed to exist, and as yet there is no general criterion for a symplectic manifold to be quantisable in this sense. In addition, demonstrating that the result of quantisation is independent of all these auxiliary choices on the symplectic manifold (M,ω) is also a very difficult task (in fact, in almost every approach to quantisation). The situation in higher quantisation is even less clear, and we do not deal with these issues in the present paper, but rather focus on prequantisation only.

This article is a companion to the longer paper [13] which, among other things, developed new categorical structures on morphisms of bundle gerbes, initiated steps towards higher geometry by introducing 2-bundle metrics, and showed how to obtain a 2-Hilbert space from the category of sections of a bundle gerbe. The present paper is partly a summary of the main constructions from [13], presented in a somewhat more informal way that we hope will be accessible to a broader readership; we refer to

[13] throughout for all technical details. Several points here are elucidated in more detail, particularly in our discussions of 2-Hilbert spaces in Sects. 7 and 8. We have also worked out some illustrative examples more thoroughly in Sect. 9, obtaining a much sharper statement concerning dimensional reductions of sections of a bundle gerbe to sections of a corresponding prequantum line bundle, as is relevant in reductions of M-theory to string theory: we show that our dimensional reduction maps descend to 2-isomorphism classes of sections of bundle gerbes (representatives of which are "constant" along the M-theory direction) and are, moreover, compatible with the respective monoidal structures and module actions. Our final example in Sect. 9.3 is new, and it nicely illustrates the construction of a 2-Hilbert space of a non-trivial bundle gerbe with torsion Dixmier–Douady class that is relevant for a certain M-theory dual of closed strings described by an asymmetric cyclic orbifold of the SU(2) WZW model; in particular, this example suggests a general prescription for our higher prequantisation that mimics the well-known situation in ordinary quantisation: given a higher prequantisation of M and a group G acting freely on M, then one obtains a higher prequantisation of the quotient M/G by taking the G-invariant part of the higher prequantisation of M.

1. 1. Glossary of notation

For the reader's convenience, we summarise here the notational conventions for categories used throughout this paper:

- Hilb,Vect: The category of finite-dimensional complex Hilbert spaces/vector spaces with linear maps.
- *H*ilb∞,Vect∞: The category of possibly infinite-dimensional complex Hilbert spaces/vector spaces.
- 2*H*ilb: The 2-category of 2-Hilbert spaces.
- *HVB*dl(M): The category of hermitean vector bundles on M with smooth fibrewise-linear maps.
- *HVB*dl$^\nabla$(M): The category of hermitean vector bundles with connection on M, where the metric is parallel with respect to the connection, with parallel morphisms of vector bundles.
- HLBdl(M),HLBdl$^\nabla$(M): The subcategory of hermitean line bundles on M without/with connection.
- BGrb(M),BGrb$^\nabla$(M): The 2-category of hermitean bundle gerbes on M without/with connection.

- For any n-category $\mathcal{C}$ and objects $a,b\in\mathcal{C}$, the $(n-1)$-category of morphisms from a to b in $\mathcal{C}$ is denoted $\mathcal{C}(a,b)$. Analogously, given 1-morphisms $f,g:a\to b$ in $\mathcal{C}$ we will write $\mathcal{C}(f,g)$ for the $(n-2)$-category of morphisms from f to g in $\mathcal{C}$.

2 GEOMETRIC QUANTISATION

Geometric quantisation starts from a symplectic manifold (M,ω), i.e. a manifold M of dimension $2n$ with a non-degenerate closed 2-form ω; this is part of the data of most classical physical systems. Such a manifold is called *prequantisable* if there exists a hermitean line bundle with connection (L,∇^L) on M whose magic flux is $F^L=2\pi\omega$; by Dirac charge quantisation this is equivalent to the statement that the de Rham class of ω lies in the image of the map $H^2(M,\mathbb{Z})\to H^2_{dR}(M)$. Sections of L then form the space of wavefunctions for the quantum theory. For compact M the space $\Gamma(M,L)$ of smooth sections carries a non-degenerate, positive-definite inner product. If we denote by h the hermitean metric on L, it is given by

$$(\psi,\phi)\longmapsto\langle\psi,\phi\rangle_{\Gamma(M,L)}=\int_M h(\psi,\phi)\,\frac{\omega^n}{n!}.$$

The pair $H_{pre}(L)=(\Gamma(M,L),\langle-,-\rangle_{\Gamma(M,L)})$ then forms a pre-Hilbert space whose Hilbert space completion we denote by $H(L)=L^2(M,L)$.[1] It is called the *prequantum Hilbert space* associated with (M,ω).

In the classical theory, observables are real-valued functions $f\in C^\infty(M,\mathbb{R})$. In the quantum theory they act on the Hilbert space as follows. First of all, to every such function there is an associated Hamiltonian vector field X_f given by $\iota_{X_f}\omega=-\mathrm{d}f$, where ι_X denotes insertion of a vector field X into the first slot of a differential form. The vector field X_f is uniquely defined because of the non-degeneracy of ω. On wavefunctions $\psi\in H_{pre}(L)$ we can now set $\mathcal{O}_f\psi=-\mathrm{i}\hbar\,\nabla^L_{X_f}\psi+2\pi\hbar\,f\,\psi$. One then has

$$[\mathcal{O}_f,\mathcal{O}_g]=-\mathrm{i}\hbar\,\mathcal{O}_{\{f,g\}}$$

where $\{-,-\}$ is the Poisson bracket on $C^\infty(M,\mathbb{R})$,

$$\{f,g\}=\iota_{X_f\wedge X_g}\omega$$

The underlying Lie algebra of the Poisson algebra $(C^\infty(M,\mathbb{R}),\{-,-\})$ is a central extension of the Lie algebra of Hamiltonian vector fields; in particular $[X_f,X_g]=-X_{\{f,g\}}$. The assignment $f\mapsto\mathcal{O}_f$ of quantum operators to classical functions is called the *Kostant–Souriau prequantisation map*, and continuously extending the action of observables to H(L) yields a representation of this Poisson algebra on the prequantum Hilbert space.

As an example, consider a 1-connected manifold M with a symplectic form $\omega\in\Omega^2(M)$ which has integer periods.[2] Fixing a base-point of M we obtain the diagram

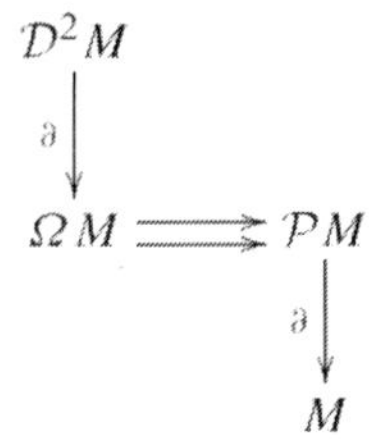

Here $\mathcal{P}M$ denotes the based path space of M,[3] ΩM denotes the based loop space, and $\mathcal{D}^2M$ denotes the space of based discs $f{:}D^2\to M$ in M.[4] The map $\mathcal{P}M\to M$ is generically a surjective submersion, but for the map $\mathcal{D}^2M\to\Omega M$ to be a surjective submersion we must require 1-connectedness of M. The horizontal arrows restrict a loop to its first and second halves, reversing the orientation of the second half in order to obtain a based map. Any disc $f{:}D^2\to M$ allows us to integrate the pullback of ω over D^2. Setting $\lambda(f)=\exp(2\pi i\int_{D^2}f^*\omega)$ defines a map λ: $\mathcal{D}^2M\to U(1)$. Two different maps $f,f'{:}D^2\to M$ which induce the same map on the boundary yield a map $f\cup_{S^1}f'{:}S^2\to M$, and because ω has integer periods, $\lambda(f)=\lambda(f')$. This implies that λ descends to a map $\hat{\lambda}{:}\Omega M\to U(1)$. By the same reasoning we see that λ defines a cocycle in the sense that for triples $(\gamma_0,\gamma_1,\gamma_2)$ of based paths with common endpoint, we have

$$\hat{\lambda}(\overline{\gamma_0}*\gamma_1)\,\hat{\lambda}(\overline{\gamma_1}*\gamma_2)=\hat{\lambda}(\overline{\gamma_0}*\gamma_2). \tag{2.1}$$

where $*$ denotes concatenation of paths and $\overline{\gamma_i}$ is the path γ_i with opposite orientation. Thus, $\hat{\lambda}$ defines a hermitean line bundle L over M, called the *tautological line bundle* of (M,ω) (see, for example, [23]). A connection on this line bundle is constructed from ωω via its transgression to $\mathcal{P}M$, $A|_\gamma(X)=-2\pi\int_{[0,1]}\gamma^*\omega(X,-)$. Its magic flux is given by $F^L=2\pi\omega$; whence, this line bundle is a prequantum line bundle for the symplectic manifold (M,ω). If M is not 1-connected, no general construction of a prequantum line bundle is known, and one has to work with the primitive statement that there exists a line bundle representative for each closed $\omega\in\Omega^2(M)$ with integer periods.

Prequantisation is only the first of two steps in the programme of geometric quantisation. The Hilbert space H(L) it produces is usually too big to properly describe the physical system. The actual physical Hilbert space is a subspace or a quotient of H(L). This redundancy can be understood as follows. In classical mechanics the symplectic manifold describing the physical system is not the configuration space Q (which might be odd-dimensional), but rather it is the phase space of the system, which is given by the cotangent bundle $M=T^*Q$ of the configuration space. This space always has twice the dimension of Q and carries a canonical symplectic structure. Moreover, all homotopy groups of T^*Q and Q are isomorphic so that T^*Q is 1-connected if and only if Q is 1-connected.[5] As an example consider a particle propagating in $Q=\mathbb{R}^n$. Then $T^*Q\cong\mathbb{R}^n\times\mathbb{R}^n$, and $\omega=\mathrm{d}p_i\wedge\mathrm{d}q^i$ where q^i are coordinates for the position space $\mathbb{R}^n$ and p_i are coordinates for the momentum space $\mathbb{R}^n$. The prequantum line bundle L in this case is trivial, $(L,\nabla^L)=((\mathbb{R}^n\times\mathbb{R}^n)\times\mathbb{C},\mathrm{d}+2\pi\mathrm{i}p_i\mathrm{d}q^i)$. Elements of H($L$) are then just square-integrable complex-valued functions on $\mathbb{R}^n\times\mathbb{R}^n$. However, H($L$) is not the Hilbert space for the quantum mechanics of a particle on RnRn. Usually, one chooses either position representation or momentum representation since the particle is already completely described by functions depending solely on either the position coordinates or the momentum coordinates. As sections of L, this would mean that these functions are covariantly constant, or parallel, in half the directions of T^*Q. This reduction of the degrees of freedom and the prequantum Hilbert space H(L) is called *polarisation*. It is usually a choice which has to be made by hand, based on physical requirements. On a generic symplectic manifold M, a splitting of M into a product does not exist, but locally a splitting of the directions on M into two halves is still possible by Darboux's theorem. In this case polarisation amounts to choosing a foliation of M which is Lagrangian for ω, and then considering sections that are parallel along the foliation. Other methods of polarisation exist as well, see, for example, [42, 44, 45] and references therein.

3 STRINGS, MEMBRANES AND FLUXES

The main physical motivation for the present paper comes from the problem of quantising closed strings and open M2-branes in the backgrounds of certain fluxes. The situation described in Sect. 2 has an analogue for open strings in a B-field background, wherein the boundaries of strings ending on a D3-brane quantise the D3-brane worldvolume M [48, 49]; the closed Kalb–Ramond field $B\in\Omega^2(M)$ serves as a magic flux on M to which the techniques of standard geometric quantisation are applicable. The situation changes, however, when we lift this configuration to M-theory via T-duality, where it describes open M2-branes ending on an M5-

brane with a constant 3-form C-field. In this case the geometry of the M5-brane worldvolume is described by replacing the Poisson bracket with a 3-bracket related to the C-field, see, for example, [14, 43, 44] and references therein. The changes incurred in the M-theory lift are the replacements of the 2-form B-field with the 3-form C-field and of the particle-like boundaries of open strings with the closed string-like boundaries of open membranes, so that the non-commutative geometry arising from quantising the Poisson bracket is replaced by a non-associative geometry quantising a 3-bracket. This kind of situation in fact arises directly in closed string theory with a supergravity background: simplifying the system by dropping the fermions, the dilaton and the metric, as well as higher form fields, this leaves us with a closed string propagating through a target space M in the presence of an NS–NS H-flux, which is described by a closed 3-form $H\in\Omega^3(M)$ such that $H/2\pi$ has integer periods by virtue of (generalised) Dirac charge quantisation. We shall write $H=2\pi\varpi$ where ϖ represents a class in $H3(M,\mathbb{Z})$.

The immediate observation at this point is that geometric quantisation does not apply to these systems, simply because we are not in the symplectic setting. We therefore seek a modification of geometric quantisation which can treat situations like this.

First of all, the notion of non-degeneracy straightforwardly carries over to 3-forms [7, 42]: we call ϖ non-degenerate if $\iota X\varpi=\varpi(X,-,-)$ is zero if and only if the vector field X is zero. A closed non-degenerate 3-form ϖ is called a *2-plectic form*, and the pair (M,ϖ) is called a *2-plectic manifold*. Thus, we have found a modification of the notion of a symplectic structure, which was the starting point in geometric quantisation.

Prequantisation then relied upon a Chern–Weil realisation of $\omega\in\Omega^2(M)$ in terms of a hermitean line bundle with connection. Again for the simple reason that we are now dealing with a 3-form, such a geometric model can no longer be given by a line bundle on M. One way out of this is to say that the actual configuration space of a closed string should be the free loop space of M, i.e. $LM=C^\infty(S^1,M)$, or the space of unparameterised oriented loops $\text{L}M=C^\infty(S^1,M)/\text{Diff}+(S^1)$, where Diff_+ denotes the orientation-preserving diffeomorphisms; we could then attempt to apply the techniques of Sect. 2 on loop space. However, on the first space the 2-form induced by ϖ via transgression is degenerate (insert a vector field tangent to the loop), and on both spaces it is extremely difficult to find good notions of square-integrability: since these spaces are not locally compact the notion of compactly supported sections does not seem appropriate, as heuristically such sections would have support on sets we expect to have measure zero. From a more fundamental perspective, string theory is a quantum theory of gravity and so is expected to quantise spacetime itself rather than its loop space. Certain features of a quantisation of M will also translate to quantum objects on its loop space, but the fundamental quantisation happens in the target space itself. The best example of this is in non-geometric string theory where non-commutative and non-associative geometries of spacetimes arise as T-duals of geometric backgrounds carrying 3-form H-fluxes (see, for instance, [2, 8, 9, 15, 30, 35]).

We therefore take a different route. Recall that in Sect. 2 we are able to give a construction of a geometric representative of a symplectic form ω with integer periods on a 1-connected manifold, taking only that form as input. The key observation there was that we can integrate the 2-form over discs, and that on 2-spheres this produces an integer. We can try to lift this construction to 3-forms and analyse the structure we obtain. In the case of a 2-plectic form $\varpi$with integer periods we could integrate over balls; analogously, on 3-spheres we then obtain integer values. Assuming this time that M is 2-connected, we thus obtain the following diagram in which vertical arrows are surjective submersions:

$$\begin{array}{ccccc} \mathcal{D}^3M \cong \mathcal{P}\Omega^2 M & & & & \\ \big\downarrow{\scriptstyle\partial} & & & & \\ S^2M \cong \Omega^2 M & \rightrightarrows & \mathcal{D}^2M \cong \mathcal{P}\Omega M & & \\ & & \big\downarrow{\scriptstyle\partial} & & \\ & & \Omega M & \rightrightarrows & \mathcal{P}M \\ & & & & \big\downarrow{\scriptstyle\partial} \\ & & & & M \end{array} \quad . \tag{3.1}$$

In each step of this stair diagram one just adds another path $\mathcal{P}$ and loop Ω iteration: since the mapping spaces are based, there are identifications $\Omega^n M \cong S^n M = C^\infty_*(S^n, M)$ and $\mathcal{P}\Omega^n M \cong \mathcal{D}^{n+1}M = C^\infty_*(D^{n+1}, M)$, where the subscript $*$ indicates that based smooth maps are used.

Using integration of pullbacks of ϖ over D^3, we now obtain a function σ: $\mathcal{D}^3M \to \mathrm{U}(1)$ by setting $\sigma(f) = \exp(2\pi \mathrm{i} \int_{D3} f^* \varpi)$ for maps $f{:}D^3 \to M$. On two balls with coinciding boundary the values of σ agree because of the integrality of ϖ. Hence, σ descends to a U(1)-valued function $\hat{\sigma}$ on S^2M. As in Sect. 2, $\hat{\sigma}$ satisfies a cocycle condition on 2-spheres with one coincident hemisphere and thus defines a hermitean line bundle L over ΩM. A connection on this hermitean line bundle is, again as in the construction of Sect. 2, given by the transgression of ϖ to $\mathcal{D}^2M, A_{|f} = -2\pi \int_{D2} f^* \varpi$. The magic flux of this connection is given by the transgression of ϖ to ΩM,

$$F^L_{|\gamma} = 2\pi \int_{S^1} \gamma^* \varpi \quad .$$

We now observe that the transgression of any differential form to a loop $\gamma = \overline{\gamma_0} * \gamma_1$ naturally splits into the difference of its transgressions to the paths γ_0 and γ_1. In our situation we get

$$F^L_{|\gamma} = 2\pi \int_{S^1} (\overline{\gamma_0} * \gamma_1)^* \varpi = 2\pi \int_{[0,1]} \gamma_1^* \varpi - 2\pi \int_{[0,1]} \gamma_0^* \varpi = B_{|\gamma_1} - B_{|\gamma_0},$$

with the 2-form $B \in \Omega^2(\mathcal{P}M)$ chosen as $B_{|\gamma} = 2\pi \int_{[0,1]} \gamma^* \varpi$.[6] Finally, recall that the magic flux of the line bundle we constructed from a 2-form in Sect. 2 was related to that 2-form. Here too we are brought full circle by observing that $dB = 2\pi \partial^* \varpi$, i.e. dB descends to M giving precisely $2\pi\varpi = H$. Thus, we say that we have represented the 3-form ϖ in a Chern–Weil manner, and we call (M,ϖ) *prequantisable.*

There is yet an additional observation we can make about this construction. The fibre of L over a loop γ is given by equivalence classes of pairs (f, z), where $z \in \mathbb{C}$ and $f: D^2 \to M$ is a based disc in M whose boundary is $\gamma = \partial f$ Two such pairs (f, z) and (f', z') are equivalent if $z' = \exp(2\pi i \int_{D^3} h_* \varpi) z$, where $h: D^3 \to M$ is any ball which is bounded by the 2-sphere defined by gluing $\overline{f}$ and f'. Consider again three based paths $\gamma_i, i=0,1,2$ in M with common endpoint. Choose three discs $f_{ij}: D^2 \to M$ such that $\partial f_{ij} = \overline{\gamma_i} * \gamma_j$ for $i<j$, i.e. the boundary of the disc f_{ij} in M is the loop defined by γ_i and γ_j. We now get two discs bounded by $\overline{\gamma_0} * \gamma_2$, namely f_{02} and $f_{01} \cup_{\gamma_1} f_{12}$. Since $\pi_2(M)=0$ there exists a ball $g: D^3 \to M$ whose boundary is composed of precisely those discs glued along their common bounding loop. We can map given elements $[f_{01}, z_{01}] \in L_{\overline{\gamma_0} * \gamma_1}$ and $[f_{12}, z_{12}] \in L_{\overline{\gamma_1} * \gamma_2}$ as

$$L_{\overline{\gamma_0} * \gamma_1} \otimes L_{\overline{\gamma_1} * \gamma_2} \ni [f_{01}, z_{01}] \otimes [f_{12}, z_{12}] \longmapsto [f_{01} \cup_{\gamma_1} f_{12}, z_{01} z_{12}] \in L_{\overline{\gamma_0} * \gamma_2}.$$

We can further modify this map by using the equivalence relation in the definition of the fibres of L to get

$$[f_{01} \cup_{\gamma_1} f_{12}, z_{01} z_{12}] = \left[f_{02}, \exp\left(2\pi i \int_{D^3} g^* \varpi \right) z_{01} z_{12} \right].$$

This yields an isomorphism

$$\mu_{\gamma_0, \gamma_1, \gamma_2} : L_{\overline{\gamma_0} * \gamma_1} \otimes L_{\overline{\gamma_1} * \gamma_2} \xrightarrow{\cong} L_{\overline{\gamma_0} * \gamma_2}. \tag{3.2}$$

This structure is analogous to the cocycle condition (2.1) and is usually referred to as a multiplication. The isomorphism (3.2) is compatible with the connection we found above.

Thus, we obtain a geometric structure which realises a closed 3-form ϖ on M with integer periods, provided that M is 2-connected. This is analogous to how the tautological line bundle in Sect. 2 realised a closed 2-form with integer periods on a 1-connected manifold. The geometric structure we have unveiled here is called the *tautological bundle gerbe* of (M,ϖ) and appeared originally in [33]. In Sect. 4 we will abstract the structures present in this example, thereby arriving at the general notion of a *bundle gerbe*.

4 BUNDLE GERBES AND B-FIELDS

In the construction of the tautological bundle gerbe it was crucial that M be 2-connected in order to ensure that the vertical maps in the diagram (3.1) are surjective. For a general bundle gerbe, we therefore start with a surjective submersion $\pi{:}Y\rightarrow M$.[7] We then get the two horizontal maps $p_i{:}Y^{[2]}\rightrightarrows Y$ automatically, where p_i is defined to forget the ith entry in $Y^{[2]}=\{(y_0,y_1)\in Y^2 \,|\, \pi(y_0)=\pi(y_1)\}$.[8] The second layer in diagram (3.1) is just the tautological line bundle on ΩM. We formalise this by taking a hermitean line bundle $L\rightarrow Y^{[2]}$. Recall the multiplication structure on L from (3.2). This is made part of the data by demanding the existence of a multiplication isomorphism of hermitean line bundles $\mu : p_2^*L \otimes p_0^*L \xrightarrow{\cong} p_1^*L$ over $Y^{[3]}$, which is associative over $Y^{[4]}$. The 3-form on M was then reconstructed from a connection on L which was compatible with μ. Hence, we demand that L be endowed with a connection ∇^L for which μ is parallel with respect to the connections induced on its source and target line bundles. This implies that its magic flux $F^L\in\Omega^2(Y^{[2]})$ satisfies $p^*_2F^L+p^*_0F^L=p^*_1F^L$ in $\Omega^2(Y^{[3]})$. In turn, this allows us to find a 2-form $B\in\Omega^2(Y)$ such that $p^*_0B-p^*_1B=F^L$, just as we were able to write down explicitly for the tautological bundle gerbe; this follows from exactness of the Čech complex $\check{C}^\bullet(\pi : Y \rightarrow M, \Omega^k)$ of sheaves of k-forms for each k (see [33]). There is some ambiguity in picking such a 2-form as the relation between B and F^L only determines B up to Čech-exact forms. A choice of such a 2-form is called a *curving* of the bundle gerbe, and a *connection on a bundle gerbe* or a *B-field* consists of a connection on L together with a curving. Finally, we see that $p^*_1dB-p^*_0dB=dF^L=0$ by the Bianchi identity for ∇^L. Again by exactness of the Čech complex of k-forms we therefore obtain a 3-form H on M such that $\pi^*H=dB$. In this case H is uniquely defined once a curving B has been chosen, since π^* is injective. The 3-form H is called the *H-flux of the bundle gerbe*. To summarise, a hermitean bundle gerbe with connection is given by data $L=(L,\mu,Y,\nabla^L,B)$. In the following the terminology bundle gerbe will always refer to a hermitean bundle gerbe.

We have already seen an example of a bundle gerbe in Sect. 3. Let us here consider a surjective submersion over M given by taking $Y=\mathcal{U}=\bigsqcup_{i\in\Lambda}U_i$ to be the total space of a good open covering of M.[9] In this case, $Y^{[2]}=\mathcal{U}^{[2]}=\bigsqcup_{i,j\in\Lambda}U_{ij}$, where $U_{ij}=U_i\cap U_j$, and a bundle gerbe defined using a surjective submersion of this kind is called a *local bundle gerbe*. A hermitean line bundle over $\mathcal{U}^{[2]}$ consists of a hermitean line bundle $L_{ij}\to U_{ij}$ for every $(i,j)\in\Lambda^2$, and as the covering is good we can choose these to be trivial as hermitean line bundles without connection. The only non-trivial input for a bundle gerbe without connection and surjective submersion given by the total space of a good cover is hence the isomorphism μ. This is now an isomorphism of trivial line bundles $\mu_{ijk}: L_{ij}\otimes L_{jk}\xrightarrow{\cong} L_{ik}$ and hence corresponds to a collection of functions $\mu_{ijk}:U_{ijk}\to \mathrm{U}(1)$ on triple intersections. The associativity constraint on μ translates to

$$\mu_{ikl}\,\mu_{ijk}=\mu_{ijl}\,\mu_{jkl},$$

for all $i,j,k,l\in\Lambda$. Thus, (μ_{ijk}) defines a $\mathrm{U}(1)$-valued Čech 2-cocycle on M. Enriching these data by the components of a connection on the bundle gerbe, one is left with a triple (μ_{ijk},A_{ij},B_i), where A_{ij} is the connection 1-form of the connection on L_{ij} and $B_i=B_{|U_i}$. A triple such as this, obtained from a bundle gerbe defined over $Y=\mathcal{U}$, defines a Čech representative of a Deligne 2-cocycle in degree 2 on M; we denote the corresponding class of the gerbe L in Deligne cohomology by $D(\mathrm{L})\in H^2(M,\mathcal{D}^\bullet(2))$. Forgetting the data of differential forms in the Deligne cocycle we recover the Čech $\mathrm{U}(1)$-cocycle (μ_{ijk}); its image in integer cohomology is called the *Dixmier–Douady class* of L and is denoted by $DD(\mathrm{L})\in H^3(M,\mathbb{Z})\cong H^2(M,\mathrm{U}(1))$.

A bundle gerbe which we will frequently encounter in the ensuing sections is the *trivial bundle gerbe*. It is defined by the data $\mathcal{I}=(M\times\mathbb{C},\cdot,M)$. Its covering of M is the identity covering $1_M:M\to M$, so that $M^{[2]}$ identifies with M. The hermitean line bundle over this space is the trivial hermitean line bundle $I=M\times\mathbb{C}\to M$. We introduce a bundle gerbe multiplication using the usual multiplication on $\mathbb{C}$, i.e. $(z,z')\mapsto z\cdot z'$. Additionally, just as the trivial line bundle can carry a non-trivial connection given by a 1-form on the base, the trivial bundle gerbe can still carry a non-trivial B-field. We define $\mathcal{I}_\rho=(M\times\mathbb{C},\cdot,M,d,\rho)$, where dddenotes the trivial connection on the trivial line bundle, and $\rho\in\Omega2(M)\rho\in\Omega2(M)$ is an arbitrary 2-form defining the curving of $\mathcal{I}_\rho$. The H-flux of $\mathcal{I}_\rho$ is given by $H=d\rho\in\Omega^3(M)$.

Another interesting example of a bundle gerbe occurs in the form of the tautological gerbe constructed in Sect. 3. Any compact, simple, simply connected Lie group G with Lie algebra ggis automatically 2-connected and has $H^3(G,\mathbb{Z})\cong\mathbb{Z}$ with generator given by the de Rham class of $\varpi = \frac{1}{3!}\langle\mu_G, [\mu_G, \mu_G]\rangle_{\mathfrak{g}}$, where $\mu_G \in \Omega^1(G,\mathfrak{g})$ denotes the Maurer–Cartan form of G [40]. Thus, we can apply the tautological bundle gerbe construction to every pair $(G,k\varpi)$ for $k\in\mathbb{Z}$. The Lie group operations are smooth; whence, we obtain group structures on the relevant spaces of smooth maps into G and also on the line bundle $L\to\Omega G$. We construct in this way the central extensions $\widehat{\Omega_k G} = L$ of the loop group ΩG, together with the additional data of a connection on the extension bundle and a curving 2-form on $\mathcal{P}G$.

5 THE 2-CATEGORY OF BUNDLE GERBES

In this section we start by summarising the 2-categorical theory of bundle gerbes as initiated in [34] and generalised in [51]. The material in Sects. 5.4 and 5.5 is new. Details and proofs can be found in [13].

5.1 Local picture

The example of the local bundle gerbe from Sect. 4 has an illuminating intuitive interpretation. The line bundles $L_{ij}\to U_{ij}$ satisfy

$$L_{ij} \otimes L_{jk} \cong L_{ik},$$

for all $i,j,k \in \Lambda$. We can compare this to the transition data of a hermitean line bundle over the same open cover, which is given by transition functions $g_{ij}:U_{ij}\to U(1)$ subject to the cocycle condition

$$g_{ij}\, g_{jk} = g_{ik},$$

for all $i,j,k\in\Lambda$.

Hence, the heuristic behind passing from hermitean line bundles to hermitean bundle gerbes is that U(1)-valued transition functions get replaced by $\mathrm{Hilb}^\times$-valued transition functions. Here HilbHilb denotes the category of finite-dimensional Hilbert spaces and linear maps. Among other structures, it carries a tensor product $\otimes$:Hilb×Hilb→Hilb. This is *symmetric* in the sense that for every $V,V'\in$Hilb there exists a distinguished isomorphism $V\otimes V'\cong V'\otimes V$ which satisfies certain coherence conditions. The *unit object* with respect to the tensor product is $\mathbb{C}$ with its natural inner product. A Hilbert space $V\in$Hilb is called *invertible* if there is another Hilbert space $V'\in$Hilb and an isomorphism $V\otimes V'\cong\mathbb{C}$. In this case we call V'

a *weak inverse* for V. We denote the subcategory of invertible objects and unitary morphisms by $\mathrm{Hilb}^\times$. For every Hilbert space V there exists a *dual* Hilbert space V^*, and it is related to V via the Riesz isomorphism

$$\theta : V \xrightarrow{\cong} \overline{V^*}, \quad \psi \longmapsto \theta(\psi) = \langle \psi, - \rangle_V.$$

If V is invertible, then its inverse is represented by V^*, which is isomorphic to the complex conjugate $\overline{V}$. In this sense the invertible objects in Hilb are also *unitary*; they are precisely the one-dimensional Hilbert spaces. The category $\mathrm{Hilb}^\times$ is closed under the tensor product, which is symmetric, and each object has a weak inverse. By construction, all morphisms in this category are invertible. We call a category with these properties a *2-group*.[10]

A function $L:Y^{[2]}\to\mathrm{Hilb}^\times$ assigns to each point $(y_0,y_1)\in Y^{[2]}$ a one-dimensional Hilbert space $L(y_0,y_1)$, and we regard such a function as smooth if the resulting family of Hilbert spaces forms a hermitean line bundle over $Y^{[2]}$. Since in $\mathrm{Hilb}^\times$ we now have isomorphisms between objects, the cocycle condition on the transition functions gets weakened from equality to the existence of an isomorphism satisfying a suitable coherence condition (which is precisely the associativity condition on μ). For connections on the line bundles, the picture is a little less clear; here, we shall endow the transition line bundles with connections in a subsequent step.[11]

Following the viewpoint that L_{ij} serve as transition functions of a *higher line bundle*, we can explore what morphisms of this object look like. Recall that for two ordinary hermitean line bundles with transition functions g_{ij} and g'_{ij}, morphisms are given by collections of maps $e_i:U_i\to\mathbb{C}$ which satisfy

$$g_{ij}\, e_j = e_i\, g'_{ij} \tag{5.1}$$

so that these functions can be glued along the transition functions g_{ij} and g'_{ij}.

The local representatives e_i are functions into the ring $\mathbb{C}$, which contains the group that the transition functions are valued in. Translating this to the framework above le us to consider smooth Hilb-valued functions $E_i:U_i\to\mathrm{Hilb}$, or in more familiar terms, hermitean vector bundles $E_i\to U_i$. Heuristically, we replace the target ring $(\mathbb{C},+,\cdot)$ by $(\mathrm{Hilb},\oplus,\otimes)$; we shall say more about this analogy in Sect. 7. The gluing relation is weakened to the existence of isomorphisms

$$\alpha_{ij} : L_{ij} \otimes E_j \xrightarrow{\cong} E_i \otimes L'_{ij}. \tag{5.2}$$

satisfying a compatibility condition with μ: we can either multiply two of the transition line bundles successively with E_i and then multiply the transition line bundles on the target side, or we can take their product first and then multiply the result with E_i, yielding

$$(1_{E_i} \otimes \mu'_{ijk}) \circ (\alpha_{ij} \otimes 1_{L'_{jk}}) \circ (1_{L_{ij}} \otimes \alpha_{jk}) = \alpha_{ik} \circ (\mu_{ijk} \otimes 1_{E_k}).$$

We can then require that E_i comes endowed with a hermitean connection such that all morphisms above are parallel with respect to the induced connections.

A crucial difference between the line bundle and bundle gerbe cases, which we have already encountered above, becomes fully visible here. Recall that the cocycle condition and the gluing condition for morphisms are only required to hold up to coherent isomorphism, i.e. since the data of a morphism of bundle gerbes are based on a hermitean vector bundle with connection, we can still have another level of morphisms between pairs of such data. Alternatively, observe that the target Hilb of the local representatives of the sections is a category rather than a set (such as $\mathbb{C}$), so that there exist non-trivial relations, or morphisms, between the objects. We call the morphisms on this higher, second-level *2-morphisms* of bundle gerbes. A 2-morphism from $(E_i,\nabla^{E_i},\alpha_{ij})$ to $(E'_i,\nabla^{E'_i},\alpha'_{ij})$ is a collection of morphisms $\psi_i:E_i\rightarrow E'_i$ of hermitean vector bundles with connection which we require to be bicovariantly constant. They are further required to be compatible with the isomorphisms in the sense that

$$\alpha'_{ij} \circ (1_{L_{ij}} \otimes \psi_j) = (\psi_i \otimes 1_{L'_{ij}}) \circ \alpha_{ij}.$$

In the following we will often make no explicit mention of the connection in the data of a 1-morphism and just write (E,α).

The data for a local bundle gerbe simplify considerably for the case of a good open covering because in that situation we can choose all hermitean line bundles to be topologically trivial, and all the information about the bundle gerbe is contained in the multiplication isomorphism μ_{ijk} and the local data A_{ij},B_i for the B-field. Similarly, in this situation we can always assume (up to 2-isomorphism) that the hermitean vector bundles that make up the morphisms of bundle gerbes are trivial over U_i. Thus, the data which are left are the isomorphism αijαij, which is now a map $\alpha_{ij}:U_{ij}\rightarrow \mathrm{U}(n)$, and 1-forms $a_i \in \Omega^1(U_i, \mathfrak{u}(n))$. The compatibility with μ and μ' becomes a *twisted cocycle condition*

$$\mu'_{ijk}\,\alpha_{ij}\,\alpha_{jk} = \alpha_{ik}\,\mu_{ijk}. \tag{5.3}$$

For equal source and target bundle gerbe, the two twists μ_{ijk} and μ'_{ijk} in this equation cancel, and we are left with transition data for a hermitean vector bundle with connection on M; 2-morphisms between such morphisms then precisely provide local representations of parallel morphisms of these hermitean vector bundles with connection. On the other hand, if the target is the trivial bundle gerbe so that $\mu'_{ijk}=1$ in (5.3), then (α_{ij},μ_{ijk}) are the data of a *rank n twisted vector bundle* [26], also known as a *bundle gerbe module of rank n* [11].

5.2 Global picture

In Sect. 5.1 we introduced morphisms of bundle gerbes which are defined over a common open covering of M. It is possible to generalise this definition of morphisms both to coverings given by general surjective submersions, and to gerbes defined over different coverings, see [51] for details. In this way, we assemble bundle gerbes on M into a 2-category $\mathcal{BG}\mathrm{rb}^{\nabla}(M)$.

The observation at the end of Sect. 5.1 that endomorphisms of local bundle gerbes are related to descent data for hermitean vector bundles with connection on M, holds as well in this 2-category. In particular, there is an equivalence of categories

$$\mathrm{R} : \mathcal{BG}\mathrm{rb}^{\nabla}(M)(\mathcal{I}_0, \mathcal{I}_0) \xrightarrow{\cong} \mathcal{HVB}\mathrm{dl}^{\nabla}(M). \tag{5.4}$$

Here, for $\mathcal{L}, \mathcal{L}' \in \mathcal{BG}\mathrm{rb}^{\nabla}(M)$, we denote by $\mathcal{BG}\mathrm{rb}^{\nabla}(M)(\mathcal{L}, \mathcal{L}')$ the category of morphisms from L to L′, and $\mathcal{HVB}\mathrm{dl}^{\nabla}(M)$ is the category of hermitean vector bundles with connection on M and morphisms given by parallel, smooth, fibrewise-linear maps. This equivalence follows from the fact that $\mathcal{HVB}\mathrm{dl}^{\nabla}$ is a stack.

As an example of 1-morphisms, consider the tautological bundle gerbe L_k over a compact simply connected Lie group G from Sect. 4. Recall that these bundle gerbes are the same as the central extensions of the loop group $\widehat{\Omega_k \mathrm{G}} \to \Omega\mathrm{G}$. Consider a representation E of $\widehat{\Omega_k \mathrm{G}}$, which is given by a map $\widehat{\Omega_k \mathrm{G}} \otimes E \to E$. Since the extension is central, the fibre U(1) acts as $\lambda \mapsto \lambda 1_E$. Thus, consider the trivial bundle $\Omega\mathrm{G}\times E\to\Omega\mathrm{G}$. Transition morphisms are given by

$$\alpha : \widehat{\Omega_k \mathsf{G}} \otimes E \longrightarrow E, \quad (\hat{a}, e)_{|b} \longmapsto (\hat{a}e)_{|ab},$$

for a lift $\hat{a} \in \widehat{\Omega_k \mathsf{G}}$ of $a \in \Omega\mathsf{G}$, and $b \in \Omega\mathsf{G}$. This is of the form of a 1-morphism $\mathcal{L}_k \to \mathcal{I}$ of bundle gerbes without connection. However, here E has infinite-rank, and bundles with infinite-dimensional fibres are not sufficiently well behaved for an interesting theory of bundle gerbes,[12] so we restrict ourselves to finite-rank hermitean vector bundles as employed above.

A 1-morphism of bundle gerbes is (weakly) invertible if and only if its underlying hermitean vector bundle is of rank 1. This stems from the analogous property of objects in (Hilb,⊗), which are invertible with respect to ⊗ if and only if they are one-dimensional. We have seen in Sect. 4 that every local bundle gerbe L defines a Deligne class $\mathrm{D}(\mathcal{L}) \in \mathrm{H}^2(M, \mathcal{D}^\bullet(2))$. With the generalised notion of 1-morphisms of bundle gerbes, it follows that every bundle gerbe is isomorphic to a local bundle gerbe. By choosing such an isomorphism and taking the Deligne class we thus obtain a Deligne class for every bundle gerbe; it is independent of the choice of isomorphism of the bundle gerbe with a local bundle gerbe. From the Čech representation of Deligne cocycles over a good open covering, it is evident that every element of $\mathrm{H}^2(M, \mathcal{D}^\bullet(2))$ arises as the Deligne class D(L) of a bundle gerbe L. Thus, we obtain a well-defined map

$$\mathrm{D} : \mathcal{BG}\mathrm{rb}^\nabla(M) \longrightarrow \mathrm{H}^2(M, \mathcal{D}^\bullet(2)),$$

which descends to a group isomorphism on 1-isomorphism classes of bundle gerbes on M.

The construction of bundle gerbes can be rephrased in terms of functors on higher Čech groupoids. For every hermitean line bundle L, there exists a covering $Y \to M$ such that L descends from the trivial line bundle over Y; these descent data are the same as a functor g: $(Y^{[2]} \rightrightarrows Y) \to (\mathbb{C} \rightrightarrows *)$, where composition in the target category is given by multiplication, while in the source category (the Čech groupoid of $Y \to M$) a pair (y_0, y_1) is understood as an isomorphism from y_0 to y_1. Morphisms of hermitean line bundles are then natural transformations of such functors. Again replacing $\mathbb{C}$ by Hilb, we can now look for functors L into Hilb$^\times$ that define transition data for a bundle gerbe. Since the target is now the 2-category Hilb$\rightrightarrows *$, we must replace the Čech groupoid by a 2-category as well. We can take this to be the Čech 2-groupoid $\mathfrak{Disc}(Y^{[2]} \rightrightarrows Y)$, which is the discrete 2-category built from the Čech groupoid $Y^{[2]} \rightrightarrows Y$ by adding an identity 2-morphism for each morphism. Thus, a bundle gerbe can be regarded as a 2-functor

$$(L, \mu) : \mathcal{D}\mathrm{isc}\big(Y^{[2]} \rightrightarrows Y\big) \longrightarrow (\mathcal{H}\mathrm{ilb} \rightrightarrows *).$$

The bundle gerbe multiplication μ arises as the part of the data of a 2-functor which establishes its weak form of compatibility with composition. Analogously to the case of line bundles, 1-morphisms and 2-morphisms arise as natural transformations of functors of 2-categories valued in Hilb, namely $(E,\alpha):(L,\mu)\Rightarrow(L',\mu')$.

5.3 Pullbacks, products and duals

Bundle gerbes are constructed from local transition or descent data. As we know from vector bundles, these data can be pulled back along smooth maps. Accordingly, any smooth map $f:N\to M$ induces a 2-functor $f^* : \mathcal{B}\mathcal{G}\mathrm{rb}^\nabla(M) \to \mathcal{B}\mathcal{G}\mathrm{rb}^\nabla(N)$, and the assignment $f\mapsto f^*$ is contravariantly functorial.[13]

Transition functions for hermitean line bundles are valued in U(1). Let $Y=\mathcal{U}$ be a good open covering of M. Given two such transition functions g,h: $\mathcal{U}^{[2]}\to$U(1) over the same good covering, their pointwise product g,h: $\mathcal{U}^{[2]}\to$U(1) also represents a hermitean line bundle over M. As every hermitean line bundle is isomorphic to one described by transition data over $\mathcal{U}$, this allows us to define a product on hermitean line bundles over M, namely the tensor product of hermitean line bundles. This product can be extended to include connections.

The product of hermitean line bundles can therefore be seen as induced or pulled back from the product structure on U(1) along the defining transition functions.[14] Carrying this viewpoint over from the abelian group U(1) to the symmetric (or abelian) 2-group (Hilb$^\times$,$\otimes$), for two transition line bundles $L,K\to\mathcal{U}^{[2]}$ we consider the transition line bundle obtained as their product, $L\otimes K\to\mathcal{U}^{[2]}$. The tensor product of their bundle gerbe multiplications makes this product into a bundle gerbe defined over $\mathcal{U}$, which we understand as the *tensor product* of the two bundle gerbes. Similarly, given two morphisms defined by functions E,F: $\mathcal{U}\to$Hilb, their tensor product yields a morphism between the tensor products of their source and target bundle gerbes. The tensor product structure in Hilb includes a tensor product on morphisms between Hilbert spaces, and this naturally pulls back to 2-morphisms of bundle gerbes. In this way, the symmetric monoidal structure on (Hilb,$\otimes$) induces a symmetric monoidal structure on $\mathcal{B}\mathcal{G}\mathrm{rb}^\nabla(M)$ which we also denote by $\otimes$.[15]

We have not yet exploited the full abelian group structure on U(1); so far, we have only made use of its commutative monoid structure. The remaining structure is the existence of inverses. For a line bundle defined by $g{:}\mathcal{U}^{[2]}\to U(1)$ we obtain new descent data by $g\mapsto g^{-1}$, the function g composed with inversion in the group U(1). The hermitean line bundle defined by these transition data are the same as the dual hermitean line bundle. The line bundle $L^*\otimes L$ is canonically isomorphic to the trivial hermitean line bundle I. For a morphism of line bundles $\psi{:}L\to L'$ with local representatives satisfying $g_{ij}\psi_j=\psi_i g'_{ij}$ over an open covering of M, we obtain the dual (or transpose) morphism $\psi^t{:}L'^*\to L^*$ by bringing the transition functions to the opposite sides, i.e. ψ^t has the same local representatives $\psi^t=(\psi_i)$, but now satisfying $g'^{-1}_{ij}\,\psi_j = \psi_i\, g^{-1}_{ij}$.[16]

As we found in Sect. 5.1, an inverse (which is unique only up to unique isomorphism) of a one-dimensional Hilbert space V is given by its dual space V . Thus, given a bundle gerbe L defined by a line bundle $L\to\mathcal{U}^{[2]}$, consider the dual bundle $L^*\to\mathcal{U}^{[2]}$. It can be regarded as the composition of the Hilb$^\times$ -valued transition function L with inversion in the 2-group Hilb$^\times$. The bundle gerbe multiplication on L^* is then given by μ^{-t}, the inverse transpose of μ, and the resulting bundle gerbe is called the *dual bundle gerbe*L*. For any bundle gerbe L there exists a canonical 1-isomorphism $\eta_{\mathcal{L}}:\mathcal{I}_0\xrightarrow{\cong}\mathcal{L}^*\otimes\mathcal{L}$. Starting from a morphism (E,α):L→L′ of bundle gerbes with isomorphisms (5.2), we obtain its dual morphism analogously to the case of line bundles: we just have to reinsert the original transition functions that get cancelled out in bringing them to the opposite sides in order to then apply α, i.e. $(E,\alpha)^t=(E,\beta){:}L'^*\to L^*$ is defined by

$$\begin{array}{ccc}
L'^*_{ij}\otimes E_j & \overset{\beta_{ij}}{\dashrightarrow} & E_i\otimes L^*_{ij}\\
{\scriptstyle 1\otimes\mu^{-1}_{iji}}\downarrow & & \uparrow{\scriptstyle \mu'^{-1}_{jij}\otimes 1}\\
L'^*_{ij}\otimes L^*_{ij}\otimes L_{ij}\otimes E_j & \xrightarrow[1\otimes\alpha_{ij}]{} & L'^*_{ij}\otimes L'_{ij}\otimes E_i\otimes L^*_{ij}
\end{array}$$

Again, this structure is straightforwardly extended to include connections at the level of bundle gerbes as well as 1-morphisms.

5.4 Direct sum

From the perspective of geometric quantisation, and also more generally, an important feature of sections and morphisms of line bundles is that they come with an operation of a sum: it would not be possible to construct a Hilbert space of sections otherwise. From the point of view of local (or descent) representatives of sections of a line bundle, the additive structure is induced by the addition in the ring $\mathbb{C}$. On the category (Hilb,⊕,⊗), by which we are replacing $\mathbb{C}$ as the ground ring, there is an additive structure

$\oplus$ present. It can readily be used to add morphisms of bundle gerbes which are defined over a mutual covering Y of M. There always exists a zero 1-morphism $(E,\alpha)=0$ given by the zero vector bundle with the zero morphism. In this case we obtain the direct sum of hermitean vector bundles over Y, which is compatible with connections. The extension to general morphisms in $\mathcal{BG}\mathrm{rb}^{\nabla}(M)$ can be found in [13].

5.5 Enrichment

To finish off the description of the 2-category of bundle gerbes, we turn to the level of 2-morphisms. As these are built from morphisms of twisted hermitean vector bundles with connection (at least for open coverings), they form a vector space over $\mathbb{C}$. The space of parallel sections of a vector bundle of rank n with connection has dimension at most n, because every covariantly constant section spans a rank 1 sub-bundle. Kernels of 2-morphisms have constant rank, which is a consequence of parallelity. Because of the hermitean structure on the vector bundles, 2-morphisms also have well-defined cokernels. With these choices of 2-morphisms, the morphism categories in $\mathcal{BG}\mathrm{rb}^{\nabla}(M)$ are *abelian categories*, just like the category $\mathcal{HVB}\mathrm{dl}^{\nabla}(M)$. Using the fact that on a hermitean vector bundle with connection ∇ and metric h one has $\nabla h=0$, together with parallelity of morphisms, we see that the hermitean metric evaluated on parallel morphisms is constant. This implies that any vector space of 2-morphisms in $\mathcal{BG}\mathrm{rb}^{\nabla}(M)$ has the structure of a finite-dimensional Hilbert space. Composition of 2-morphisms is compatible with this structure, as are composition and direct sum of 1-morphisms as well as the tensor product of bundle gerbes: we say that $\mathcal{BG}\mathrm{rb}^{\nabla}(M)$c is *2-enriched in* Hilb.

6 BUNDLE GERBES AS HIGHER LINE BUNDLES

6.1 Module categories of sections

From the perspective of higher geometric quantisation, and also of higher geometry itself, we wish to find a notion of section of a bundle gerbe. For a hermitean line bundle L these are usually defined as maps $M\to L$ splitting the projection $\pi:L\to M$. However, this definition is not suitable for bundle gerbes in the language employed here. If one desires to stay in the setting of maps to a total space, the notion of space has to be generalised to incorporate total spaces of higher bundles. A detailed treatment of this idea can be found in [36, 37].

There is a different, purely categorical perspective on sections of a line bundle. Let us denote the monoidal category of hermitean line bundles on *M* without connection by $\mathcal{HLB}\mathrm{dl}(M)$; its morphisms are given by smooth fibrewise-linear maps. The category of hermitean line bundles with connection and parallel morphisms of line bundles is denoted $\mathcal{HLB}\mathrm{dl}(M)$. As is evident from (5.1), sections of a hermitean line bundle $L \in \mathcal{HLB}\mathrm{dl}(M)$ can equivalently be viewed as morphisms from the trivial hermitean line bundle *I* to *L*,

$$\Gamma(M, L) \cong \mathcal{HLB}\mathrm{dl}(M)(I, L).$$

The trivial line bundle *I* can in fact be defined to be the monoidal unit in $(\mathcal{HLB}\mathrm{dl}(M), \otimes)$.

Hence, for a bundle gerbe L we define the *category of sections of* L via

$$\Gamma(M, \mathcal{L}) := \mathcal{BG}\mathrm{rb}^{\nabla}(M)(\mathcal{I}_0, \mathcal{L}) .$$

The idea for this definition of sections of bundle gerbes is not new (cf. [50]), but, to our knowledge, has not appeared in the literature before in this form. Here we write the category with connections, since parallelity is imposed at the level of 2-morphisms, whereas 1-morphisms are unconstrained; we shall require bicovariantly constant 2-morphisms in Sect. 8. Hence, instead of the trivial bundle gerbe we must use the trivial bundle gerbe with the trivial connection for consistency. Note that $\mathcal{BG}\mathrm{rb}^{\nabla}(M)(\mathcal{I}_0, \mathcal{I}_0)$ carries both direct sum and tensor product, and that the tensor product distributes over the direct sum in a categorical sense. Such a category is commonly referred to as a *rig category*, since it has the structure of a categorified ring but is missing the additive inverses of objects. Other important examples of rig categories include $(\mathcal{HVB}\mathrm{dl}^{\nabla}(M), \oplus, \otimes)$ and $(\mathcal{H}\mathrm{ilb}, \oplus, \otimes)$.

Every morphism category $\mathcal{BG}\mathrm{rb}^{\nabla}(\mathcal{L}, \mathcal{L}')$ is an abelian monoidal category under direct sum of morphisms and enriched in HilbHilb as discussed in Sect. 5. Under the natural isomorphisms $l_{\mathcal{L}} : \mathcal{I}_0 \otimes \mathcal{L} \xrightarrow{\cong} \mathcal{L}$ and $r_{\mathcal{L}} : \mathcal{L} \otimes \mathcal{I}_0 \xrightarrow{\cong} \mathcal{L}$, we obtain an action of the rig category $\mathcal{BG}\mathrm{rb}^{\nabla}(M)(\mathcal{I}_0, \mathcal{I}_0)$ on every morphism category. It is given (as a right-action) by

$$(E, \alpha) \lhd (F, \beta) = r_{\mathcal{L}'} \circ \big((E, \alpha) \otimes (F, \beta)\big) \circ r_{\mathcal{L}}^{-1}. \qquad (6.1)$$

Hence, such actions exist in particular on the categories of sections of any bundle gerbe L. (Note that this module action extends the one obtained from composition in [50] to morphism categories between any two bundle gerbes, rather than just being defined on sections.) We may use the equivalence from (5.4) to translate this into an action of $\mathcal{HVB}\mathrm{dl}^{\nabla}(M)$ on any morphism category.

The action defined by (6.1) is built from the tensor product of vector bundles and therefore distributes over the direct sum. Hence, the morphism categories have the structure of *rig-module categories over* $\mathcal{HVB}\mathrm{dl}^{\nabla}(M)$. The term 'module category' usually refers to a category with an action of a monoidal category. Here we actually have much more: the acting category is a rig category, the categories acted upon are monoidal, and the action respects both the rig structure on the right as well as the monoidal structure on the left. Hence, this kind of action of categories is as closely related to the algebraic notion of a module over a ring as we can get.

6.2 Hermitean 2-bundle metrics

In this section we give a non-technical account of what we term a hermitean 2-bundle metric on a hermitean bundle gerbe. This extends the common point of view that a bundle gerbe is a higher line bundle to the statement that a *hermitean* bundle gerbe as considered here is a higher *hermitean* line bundle. A full treatment can be found in [13]. The existence of a hermitean structure on the prequantum line bundle $L \to M$ is crucial in geometric quantisation, as it is this structure which allows one to naturally promote the vector space structure on sections of L to a Hilbert space structure. That is, the notion of amplitudes and probabilities, and hence, the possibility of making contact with quantum mechanics in geometric quantisation relies on the existence of a hermitean metric on L. Thus, it will be necessary to have a higher analogue of such a hermitean metric on the prequantum bundle gerbe L.

A hermitean line bundle is a complex line bundle with a fibrewise hermitean metric. On morphisms $\psi: L \to L'$ there is an induced hermitean metric. Using the identification $C^\infty(M, \mathbb{C}) \cong \mathcal{HLB}\mathrm{dl}(M)(I, I)$, this turns out to be an operation in the category of line bundles on M,

$$h : \overline{\mathcal{HLB}\mathrm{dl}(M)(L, L')} \times \mathcal{HLB}\mathrm{dl}(M)(L, L') \longrightarrow \mathcal{HLB}\mathrm{dl}(M)(I, I),$$

where the bar over the first argument indicates that h is $C^\infty(M, \mathbb{C})$ -antilinear in this argument. To see this, we first use the hermitean structures on the objects of $\mathcal{HLB}\mathrm{dl}(M)$ to obtain an antilinear isomorphism

$$\theta : \overline{\mathcal{HLB}\mathrm{dl}(M)} \xrightarrow{\cong} \mathcal{HLB}\mathrm{dl}(M), \quad \theta\big(L \xrightarrow{\psi} L'\big) \longmapsto \big(L^* \xrightarrow{\psi^{\mathrm{t}*}} L'^*\big),$$

where the superscript 't' refers to the fibrewise transpose of a morphism and the star operation takes the fibrewise adjoint with respect to the hermitean structures. Given a second morphism ϕ:$L \to L'$ we can then use the tensor product of line bundles to obtain a morphism

$$\theta(\psi) \otimes \phi : L^* \otimes L \longrightarrow L'^* \otimes L'.$$

Recalling that $L^* \otimes L$ is canonically isomorphic to I in $\mathcal{HLB}\mathrm{dl}(M)$, we obtain an endomorphism of the trivial bundle as the composition

$$\eta_{L'}^{-1} \circ \big(\theta(\psi) \otimes \phi\big) \circ \eta_L \ \in \ \mathcal{HLB}\mathrm{dl}(M)(I, I),$$

where $\eta_L : I \xrightarrow{\cong} L^* \otimes L$ is the canonical isomorphism given by $\eta_L(z) = z\,1_L$. One then has

$$\eta_{L'}^{-1} \circ \big(\theta(\psi) \otimes \phi\big) \circ \eta_L = h(\psi, \phi). \tag{6.2}$$

under the identification $\mathcal{HLB}\mathrm{dl}(M)(I, I) \cong C^\infty(M, \mathbb{C})$. Thus, the hermitean metrics on line bundles can equivalently be obtained via the Riesz-type equivalence θ which acts $C^\infty(M, \mathbb{C})$-antilinearly on morphisms.

This functorial version of a hermitean bundle metric is easier to lift to the higher categorical world of bundle gerbes. We have already found a dual (or transpose) of a morphism between bundle gerbes in Sect. 5.3, but the ordinary dual functor produces morphisms going in the wrong direction and is linear rather than antilinear. However, given a bundle gerbe morphism (E,α):$\mathcal{L} \to \mathcal{L}'$, observe that

$$(E^*, \alpha^{-\mathrm{t}}) = (E^*, \alpha^{\mathrm{t}*}) : \mathcal{L}^* \longrightarrow \mathcal{L}'^*,$$

just as desired. From the point of view of local representations of a section this amounts to replacing the local representatives by their complex conjugates. In particular, if we just consider a hermitean vector bundle $E \to M$ as an endomorphism $(E, 1_E) : \mathcal{I}_0 \to \mathcal{I}_0$ of the trivial bundle gerbe on M, this construction yields the dual vector bundle.[17] Thus, we can define a 2-functor

$$\Theta : \mathcal{BG}\mathrm{rb}^\nabla(M) \longrightarrow \mathcal{BG}\mathrm{rb}^\nabla(M)$$

acting on 1-morphisms as

$$\Theta\big((E,\alpha):\mathcal{L}\to\mathcal{L}'\big)=\big((E^*,\alpha^{t*}):\mathcal{L}^*\to\mathcal{L}'^*\big)$$

On 2-morphisms we could just have it take the transpose, thereby acting contravariantly on 2-morphisms. However, as we wish to employ the hermitean structure we define the action on $\psi:(E,\alpha)\Rightarrow(F,\beta)$ by

$$\Theta(\psi)=\psi^{t*}=\psi^{*t}:\Theta(E,\alpha)\Longrightarrow\Theta(F,\beta).$$

With this convention, Θ becomes a covariant 2-functor,[18] which is compatible with tensor products and direct sums of morphisms, and it acts antilinearly in the sense that $\Theta((E,\alpha)\lhd F)=\Theta(E,\alpha)\lhd F^*$ for any hermitean vector bundle $F\in\mathcal{HVB}\mathrm{dl}^\nabla(M)$ and morphism of gerbes (E,α).

Given a pair of morphisms $(E,\alpha),(F,\beta):\mathrm{L}\to\mathrm{L}'$ we can now mimic (6.2): we define

$$\mathfrak{h}\big((E,\alpha),(F,\beta)\big):=\mathrm{R}\big(\eta_{\mathcal{L}'}^{-1}\circ\big(\Theta(E,\alpha)\otimes(F,\beta)\big)\circ\eta_{\mathcal{L}}\big)\in\mathcal{HVB}\mathrm{dl}^\nabla(M).$$

We included the equivalence RR from (5.4) in order to end up in the more familiar category of hermitean vector bundles on *M*, which corresponds directly to functions on *M* valued in HilbHilb; they can be regarded as *higher functions*. Without RR in the definition of hh we would only map to the endomorphisms of the trivial bundle gerbe.[19] Via the above module actions, morphisms of bundle gerbes, and in particular sections of bundle gerbes, carry a module structure over these higher functions, just as morphisms and sections of vector bundles are modules over the ring of functions $C^\infty(M,\mathbb{C})$. The functor $\mathfrak{h}$ is sesquilinear with respect to this module action and complex conjugation of hermitean vector bundles as defined by Θ. Thus, we obtain a *2-bundle metric* on $\mathcal{BG}\mathrm{rb}^\nabla(M)(\mathcal{L},\mathcal{L}')$ as the composition of functors and natural 1-isomorphisms inherent to the category of bundle gerbes with connection on *M*. It has the useful property[20]

$$\Gamma_{\mathrm{par}}\big(M,\ \mathfrak{h}\big((E,\alpha),(F,\beta)\big)\big)\cong\mathcal{BG}\mathrm{rb}^\nabla(M)\big((E,\alpha),(F,\beta)\big).\qquad(6.3)$$

where $\Gamma_{\mathrm{par}}(M,-):\mathcal{HVB}\mathrm{dl}^\nabla(M)\to\mathcal{H}\mathrm{ilb}$ is the functor that assigns to a hermitean vector bundle on *M* with connection its Hilbert space of covariantly constant sections; this follows from the fact that $\Theta(E,\alpha)\otimes(F,\beta)$ is descent data for a hermitean vector bundle with connection on *M*, while the 2-morphisms between these bundles precisely correspond to sections of this descent bundle. With our choice of Θ, however, the isomorphism (6.3) holds on objects only. A natural isomorphism exists only in the form

$$\eta_{(E,\alpha),(F,\beta)} : \mathcal{B}\mathcal{G}\mathrm{rb}^{\nabla}(M)\big((E,\alpha),(F,\beta)\big) \xRightarrow{\cong} \mathfrak{h} \circ \big(\vartheta(E,\alpha) \times (F,\beta)\big),$$

where $\vartheta : \mathcal{B}\mathcal{G}\mathrm{rb}^{\nabla}(M) \to \mathcal{B}\mathcal{G}\mathrm{rb}^{\nabla}(M)$is the 2-functor which is the identity on objects and 1-morphisms, but sends 2-morphisms to their adjoints, i.e. $\vartheta(\psi)=\psi^*$.

7 2-Hilbert spaces

The key idea of the preceding sections has been that the ground ring $(\mathbb{C},+,\cdot)$ with its natural structure of a finite-dimensional Hilbert space should be replaced by a categorical object. This object should be endowed with categorical analogues of the algebraic structures of $\mathbb{C}$, including the inner product. General accounts can be found, for example, in [5, 24].

There are (at least) two candidate categories which provide categorifications of $\mathbb{C}$. The complex numbers form a set, or a 0-category. From any set S one can canonically construct a category by adding to the data an identity morphism for each element of S. This is called the *discrete category* $\mathcal{D}$isc(S) *of the set* S.[21] In our case, $\mathcal{D}\mathrm{isc}(\mathbb{C})=(\mathbb{C}\rightrightarrows\mathbb{C})$; source and target of $z\in\mathbb{C}$ are both z. Composition in this category is trivial since all morphisms are identities. The set of morphisms inherits the ring structure from the objects. From this construction we therefore obtain a category in Hilb. This means that both collections of objects and morphisms form finite-dimensional Hilbert spaces, and the structure maps are compatible with this structure.

The category $\mathcal{D}\mathrm{isc}(\mathbb{C})$Disc(C) carries a natural action of the discrete category $\mathcal{D}\mathrm{isc}(\mathrm{U}(1))=(\mathrm{U}(1)\rightrightarrows\mathrm{U}(1))$. However, this is not the categorification of the group U(1) which appears in the definition of a bundle gerbe, where the pertinent 2-group is $(\mathrm{Hilb}^{\times},\otimes)$. For every one-dimensional Hilbert space there exists a unitary isomorphism to $\mathbb{C}$, and endomorphisms of any such Hilbert space identify canonically with $\mathbb{C}$. Thus, the inclusion

$$\mathcal{B}\mathsf{U}(1) = \big(\mathsf{U}(1) \rightrightarrows *\big) \hookrightarrow (\mathcal{H}\mathrm{ilb}^{\times}, \otimes) \tag{7.1}$$

sending the single object of the suspension $\mathcal{B}$U(1) of U(1) to $\mathbb{C}$, is an equivalence of categories. It is compatible with the tensor structure and therefore is an equivalence of 2-groups. This relates our definition of bundle gerbes to the common statement that they are principal 2-bundles with structure 2-group $\mathcal{B}U(1)$. From a topological perspective, this structure 2-group is big enough to capture all bundle gerbes up to stable isomorphism.[22] However, it does not capture important examples such as the central extensions of loop groups discussed in Sect. 4. For this reason we stick to our choice of $(\mathrm{Hilb}^\times,\otimes)$ as the structure 2-group.[23]

The most natural object to represent $(\mathrm{Hilb}^\times,\otimes)$ on is the rig category $(\mathrm{Hilb},\oplus,\otimes)$. It acts on itself via the tensor product in the same way that a ring acts on itself via multiplication. In this way, Hilb is a module category over itself generated by a single object. As in the case of the inclusion (7.1), there exist unitary isomorphisms for every Hilbert space of dimension n to $\mathbb{C}^n$ with its natural inner product. A choice of two such isomorphisms establishes a bijection between morphisms $V\to W$ and spaces of complex n×mn×m-matrices $\mathbf{Mat}(n\times m,\mathbb{C})$. Hence, the inclusion

$$(\mathrm{Mat}(-\times -,\mathbb{C}) \rightrightarrows \mathbb{N}_0) \hookrightarrow \mathcal{H}\mathrm{ilb} \tag{7.2}$$

is an equivalence of rig categories, where the category on the left has objects given by numbers $n\in\mathbb{N}_0$ and morphisms $n\to m$ given by $n\times m$ -matrices. Using the smaller category in the associated construction of morphisms from Sect. 5 would only enable us to obtain trivial hermitean vector bundles $E\to Y$, and thus only a restricted class of morphisms of bundle gerbes. If we would like to understand a hermitean vector bundle on Y as a map into a generalised ring, then this generalised ring has to be the bigger category Hilb because although the fibres of any rank n hermitean vector bundle are all isomorphic to $\mathbb{C}^n$, they are not $\mathbb{C}^n$ on the nose: we can find a consistent simultaneous identification of all fibres with $\mathbb{C}^n$ if and only if the vector bundle is trivialisable. This justifies our use of $(\mathcal{H}\mathrm{ilb},\oplus,\otimes)$ as the ground rig.

The space of smooth sections of a line bundle gives rise to a pre-Hilbert space, which is algebraically a $\mathbb{C}$-module together with a sesquilinear $\mathbb{C}$-valued inner product. If the underlying vector space of such a module is of finite dimension, then a pre-Hilbert space is automatically a Hilbert space. Translating this to the setting where the ground rig is $\mathcal{H}\mathrm{ilb}$, we are interested in rig-module categories over $\mathcal{H}\mathrm{ilb}$. Such a category is a symmetric monoidal category $(\mathcal{C},\oplus)$ together with a module action of $\mathcal{H}\mathrm{ilb}$,[24] that we may choose to be from the right, which is a functor $\triangleleft:\mathcal{C}\times\mathcal{H}\mathrm{ilb}\to\mathcal{C}$ satisfying the categorified properties of a ring-module action up to coherent isomorphisms. Additionally, we would like the action of the functor on morphisms to be compatible with the vector space structures on the morphism spaces in $\mathcal{H}\mathrm{ilb}$. The action $C\mapsto C\triangleleft\mathbb{C}\cong C$ of the unit element in $\mathcal{H}\mathrm{ilb}$ then forces all the morphism sets in $\mathcal{C}$ to be vector spaces, i.e. we demand that $\mathcal{C}$ be enriched in $\mathcal{V}\mathrm{ect}$, the category of finite-dimensional complex vector spaces.[25] Sometimes $\mathcal{C}$ is required to be enriched in $\mathcal{H}\mathrm{ilb}$, but it seems more natural to just require enrichment in $\mathcal{V}$ect for the general case.[26] A *2-vector space* is a monoidal $\mathcal{V}$ect-enriched category $\mathcal{C}$ with a $\mathcal{H}$ilb rig-module action that factors through the forgetful functor $\mathcal{H}$ilb→$\mathcal{V}$ect which drops the inner product.

We are interested in *2-Hilbert spaces*. In order to obtain a Hilbert-like structure we need an inner product. Since our ground rig is $\mathcal{H}$ilb, this should be the target of inner products. Thus, we define an inner product on a 2-vector space $\mathcal{C}$ to be a sesquilinear, non-degenerate functor $\langle -,- \rangle_{\mathcal{C}} : \mathcal{C} \times \mathcal{C} \to \mathcal{H}\mathrm{ilb}$. Sesquilinearity is to be understood with respect to the rig-module action of $\mathcal{H}$ilb and the involution on $\mathcal{H}$ilb given by

$$\Theta : \mathcal{H}\mathrm{ilb} \longrightarrow \mathcal{H}\mathrm{ilb}, \quad \left(V \xrightarrow{\psi} V'\right) \longmapsto \left(V^* \xrightarrow{\psi^{t*}} V'^*\right),$$

whereas in Sect. 6.2 we use $\Theta(\psi)=\psi^{t*}$. Non-degeneracy here means that $\langle V,V \rangle_{\mathcal{C}}=0$ if and only if $V=0$ in $\mathcal{C}$.[27] The category $\mathcal{H}$ilb itself carries such an inner product given by

$$\langle V, W \rangle_{\mathcal{H}\mathrm{ilb}} = \Theta V \otimes W = V^* \otimes W \cong \mathcal{H}\mathrm{ilb}(V, W),$$

where the last identification follows since we are considering finite-dimensional Hilbert spaces. Given two morphisms $\psi: V \to V'$ and $\phi: W \to W'$, we obtain

$$\langle \psi, \phi \rangle_{\mathcal{H}\mathrm{ilb}} = \Theta(\psi) \otimes \phi : \langle V, W \rangle_{\mathcal{H}\mathrm{ilb}} \longrightarrow \langle V', W' \rangle_{\mathcal{H}\mathrm{ilb}}.$$

This is the same as the 2-bundle metric from Sect. 6.2 on the category $\mathcal{HVB}\mathrm{dl}^{\nabla}(\mathrm{pt}) = \mathcal{H}\mathrm{ilb}$. With the definition $\langle -,- \rangle_{\mathcal{H}\mathrm{ilb}} = \Theta(-) \otimes (-)$, it is not true that $\langle -,- \rangle_{\mathcal{H}\mathrm{ilb}}$ is the same as $\mathcal{H}\mathrm{ilb}(-,-)$ for the same reasons as explained at the end of Sect. 6.2: this identity holds on objects only. The inner product functor is covariant in both arguments, whereas the hom-functor is contravariant in its first argument. In particular, for $\psi,\phi \in \mathcal{H}\mathrm{ilb}(\mathbb{C},\mathbb{C})$ we have $\langle \psi,\phi \rangle_{\mathcal{H}\mathrm{ilb}} = \langle \psi,\phi \rangle_{\mathbb{C}}$ under the identification $\mathcal{H}\mathrm{ilb}(\mathbb{C},\mathbb{C}) \cong \mathbb{C}$. Thus, the original Hilbert space $\mathbb{C}$ (also as an algebra with inner product) sits inside $\mathcal{H}$ilb as the endomorphism space of the monoidal unit. Using $\mathcal{H}\mathrm{ilb}(-,-)$ as the inner product would only yield $\mathcal{H}\mathrm{ilb}(\psi,\phi)=\psi\phi$.

Morphisms of 2-Hilbert spaces are functors of $\mathcal{H}$ilb-module categories, i.e. functors compatible with the module actions and the direct sums. As morphisms are certain functors, there exists a notion of 2-morphism of 2-Hilbert spaces given by natural transformations of HilbHilb-module functors. By including morphisms and 2-morphisms of 2-Hilbert spaces, we obtain a 2-category $2\mathcal{H}\mathrm{ilb}$ of 2-Hilbert spaces.

A morphism of 2-Hilbert spaces is *unitary* if it preserves the inner product up to a well-defined natural isomorphism. The simplest type of 2-Hilbert spaces is free 2-Hilbert spaces, i.e. those of the form $\mathcal{H}\mathrm{ilb}^n$. Given a generic 2-Hilbert space $\mathcal{C}$, there might be unitary isomorphisms $U\colon \mathcal{C} \xrightarrow{\cong} \mathcal{H}\mathrm{ilb}^n$ for some $n \in \mathbb{N}_0$. On objects only, we then have

$$\mathcal{C}(C, D) \cong \mathcal{H}\mathrm{ilb}^n(UC, UD) = \bigoplus_{i,j=1}^{n} \mathcal{H}\mathrm{ilb}\big((UC)_i, (UD)_j\big) \in \mathrm{Mat}(n \times n, \mathcal{H}\mathrm{ilb})$$
.

Because of the unitarity of U, we therefore have

$$\begin{aligned}\langle C, D\rangle_{\mathcal{C}} \cong \langle UC, UD\rangle_{\mathcal{H}\mathrm{ilb}^n} &= \bigoplus_{i=1}^{n} \big\langle (UC)_i, (UD)_i\big\rangle_{\mathcal{H}\mathrm{ilb}} \\ &= \mathrm{tr}\big(\mathcal{H}\mathrm{ilb}^n(UC, UD)\big) \in \mathcal{H}\mathrm{ilb}\end{aligned}$$
.

This is completely analogous to what happens in an ordinary Hilbert space V: given two vectors $\psi, \phi \in V$, we can view them as morphisms $\psi, \phi \in \mathcal{H}\mathrm{ilb}(\mathbb{C}, V)$, and combine them into a morphism $\phi \circ \psi^* \in \mathcal{H}\mathrm{ilb}(V, V)$. With this identification of vectors, given any unitary transformation $u\colon V \to \mathbb{C}^n$ we obtain an endomorphism of $\mathbb{C}^n$ as

$$u \circ (\phi \circ \psi^*) \circ u^{-1} = u(\phi) \circ u(\psi)^* \in \mathcal{H}\mathrm{ilb}(\mathbb{C}^n, \mathbb{C}^n) = \mathrm{Mat}(n \times n, \mathbb{C})$$

,where $u(\phi) = u \circ \phi$, and finally

$$\langle \psi, \phi\rangle_V = \mathrm{tr}\big(u(\phi) \circ u(\psi)^*\big) = \mathrm{tr}(\phi \circ \psi^*)$$
.

The transformation and invariance laws for these objects translate to 2-Hilbert spaces with equalities weakened to canonical isomorphisms.

To summarise, we take a 2-Hilbert space to be a $\mathcal{V}$ect-enriched rig-module category $\mathcal{C}$ over $\mathcal{H}\mathrm{ilb}$ with a covariant inner product functor $\langle -,- \rangle_{\mathcal{C}}\colon \mathcal{C} \times \mathcal{C} \to \mathcal{H}\mathrm{ilb}$. An object C in a 2-Hilbert space $\mathcal{C}$ is called *simple* if $\mathcal{C}(C, C) \cong \mathbb{C}1_C$, i.e. its endomorphism algebra is one-dimensional, and *normalised* or *of unit length* if $\langle C, C\rangle_{\mathcal{C}} \cong \mathbb{C}$. If we are in the situation that $\langle -,- \rangle_{\mathcal{C}}$ and $\mathcal{C}(-,-)$ agree on objects, the notions of simplicity and unit length coincide. Particularly nice 2-Hilbert spaces are those which are *semisimple*, i.e. where every object decomposes into direct sums of simple objects (up to isomorphism) and where we additionally assume that the inner product agrees with the hom-functor on objects. In such a decomposition, several isomorphic copies of the same simple object may occur. Fixing one simple object C_i in each isomorphism class of simple objects, each of the C_i will occur n_i times for some $n_i \in \mathbb{N}_0$. Hence, we can write $C = \oplus_i (C_i \lhd \mathbb{C}^{n_i})$. Therefore, any semisimple 2-Hilbert space is unitarily equivalent to a free 2-Hilbert space via the choice of an orthonormal basis, which in this case amounts to picking one object in each isomorphism class of simple objects.

There does not yet exist a generally accepted notion of a 2-Hilbert space. The definition we have presented here may be altered in several ways. To point out just a few variations, sometimes the inner product is assumed to be given by the hom-functor, sometimes the underlying category is assumed to be abelian or semisimple, or $\mathcal{H}$ilb-enrichment may be required instead of $\mathcal{V}$ect-enrichment. An extensive treatment can be found in [5]. Another interesting question is of the necessity of a 2-Hilbert space completion. This is completely open as of yet; some approaches to parameterised families of Hilbert spaces can be found in [16, 17], and a self-contained account in view of general continuous 2-group representations is given in [6].

The prototypical 2-Hilbert space, which is semisimple, is $\mathcal{H}$ilb itself, just as $\mathbb{C}$ is the prototypical Hilbert space. The question is which of the features of $\mathbb{C}$ to take over to the general setting since $\mathbb{C}$ is an extremely special object with a particularly rich structure. Presumably, the first step of categorification from objects based on sets to objects based on categories is the hardest to find, whereas one might hope that once the first additional layer and weakening of structure has been found it is easier to see how even higher n-Hilbert spaces should be defined. In Sect. 8 we will encounter an example of a 2-Hilbert space, in the above sense, arising from the geometric data of a bundle gerbe.

8 THE PREQUANTUM 2-HILBERT SPACE OF A BUNDLE GERBE

In this section we will put to use the properties and structures of bundle gerbes that we have accumulated in earlier sections to the goal of obtaining a 2-Hilbert space of sections of a bundle gerbe on a manifold M. This extends the statement in [50] that $\Gamma(M,\mathrm{L})$ is a 2-vector space. If the category of sections of L is to be the space of states of a quantum theory, there has to be a pairing on these states in order to define amplitudes (whose interpretation in physical terms is yet to be understood in the higher settings) and hermitean observables. Using sections of a bundle gerbe as higher quantum states has been proposed in [41, 42] and also in [46], though the hermitean and inner product structures have not been present there, and the treatment was restricted to local bundle gerbes.

Recall from Sect. 2 that in geometric quantisation the pre-Hilbert space of interest was the space of sections $\Gamma(M,L)$ with the inner product given by integrating the evaluation of the hermitean metric of L on a pair of sections over M. In 2-plectic quantisation we would now like to replace the hermitean line bundle L by a hermitean bundle gerbe L. In Sect. 6.1 we defined the category of sections of L as $\Gamma(M,\mathrm{L})=\mathcal{BG}\mathrm{rb}^\nabla(M)(\mathcal{I}_0,\mathcal{L})$. We also elaborated that this forms a module category over the rig category $\mathcal{HVB}\mathrm{dl}^\nabla(M)$ of higher functions on M. From here on, we assume M to be connected.

For the Hilbert space structure of $\Gamma(M,L)$ the relevant module structure is that over $\mathbb{C}$rather than that over $C^\infty(M,\mathbb{C})$. However, the former module structure is actually contained in the latter as follows. We simply note that the action of a complex number z∈$\mathbb{C}$ on a section $\psi\in\Gamma(M,L)$ factors through the inclusion of complex numbers as constant functions,

$$c:\mathbb{C}\hookrightarrow C^\infty(M,\mathbb{C}),\quad c(z)(x)=z,$$

for all $x\in M$. For a 2-Hilbert space we need a module structure over $\mathcal{H}$ilb. The above picture of the ground ring sitting inside the ring of functions as constant functions generalises appropriately: given a Hilbert space $V\in\mathcal{H}\mathrm{ilb}$ we can form the trivial hermitean vector bundle with fibre V, i.e. $M\times V\in\mathcal{HVB}\mathrm{dl}(M)$. This becomes even clearer from the perspective that hermitean vector bundles are functions into $\mathcal{H}$ilb: the bundle $M\times V$ is precisely the constant function $x\mapsto V$. However, this is not yet a higher function as was defined in Sect. 6.2 since we are missing a connection. Again there is a constant choice for a connection on a trivial bundle, namely the trivial connection given by just the exterior derivative. Thus, the inclusion of higher numbers into higher functions re as

$$c:\mathcal{H}\mathrm{ilb}\hookrightarrow\mathcal{HVB}\mathrm{dl}^\nabla(M),\quad V\longmapsto\big(M\times V,\mathrm{d}\big).$$

This is a functor, and it is compatible with the structure functors ⊕,⊗ and θ on $\mathcal{H}$ilb. Composing the module action of $\mathcal{HVB}\mathrm{dl}^\nabla(M)$ with this inclusion yields a module action of $\mathcal{H}$ilb on $\Gamma(M,\mathrm{L})$ for any $\mathcal{L}\in\mathcal{BG}\mathrm{rb}^\nabla(M)$.

For the inner product, we have already constructed a 2-bundle metric on L in the algebraic sense in Sect. 6.2. This was defined on sections as a functor $\mathfrak{h}:\Gamma(M,\mathcal{L})\times\Gamma(M,\mathcal{L})\to\mathcal{HVB}\mathrm{dl}^\nabla(M)$. In the case of a hermitean line bundle, the inner product of two sections is given by first inserting them into the metric and then integrating the resulting function over M.

Inserting two sections into the 2-bundle metric yields a hermitean vector bundle with connection. We would now like to integrate this over *M*, but first we have to settle on a suitable notion of integration.

First of all, viewing the bundle as a function to $\mathcal{H}$ilb would suggest adding up all the fibres, possibly weighted with a measure. The measure should then be valued in HilbHilb as well, but such a measure would have that be of rank 1 at least if it is non-vanishing, leading inevitably to infinite-dimensional Hilbert spaces which are not separable. Another approach is known from harmonic analysis (see, for instance, [21]), where the *direct integral* of a family of Hilbert spaces over *M* (e.g. a hermitean vector bundle with possibly infinite-dimensional fibres) is defined to be the space of L^2-sections of the family. We could technically apply the direct integral to our finite-rank hermitean vector bundles, but this would again produce infinite-dimensional Hilbert spaces, which are now separable. From the point view of our ground rig $\mathcal{H}$ilb, an infinite-dimensional separable Hilbert space $\mathcal{H}$ would be the same as ∞ is for $\mathbb{C}$; for example, $V \oplus \mathcal{H} \cong \mathcal{H}$ and $V \otimes \mathcal{H} \cong \mathcal{H}$ for any Hilbert space V (including $V = \mathcal{H}$).

A modification of the direct integral which allows us to stay in the finite-dimensional world is the functor

$$\int_M^{\mathrm{par}} = \Gamma_{\mathrm{par}}(M, -) : \mathcal{HVB}\mathrm{dl}^{\nabla}(M) \longrightarrow \mathcal{H}\mathrm{ilb}, \quad E \longmapsto \int_M^{\mathrm{par}} E = \Gamma_{\mathrm{par}}(M, E).$$

This respects the direct sum and involution in $\mathcal{HVB}\mathrm{dl}^{\nabla}(M)$, and as a consequence is compatible with the module action of $\mathcal{H}$ilb, i.e.

$$\begin{aligned} \int_M^{\mathrm{par}} \big((E \lhd V) \oplus (F^* \lhd W)\big) &= \Gamma_{\mathrm{par}}\big(M, (E \lhd V) \oplus (F^* \lhd W)\big) \\ &= \big(\int_M^{\mathrm{par}} E\big) \otimes V \ \oplus \ \big(\int_M^{\mathrm{par}} F\big)^* \otimes W. \end{aligned}$$

The inner product on $\Gamma_{\mathrm{par}}(M,E) \in \mathcal{H}$ilb is canonically obtained from the hermitean metric *h* on *E*: for $\psi, \phi \in \Gamma_{\mathrm{par}}(M,E)$ define their inner product to be $h_{|x}(\psi, \phi) \in \mathbb{C}$ for any $x \in M$. The independence of the choice of point *x* follows from the fact that the sections and the hermitean metric are (bi)covariantly constant, and the assumption that *M* be connected.

Therefore, to a bundle gerbe $\mathcal{L} \in \mathcal{BG}\mathrm{rb}^{\nabla}(M)$ with connection on *M*, we obtain a 2-Hilbert space of sections given as

$$\mathcal{H}(\mathcal{L}) = \big(\Gamma(M, \mathcal{L}), \int_M^{\mathrm{par}} \circ \mathfrak{h}\big) \in 2\mathcal{H}\mathrm{ilb}.$$

It follows immediately from the functorial nature of this assignment that we have in fact found a 2-functor

$$\mathcal{H} : \mathcal{BG}\mathrm{rb}^{\nabla}(M) \longrightarrow 2\mathcal{H}\mathrm{ilb}.$$

It is merely the global section functor $\Gamma(M,-) = \mathcal{BG}\mathrm{rb}^{\nabla}(M)(\mathcal{I}_0,-)$ with the additional information of the inner product; this matches the situation for line bundles precisely. This inner product is non-degenerate since for every vector bundle with connection, the identity morphism is parallel: Hence, $\langle(E,\alpha),(E,\alpha)\rangle_{\Gamma(M,L)}=0$ if and only if $(E,\alpha)=0$. Thus, HHrefines the global section functor $\Gamma(M,-)$ to a functor from bundle gerbes to 2-Hilbert spaces. This is in analogy to how sections of a hermitean line bundle define not just a vector space, but even a Hilbert space of sections, and it is this *Hilbert* space structure that makes the space of sections of a prequantum line bundle into a viable habitat for a quantum theory as it allows to define transition amplitudes and self-adjoint observables.

Let us point out a few more features of the parallel section functor. If $E \in \mathcal{HVB}\mathrm{dl}^{\nabla}(M)$, the assignment $U \mapsto \Gamma_{\mathrm{par}}(U,E|U)$ produces a sheaf of Hilbert spaces on M. We furthermore get a functor

$$\Gamma_{\mathrm{par}} : \mathcal{HVB}\mathrm{dl}^{\nabla}(M) \longrightarrow \mathrm{Sh}_{\mathcal{H}\mathrm{ilb}}(M)$$

from the category of hermitean vector bundles with connection on M to the category $\mathrm{Sh}_{\mathcal{H}\mathrm{ilb}}(M)$ of $\mathcal{H}$ilb-valued sheaves on M. Every $\Gamma_{\mathrm{par}}(-,E)$ is a sheaf of modules over the sheaf of constant functions $\mathbb{C}_M \cong \Gamma_{\mathrm{par}}(-,c(\mathbb{C}))$. There is a generalised Mayer–Vietoris property for open sets $U,U' \subset M$, as there is an exact sequence

$$0 \longrightarrow \int_{U\cup U'}^{\mathrm{par}} E \longrightarrow \int_{U}^{\mathrm{par}} E \oplus \int_{U'}^{\mathrm{par}} E \longrightarrow \int_{U\cap U'}^{\mathrm{par}} E. \qquad (8.1)$$

This sequence does not extend to a short exact sequence for general U and U'. For example, consider U and U' with non-trivial fundamental group, but such that their intersection is contractible. Let E be a flat hermitean line bundle with non-trivial holonomy around the cycles in U and U'. Then the first and second Hilbert spaces of the exact sequence (8.1) are zero, while the rightmost space is isomorphic to $\mathbb{C}$. Nevertheless, this exact sequence resembles the familiar property of the Lebesgue integral:

$$\int_{U\cup U'} f\,\mathrm{d}\mu = \int_U f\,\mathrm{d}\mu + \int_{U'} f\,\mathrm{d}\mu - \int_{U\cap U'} f\,\mathrm{d}\mu.$$

Recall that there are two ways to obtain a Hilbert space from a pair (E,α) and (F,β) of sections of L. We can either use the inner product on the 2-Hilbert space $\mathcal{H}$(L) of sections of L, giving $\Gamma_{\mathrm{par}}(M,\mathfrak{h}((E,\alpha),(F,\beta)))$Γpar, or we can consider the space of 2-morphisms $\mathcal{BG}\mathrm{rb}^{\nabla}(M)((E,\alpha)$ between the two sections in the 2-category of bundle gerbes. As pointed out in Sect. 6.2 these two Hilbert spaces are canonically isomorphic by construction. The corresponding isomorphisms assemble into a natural isomorphism

$$\eta'_{(E,\alpha),(F,\beta)} : \mathcal{B}\mathcal{G}\mathrm{rb}^{\nabla}(M)\big((E,\alpha),(F,\beta)\big) \overset{\cong}{\Longrightarrow} \big\langle \vartheta(E,\alpha),(F,\beta)\big\rangle_{\Gamma(M,\mathcal{L})}$$

From the fact that this natural isomorphism is the identity on objects and since the morphism categories are abelian, we infer that $\Gamma(M,L)$ is a semisimple 2-Hilbert space.

We summarise the most important structures and analogies in higher prequantisation with the following table (see [13] and Sect. 10 for further discussion):

Structure	Geometric quantisation	2-Plectic quantisation
Geometric data	Symplectic form ω	2-Plectic form ϖ
Chern–Weil object	Line bundle L	Bundle gerbe $\mathcal{L}$
Ground ri(n)g	$(\mathbb{C},+,\cdot)$	$(\mathcal{H}\mathrm{ilb},\oplus,\otimes)$
Functions on M	$C^\infty(M,\mathbb{C})$	$\mathcal{H}\mathcal{V}\mathcal{B}\mathrm{dl}^{\nabla}(M)$
Algebra	Poisson algebra	Unknown for full category
Sections $\Gamma(M,-)=$	$\mathcal{H}\mathcal{L}\mathcal{B}\mathrm{dl}(M)(I,L)$	$\mathcal{B}\mathcal{G}\mathrm{rb}^{\nabla}(M)(\mathcal{I}_0,\mathcal{L})$
Sections form a	$\mathbb{C}$-Hilbert space $\mathcal{H}(L)$	$\mathcal{H}\mathrm{ilb}$-module category $\mathcal{H}(\mathcal{L})$
Inner product	$\int_M \frac{\omega^n}{n!}\circ h : \mathcal{H}(L)^2 \to \mathbb{C}$	$\int_M^{\mathrm{par}} \circ\, \mathfrak{h} : \mathcal{H}(\mathcal{L})^2 \to \mathcal{H}\mathrm{ilb}$
Prequantisation map	Kostant–Souriau map	Unknown
Polarisation	Lagrangian foliation	Unknown

9 EXAMPLES

9.1 *H*-flux on $\mathbb{R}^3$

In Type II string theory, the prototypical example of a non-geometric flux background arises in the T-duality chain that starts from a 3-torus with constant geometric *H*-flux. In many situations local considerations suffice, so that one can take the decompactification limit and work on $\mathbb{R}^3$ instead, see, for example, [2, 8, 35]. This setting is appropriate for our needs as well: the constructions of this paper apply to bundle gerbes on $\mathbb{R}^3$ and we can explicitly construct the pertinent 2-Hilbert space in 2-plectic quantisation.

On $M=\mathbb{R}^3$ we can consider the trivial bundle gerbe $\mathcal{L}=\mathcal{I}_\rho$ with non-trivial curving given by $\rho = \frac{\pi}{3}\,\epsilon_{ijk}\,x^i\,\mathrm{d}x^j\wedge\mathrm{d}x^k$. It has *H*-flux

$H = \mathrm{d}\rho = 2\pi\, \mathrm{d}x^1 \wedge \mathrm{d}x^2 \wedge \mathrm{d}x^3$. Technically, any curving would be admissible at this stage, but the 3-form H-flux we obtain from this choice shows that $\mathcal{I}_\rho$ is a geometric representative of the canonical 2-plectic form $\varpi = \mathrm{d}x^1 \wedge \mathrm{d}x^2 \wedge \mathrm{d}x^3$ on $\mathbb{R}^3$; the corresponding classical 3-brackets define the standard Nambu–Poisson structure on $\mathbb{R}^3$and hence our construction will produce the prequantum 2-Hilbert space for the elusive quantised Nambu–Heisenberg algebra. As the bundle gerbe is trivial, its category of sections is

$$\Gamma(\mathbb{R}^3, \mathcal{I}_\rho) \cong \mathcal{B}\mathcal{G}\mathrm{rb}^\nabla(\mathbb{R}^3)(\mathcal{I}_0, \mathcal{I}_0) \cong \mathcal{H}\mathcal{V}\mathcal{B}\mathrm{dl}^\nabla(\mathbb{R}^3).$$

The first step makes use of the fact that morphisms of bundle gerbes do not see their curvings,[28], while the second step is the, by now, familiar equivalence of categories coming from the fact that vector bundles with connection form a stack.

Here we are in the special situation that $\mathbb{R}^3$ is contractible so that every hermitean vector bundle is trivialisable. This is true, however, only at the level without connections. Thus, every hermitean vector bundle of rank n with connection is isomorphic, as a vector bundle with connection, to $(\mathbb{R}^3 \times \mathbb{C}^n, \mathrm{d} + \mathrm{i}\, A)$ for some 1-form $A \in \Omega^1(\mathbb{R}^3, \mathfrak{u}(n))$; we can use just A to label a section of this form. A 2-morphism $\psi: A \Rightarrow A'$ is a function $f: \mathbb{R}^3 \to \mathsf{Mat}(n \times n', \mathbb{C})$which is bicovariantly constant: $\mathrm{d}f + \mathrm{i}A'f - \mathrm{i}fA = 0$. Denoting by $\Omega^1(\mathbb{R}^3, \mathfrak{u})$ the category with objects given by 1-forms on $\mathbb{R}^3$ valued in $\mathfrak{u}(n)$ for some $n \in \mathbb{N}_0$ and morphisms given by bicovariantly constant functions on $\mathbb{R}^3$ valued in $\mathsf{Mat}(n \times n', \mathbb{C})$ for $n, n' \in \mathbb{N}_0$, we have a further equivalence of categories

$$\Gamma(\mathbb{R}^3, \mathcal{I}_\rho) \cong \Omega^1(\mathbb{R}^3, \mathfrak{u}).$$

Direct sum and tensor product on $\Omega^1(\mathbb{R}^3, \mathfrak{u})$ are induced by those of connection 1-forms, so that $\oplus$ sends (A, A') to $A \oplus A'$, while $\otimes$ sends (A, A') to $A \otimes 1 + 1 \otimes A'$. The inner product on the 2-Hilbert space $\Omega^1(\mathbb{R}^3, \mathfrak{u})$ re as

$$\langle (A, A' \rangle_{\Omega^1(\mathbb{R}^3, \mathfrak{u})} = \{ g : \mathbb{R}^3 \to \mathsf{Mat}(n \times n', \mathbb{C}) \mid \mathrm{d}g + \mathrm{i}\, A'\, g - \mathrm{i}\, g\, A = 0 \}$$

on objects, whereas on morphisms we find

$$\langle f, f' \rangle_{\Omega^1(\mathbb{R}^3, \mathfrak{u})} = \overline{f} \otimes f'.$$

The inner product on the Hilbert space $\langle A, A' \rangle_{\Omega^1(\mathbb{R}^3,\mathfrak{u})} \in \mathcal{H}\mathrm{ilb}$ is given by

$$(g, g') \longmapsto \mathrm{tr}(g^* g')$$

This is easily seen to be constant via the calculation

$$\begin{aligned} \mathrm{d}\,\mathrm{tr}(g^* g') &= \mathrm{tr}(\mathrm{d}(g^* g')) \\ &= \mathrm{tr}(\mathrm{d}(g^* g') + \mathrm{i}\, A\, g^* g' - \mathrm{i}\, g^* g' A) \\ &= \mathrm{tr}((\mathrm{d}g^* + \mathrm{i}\, A\, g^* - \mathrm{i}\, g^* A')\, g' + g^* (\mathrm{d}g' + \mathrm{i}\, A' g' - \mathrm{i}\, g' A)) \\ &= 0 \end{aligned}$$

by the condition on 2-morphisms.

This category bears a remarkable similarity to the Lie 2-algebras of classical observables in 2-plectic geometry considered by [7, 42] (see [13] for further discussion of this point). It would be interesting to find an extension of this Lie 2-algebra structure to the entire category of higher functions in the general case, but even an extension to the simpler category $\Omega^1(M,\mathfrak{U})$ would give valuable insight into the quantised version of the rig category of higher functions. Understanding the induced non-commutative and non-associative structure on the complete higher functions would be a huge step towards full higher geometric quantisation.

To finish this section, let us point out a slight modification of the above example which has an interpretation in terms of M-theory. The M-theory *C*-field serves as an *H*-flux on the six-dimensional worldvolume of an M5-brane on which open membranes end. On $M = \mathbb{R}^6 = \mathbb{R}^3 \times \mathbb{R}^3$ the *C*-field can be modelled on a trivial bundle gerbe $\mathcal{L} = \mathcal{I}_\rho$ on $\mathbb{R}^3 \times \mathbb{R}^3$ with *H*-flux [14] $H = 2\pi\,(\mathrm{d}x^1 \wedge \mathrm{d}x^2 \wedge \mathrm{d}x^3 + \mathrm{d}x^4 \wedge \mathrm{d}x^5 \wedge \mathrm{d}x^6)$, which is a geometric realisation of the 2-plectic form on $\mathbb{R}^3 \times \mathbb{R}^3$ given by the sum of the canonical 2-plectic forms on each of the transverse $\mathbb{R}^3$ subspaces. The construction of the corresponding prequantum 2-Hilbert space proceeds verbatum to that above, yielding again an equivalence of rig-module categories

$$\Gamma(\mathbb{R}^3 \times \mathbb{R}^3, \mathcal{I}_\rho) \cong \Omega^1(\mathbb{R}^3 \times \mathbb{R}^3, \mathfrak{u}).$$

9.2 M-theory lift of $\mathbb{R}^2$

A bound D2–D4-brane configuration in a *B*-field background of Type IIA string theory lifts to a bound M2–M5-brane configuration in a *C*-field background of M-theory. If the transverse space to the D2-brane worldvolume is $\mathbb{R}^2$ with its canonical symplectic form $\omega = dx^1 \wedge dx^2$ fluxed by the *B*-field, then the transverse space to the membrane worldvolume is the total space of a circle bundle *M* over $\mathbb{R}^2$ with a 2-plectic form that is fluxed by the *C*-field. Since the base $\mathbb{R}^2$ is contractible any such fibration is trivial, so that $M = \mathbb{R}^2 \times S^1$ and it comes endowed with the natural 2-plectic form given by $\varpi = \omega \wedge d\tau = dx^1 \wedge dx^2 \wedge d\tau$, where $\tau \in [0,1)$ is the coordinate along the M-theory direction S^1. Since *M* is homotopy equivalent to S^1, we see that $\mathrm{H}^3(M, \mathbb{Z}) = 0$ and therefore every bundle gerbe on $\mathbb{R}^2 \times S^1$ is trivial. A geometric representative of the 2-plectic form is given by $\mathcal{I}_\rho$ with $\rho = \pi\, \epsilon_{ij}\, x^i\, dx^j \wedge d\tau$.

The similarity to the example of Sect. 9.1 goes even further. Isomorphism classes of rank *n* hermitean vector bundles (without connection) on *M* are classified by $[M, \mathrm{BU}(n)]$, the homotopy classes of based maps from *M* to the classifying space $\mathrm{BU}(n)$.[29] We can then use the homotopy equivalence $\mathbb{R}^2 \times S^1 \simeq S^1$ to obtain

$$[\mathbb{R}^2 \times S^1, \mathsf{BU}(n)] \cong [S^1, \mathsf{BU}(n)] \cong [\Sigma S^0, \mathsf{BU}(n)] \cong [S^0, \Omega\mathsf{BU}(n)] = \pi_0(\Omega\mathsf{BU}(n)),$$

where Σ denotes the suspension of a topological space. Using the equivalence $\Omega\mathrm{BU} \simeq \mathrm{G}$ for any topological group G we find that $[\mathbb{R}^2 \times S^1, \mathrm{BU}(n)]$ is trivial. Therefore, just as on $\mathbb{R}^3$, every hermitean vector bundle (without connection) is isomorphic to a trivial hermitean vector bundle. Hence, we have once again an equivalence of rig categories

$$\mathcal{HVB}\mathrm{dl}^\nabla(M) \cong \Omega^1(\mathbb{R}^2 \times S^1, \mathfrak{u}),$$

and of $\Omega^1(\mathbb{R}^2 \times S^1, \mathfrak{u})$-module categories

$$\Gamma(\mathbb{R}^2 \times S^1, \mathcal{I}_\rho) \cong \Gamma(\mathbb{R}^2 \times S^1, \mathcal{I}_0) \cong \mathcal{HVB}\mathrm{dl}^\nabla(M) \cong \Omega^1(\mathbb{R}^2 \times S^1, \mathfrak{u}).$$

We would now like to check how this higher prequantisation behaves under dimensional reduction of M-theory back to string theory along the S^1-direction, wherein the M2–M5-brane system reduces to the D2–D4-brane system for which ordinary geometric prequantisation should apply. Dimensional reduction is implemented by integrating over the S1S1-fibres. First of all, the 2-plectic form $\varpi = \omega \wedge d\tau$ nicely reduces to the symplectic form $\omega = \iota_{\partial_\tau}\varpi$ on $\mathbb{R}^2$. It is represented by the trivial hermitean line bundle $I \to \mathbb{R}^2$ with connection $a = \iota_{\partial_\tau}\rho = \pi\, \epsilon_{ij}\, x^i\, dx^j$; indeed $da = 2\pi\omega$.

Next, given a section $A \in \Omega^1(\mathbb{R}^2 \times S^1, \mathfrak{U})$, we obtain a function on $\mathbb{R}^2$ given by the path-ordered exponential

$$(x^1, x^2) \longmapsto \mathrm{W}(A)(x^1, x^2) = \mathrm{tr}\, \mathrm{P} \exp\left(\mathrm{i} \int_0^1 A_{|(x^1, x^2, \tau)}(\partial_\tau)\, \mathrm{d}\tau\right) \quad (9.1)$$

The value of $\mathrm{W}(A)$ at $(x^1, x^2) \in \mathbb{R}^2$ is just the Wilson loop of the hermitean vector bundle with connection obtained as the restriction of $(\mathbb{R}^2 \times S^1 \times \mathbb{C}^n, \mathrm{d} + \mathrm{i}\, A)$ to the S^1 fibre of $\mathbb{R}^2 \times S^1$over the point (x^1, x^2). One can see that 2-isomorphic sections of $\mathcal{I}_\rho$ reduce to identical functions on $\mathbb{R}^2$. Hence, the dimensional reduction descends to 2-isomorphism classes of sections of $\mathcal{I}_\rho$. The fact that the trace converts direct sums into sums and tensor products into products implies that this prescription is compatible with the module action of higher functions in the following sense. Higher functions on $\mathbb{R}^2 \times S^1$ are given by the category $\Omega^1(\mathbb{R}^2 \times S^1, \mathfrak{U})$ up to canonical equivalence, since the prequantum bundle gerbe $\mathcal{I}_\rho$ is topologically trivial. Therefore, the action of higher functions on sections is, in this situation, simply the tensor product of higher functions. As the holonomy of a tensor product of vector bundles is the tensor product of the holonomies, we obtain $\mathrm{W}(A \triangleleft A') = \mathrm{W}(A)\mathrm{W}(A')$.

The hermitean 2-bundle metric evaluated on a pair of sections $A, A' \in \Omega^1(\mathbb{R}^2 \times S^1, \mathfrak{U})$ yields $\mathfrak{h}(A, A') = -A^{\mathrm{t}} \otimes 1 + 1 \otimes A'$. Since the occurring 1-forms are self-adjoint we can rewrite this as $\mathfrak{h}(A, A') = -\overline{A} \otimes 1 + 1 \otimes A'$, so that

$$\mathrm{W}\big(\mathfrak{h}(A, A')\big) = \overline{\mathrm{W}(A)}\, \mathrm{W}(A') = h\big(\mathrm{W}(A), \mathrm{W}(A')\big)$$

where h is the hermitean metric on the trivial hermitean line bundle with connection. Thus, dimensional reduction in this example maps the hermitean 2-bundle metric on $\mathcal{I}_\rho$ to the hermitean bundle metric on $(I, \mathrm{d} + \mathrm{i}a)$. However, while dimensional reduction works out nicely at the level of sections and the hermitean structures, it is not evident how to relate the inner products $\langle A, A' \rangle_{\Omega^1(\mathbb{R}^2 \times S^1, \mathfrak{u})}$ and $\langle \mathrm{W}(A), \mathrm{W}(A') \rangle_{C^\infty(\mathbb{R}^2, \mathbb{C})}$ on the respective categories and spaces of sections.

9.3 M-theory lifts of lens spaces

The next natural example we would like to consider is the case wherein the transverse space of the D2–D4-brane system is a 2-sphere S^2 with its canonical symplectic structure given by the area form; quantisation in the

T-dual picture then describes the polarisation of open D1-branes into the usual fuzzy 2-spheres. In the uplift to M-theory, this configuration becomes an M2–M5-brane system with transverse space a 3-sphere S^3, which is an S^1-bundle over S^2 via the Hopf fibration $S^3 \to S^2$, and with its canonical 2-plectic structure given by the volume form. Quantisation would then describe the polarisation of open M2-branes into what should be analogously called a 'fuzzy 3-sphere', which is also relevant in non-geometric flux compactifications of closed string theory originating via T-duality from S^3 with geometric H-flux (see, for example, [9]); this quantum geometry has been thus far elusive despite various attempts. Unfortunately, this example also lies outside our scope of quantisation, as the cohomology ring of the 3-sphere S^3 is torsion-free so that the interesting prequantum bundle gerbes on S^3 which geometrically realise the canonical 2-plectic form do not have non-trivial sections in our framework.[30]

Hence, we shall instead consider a modification of this example, which has nonzero torsion in its third integer cohomology, by taking quotients of S^3; this will provide illuminating modifications of the local 2-Hilbert spaces $\Omega^1(M,\mathfrak{U})$ associated with trivial prequantum bundle gerbes. We employ an explicit construction from [23] of bundle gerbes realising cup product Dixmier–Douady classes, which then allows us to reformulate the category of sections of such bundle gerbes in terms of more familiar equivariant vector bundles. The twist introduced by the gerbe is thereby translated into a twist between two group actions of $\mathbb{Z}p$ and $\mathbb{Z}$. This way, we can explore higher geometric structures in terms of ordinary geometry. To finish the section with, we examine the dimensional reduction of such sections along the circle. Using the aforementioned view on sections of the bundle gerbe, we see that these nicely reduce to sections of a prequantum line bundle on the lens space which realises its second cohomology.

Geometric set-ups as outlined above arise in string compactifications which are described by an asymmetric orbifold of the SU(2) WZW model by a left-acting cyclic subgroup $\mathbb{Z}p \subset \mathrm{U}(1) \subset \mathrm{SU}(2)$ for $p \in \mathbb{N}$, see, for example, [31]. Geometrically, this model describes a closed string propagating on a three-dimensional lens space $\mathbb{L}_p = \mathbb{L}(p;1) = S^3/\mathbb{Z}_p$, the quotient of the 3-sphere $S^3 \subset \mathbb{C}^2$ by the free action of the cyclic group acting in the fundamental representation $\mathbb{C}^2$ of $\mathrm{SU}(2)$. Then $\mathbb{L}_p$ is connected with $\pi_1(\mathbb{L}_p) = \pi_0(\mathbb{Z}_p) = \mathbb{Z}_p$ as S^3 is simply connected, and it is orientable (and spin) if $p \geqslant 3$. The generator of the fundamental group $[f] \in \pi_1(\mathbb{L}_p)$ may be taken to be any loop $f: S^1 \to \mathbb{L}_p$ obtained by projecting a path on the universal cover $q_p: S^3 \to \mathbb{L}_p$ which connects two points on S3S3 in the same orbit of the $\mathbb{Z}_p$-action.

The lens space $\mathbb{L}_p$ has a CW-complex decomposition with one k-cell e_k in each dimension k=0,1,2,3, and a p-fold covering map $e_2 \to e_1$ [18]. Its singular homology is thus $\mathrm{H}_0(\mathbb{L}_p, \mathbb{Z}) = \mathrm{H}_3(\mathbb{L}_p, \mathbb{Z}) = \mathbb{Z}$, $\mathrm{H}_1(\mathbb{L}_p, \mathbb{Z}) = \mathbb{Z}_p$ and $\mathrm{H}_2(\mathbb{L}_p, \mathbb{Z}) = 0$, and hence by the universal coefficient theorem the cohomology ring of $\mathbb{L}_p$ is given by

$$\mathrm{H}^0(\mathbb{L}_p, \mathbb{Z}) = \mathrm{H}^3(\mathbb{L}_p, \mathbb{Z}) = \mathbb{Z}, \quad \mathrm{H}^2(\mathbb{L}_p, \mathbb{Z}) = \mathbb{Z}_p, \quad \mathrm{H}^1(\mathbb{L}_p, \mathbb{Z}) = 0$$

The generator of $\mathrm{H}^2(\mathbb{L}_p, \mathbb{Z}) = \mathbb{Z}_p$ can be described geometrically in the following way [27]. Let $K \to S^2$ be the hermitean line bundle canonically associated with the Hopf fibration $S^3 \to S^2$of degree 1. Then $c_1(K)$ is a generator of $\mathrm{H}^2(S^2, \mathbb{Z}) = \mathbb{Z}$, and the lens space can be identified as the total space $\mathbb{L}_p \cong S(K^{\otimes p})$ of the circle bundle $\Pi_p : S(K^{\otimes p}) \to S^2$ of the hermitean line bundle $K^{\otimes p}$ with degree p. With $J := \Pi_p^* K$, the Chern class $c_1(J) = \Pi_p^* c_1(K)$ generates $\mathrm{H}^2(\mathbb{L}_p, \mathbb{Z}) = \mathbb{Z}_p$.

The third cohomology here is still torsion-free, so we consider an M-theory lift of the lens space to an oriented S^1-bundle $\pi_p : M_p \to \mathbb{L}_p$. The cohomology groups of M_p are related to those of $\mathbb{L}_p$ through the Gysin exact sequence

$$\cdots \longrightarrow \mathrm{H}^k(M_p, \mathbb{Z}) \xrightarrow{\pi_{p!}} \mathrm{H}^{k-1}(\mathbb{L}_p, \mathbb{Z}) \xrightarrow{\smile e} \mathrm{H}^{k+1}(\mathbb{L}_p, \mathbb{Z}) \xrightarrow{\pi_p^*} \mathrm{H}^{k+1}(M_p, \mathbb{Z}) \longrightarrow \cdots$$

where $\pi_{p!}$ is the Gysin pushforward map on cohomology, π^*_p is the usual pullback on cohomology, and $\smile e$ denotes cup product with the Euler class $e \in \mathrm{H}^2(\mathbb{L}_p, \mathbb{Z})$ of the fibration. In particular, putting k=3 shows that the third cohomology of M_p sits in a short exact sequence

$$\begin{array}{ccccccccc} 0 & \longrightarrow & \mathrm{H}^3(\mathbb{L}_p, \mathbb{Z}) & \xrightarrow{\pi_p^*} & \mathrm{H}^3(M_p, \mathbb{Z}) & \xrightarrow{\pi_{p!}} & \mathrm{H}^2(\mathbb{L}_p, \mathbb{Z}) & \longrightarrow & 0 \\ & & \| & & & & \| & & \\ & & \mathbb{Z} & & & & \mathbb{Z}_p & & \end{array}$$

indicating that $\mathrm{H}^3(M_p, \mathbb{Z})$ can contain torsion. In particular, for the trivial bundle $M_p = \mathbb{L}_p \times S^1$ with e=0 and π_p the projection to the first factor, this sequence has a canonical splitting via the Kunh theorem to give

$$\mathrm{H}^3(\mathbb{L}_p \times S^1, \mathbb{Z}) \cong \mathbb{Z} \oplus \mathbb{Z}_p.$$

Writing $[1_{S^1}]$ for the generator of $\mathrm{H}^1(S^1,\mathbb{Z}) \cong \mathrm{H}^0(S^1,\mathsf{U}(1)) = \mathbb{Z}$, the class[31]

$$\delta_p = c_1(J) \smile \big[1_{S^1}\big]. \tag{9.2}$$

is a generator for the torsion subgroup $\mathrm{Tor}\big(\mathrm{H}^3(\mathbb{L}_p \times S^1,\mathbb{Z})\big) = \mathbb{Z}_p$. In the following we fix a connection ∇^J on the hermitean line bundle J. Since the fibration $\Pi_p : S(K^{\otimes p}) \to S^2$ has degree p, we may choose ∇^J so that its magic flux is given by

$$F^J = 2\pi\, p\, \Pi_p^*(\omega). \tag{9.3}$$

for a normalised symplectic form ω on S^2, i.e. $\int_{S^2}\omega = 1$. Then F^J has trivial de Rham class in $\mathrm{H}^2_{\mathrm{dR}}(\mathbb{L}_p \times S^1)$ because $\mathrm{H}^2_{\mathrm{dR}}(\mathbb{L}_p) = 0$.

To construct a bundle gerbe $\mathcal{L}_p$ with connection representing the torsion class δ_p, we use the universal covering map $\pi : \mathbb{R} \to S^1 = \mathbb{R}/\mathbb{Z}$ to define the $\mathbb{Z}$-fibration $1 \times \pi : Y = \mathbb{L}_p \times \mathbb{R} \to \mathbb{L}_p \times S^1$. Then the fibre product $Y^{[2]}$ can be identified with $(\mathbb{L}_p \times \mathbb{R}) \times \mathbb{Z}$, and the groupoid structure maps $p_i : Y^{[2]} \rightrightarrows Y$ as $p_0 : (x,r,n) \mapsto (x,r)$ and $p_1 : (x,r,n) \mapsto (x,r+n)$ for $(x,r,n) \in \mathbb{L}_p \times \mathbb{R} \times \mathbb{Z}$. Omitting pullbacks to $\mathbb{L}_p \times \mathbb{R}$, for the hermitean line bundle $L \to Y^{[2]}$ we take $J^{\otimes \mathbb{Z}} \to (\mathbb{L}_p \times \mathbb{R}) \times \mathbb{Z}$ with fibres $J^{\otimes \mathbb{Z}}_{|(x,r,n)} = J^{\otimes n}_{|(x,\pi(r))}$; since $J^{\otimes p}$ is trivialisable as a line bundle without connection, there is a p-periodicity $L_{|(x,r,n+m\,p)} \cong L_{|(x,r,n)}$ for all $m \in \mathbb{Z}$ reflecting the torsion. The bundle gerbe multiplication

$$\mu_{|(x,r,n,m)} : L_{|(x,r,n)} \otimes L_{|(x,r+n,m)} \xrightarrow{\ \cong\ } L_{|(x,r,n+m)}$$

is given by the canonical isomorphism $J^{\otimes n} \otimes J^{\otimes m} \to J^{\otimes (n+m)}$. In this way we obtain the topological data (L,μ,Y) of a bundle gerbe $\mathcal{L}_p$ on $\mathbb{L}_p \times S^1$ with Dixmier–Douady invariant $\mathrm{DD}(\mathcal{L}_p) = \delta_p$, see, for example, [12, 23].

It remains to define a connection on the bundle gerbe (L,μ,Y). This is achieved by extending this geometric realisation of the cup product $\smile : \mathrm{H}^2(\mathbb{L}_p,\mathbb{Z}) \otimes_{\mathbb{Z}} \mathrm{H}^1(S^1,\mathbb{Z}) \to \mathrm{H}^3(\mathbb{L}_p \times S^1,\mathbb{Z})$, which defines the cohomology class (9.2), to Deligne cohomology. Since the connection ∇^J on the line bundle $J \to \mathbb{L}_p \times S^1$ induces connections on tensor products $J^{\otimes n}$, we obtain a connection ∇^L on the hermitean line bundle L which is compatible with μ and has magic flux $F^L \in \Omega^2(\mathbb{L}_p \times \mathbb{R} \times \mathbb{Z})$ given by $F^L_{|(x,r,n)} = n\, F^J_{|(x,\pi(r))}$.

The curving $B \in \Omega^2(\mathbb{L}_p \times \mathbb{R})$ given by $B_{|(x,r)} = r\, F^J_{|(x,\pi(r))}$ easily obeys $B_{|(x,r+n)} - B_{|(x,r)} = F^L_{|(x,r,n)}$, and by the Bianchi identity for ∇^J we obtain $\mathrm{d}B = \mathrm{d}r \wedge (1\times\pi)^* F^J = (1\times\pi)^*(\mathrm{d}\tau \wedge F^J)$ with $\tau \in [0,1)$, which identifies the H-flux

$$H=F^J\wedge \mathrm{d}\tau \tag{9.4}$$

having trivial de Rham class in $\mathrm{H}^3_{\mathrm{dR}}(\mathbb{L}_p \times S^1)$ because the magic flux (9.3) has trivial class $[F^J] = 0$ in $\mathrm{H}^2_{\mathrm{dR}}(\mathbb{L}_p \times S^1)$. In this way we obtain the desired bundle gerbe with connection $\mathcal{L}_p = (L, \mu, Y, \nabla^L, B)$.

The physical meaning of this torsion H-flux can be understood as follows. A closed string winding around the non-contractible cycle of the lens space $\mathbb{L}_p$ can be regarded as an open string on the universal cover S^3 (with endpoints in the same orbit of the $\mathbb{Z}_p$ -action), and the pullback $q_p^* F^J$ of the magic flux serves as a B-field source for open strings on the worldvolume of a fractional D3-brane wrapping S^3 [31]. Projecting back thus produces a fundamental string charge (holonomy) in $\mathbb{L}_p$ which is valued in $\mathrm{H}_1(\mathbb{L}_p, \mathbb{Z}) = \mathbb{Z}_p$ [22]. In the lift to M-theory, fundamental string winding numbers become membrane wrapping numbers, so that the membrane charge is valued in $\mathrm{H}_2(\mathbb{L}_p \times S^1, \mathbb{Z}) = \mathbb{Z}_p$ [1]. In this way the torsion magic flux F^J on $\mathbb{L}_p$ induces a torsion H-flux on $\mathbb{L}_p \times S^1$, regarded as a C-field source for open M2-branes on the worldvolume of a fractional M5-brane wrapping the cover $S^3 \times \mathbb{R}$ (with the boundary lines of the membranes in the same orbit of the $\mathbb{Z}_p \times \mathbb{Z}$-action).

This physical picture appears in the description of the prequantum 2-Hilbert space of sections $\Gamma(\mathbb{L}_p \times S^1, \mathcal{L}_p) = \mathcal{BG}\mathrm{rb}^\nabla(\mathbb{L}_p \times S^1)(\mathcal{I}_0, \mathcal{L}_p)$, which may be described as bundle gerbe modules [51]. A section is a triple (E, ∇^E, α), where $E \to \mathbb{L}_p \times \mathbb{R}$ is a hermitean vector bundle with connection ∇^E together with parallel isomorphisms

$$\alpha_{|(x,r,n)} : E_{|(x,r+n)} \xrightarrow{\cong} E_{|(x,r)} \otimes J^{\otimes n}_{|(x,\pi(r))} \tag{9.5}$$

which by construction are automatically compatible with μ. The projection $q_p : S^3 \to \mathbb{L}_p$ does not commute with the U(1) -actions, but $q_p^* J = (\Pi_p \circ q_p)^* K \cong S^3 \times \mathbb{C} = I$ which implies that the trivial hermitean line bundle $S^3 \times \mathbb{C}$ can be used as descent data for J: Identifying $S^{3[2]}$ with $S^3 \times \mathbb{Z}_p$, the morphism $p_1^* I \to p_0^* I$ is given by $(\hat{x}, z, \zeta) \mapsto (\hat{x} \cdot \zeta, \zeta^{-1}$ for $(\hat{x}, z, \zeta) \in S^3 \times \mathbb{C} \times \mathbb{Z}_p$. Defining $I_p \to S^3$ to be the trivial hermitean line bundle with this $\mathbb{Z}_p$-action, it follows that J is the descent bundle of (I_p, q_p). The connection ∇^J on J is represented in this descent by the connection d+iκ on I_p for some $\mathbb{Z}_p$-invariant 1-form κ on S^3. Since $\mathcal{HVBdl}^{\nabla}$ is a stack, it also follows that the category $\mathcal{HVBdl}^{\nabla}(\mathbb{L}_p \times \mathbb{R})$ of hermitean vector bundles with connection on $\mathbb{L}_p \times \mathbb{R}$ is equivalent to the category $\mathcal{HVBdl}^{\nabla}_{\mathbb{Z}_p}(S^3 \times \mathbb{R})$-of $\mathbb{Z}_p$ equivariant hermitean vector bundles with connection on $S^3 \times \mathbb{R}$with respect to the $\mathbb{Z}_p$-action $(\hat{x}, r) \mapsto (\hat{x} \cdot \zeta, r)$. Hence, $E \to \mathbb{L}_p \times \mathbb{R}$lifts to a pair $(\hat{E}, \phi^{\hat{E}})$, where $\hat{E} \to S^3 \times \mathbb{R}$ is a hermitean vector bundle with connection and $\phi^{\hat{E}}$ is a parallel isomorphism with

$$\phi^{\hat{E}}_{\zeta|(\hat{x},r)} : \hat{E}_{|(\hat{x}\cdot\zeta,r)} \xrightarrow{\cong} \hat{E}_{|(\hat{x},r)}$$

for all $\zeta \in \mathbb{Z}_p$. As the $\mathbb{Z}$-action and the $\mathbb{Z}_p$-action on $S^3 \times \mathbb{R}$ commute, the isomorphism (9.5) correspondingly lifts to an isomorphism

$$\hat{\alpha}_{|(\hat{x},r,n)} : \hat{E}_{|(\hat{x},r+n)} \xrightarrow{\cong} \hat{E}_{|(\hat{x},r)} \otimes I^{\otimes n}_{p|\hat{x}}$$

of $\mathbb{Z}_p$-equivariant hermitean vector bundles with connection.

Thus far, we have found an equivalence between the category of sections $\Gamma(\mathbb{L}_p \times S^1, \mathcal{L}_p)$and the category whose objects are triples $(\hat{E}, \phi^{\hat{E}}, \hat{\alpha})$, with $(\hat{E}, \phi^{\hat{E}}) \in \mathcal{HVBdl}^{\nabla}_{\mathbb{Z}_p}(S^3 \times \mathbb{R})$ and $\hat{\alpha} \in \mathcal{HVBdl}^{\nabla}_{\mathbb{Z}_p}(S^3 \times \mathbb{R})(\hat{p}_1^*\hat{E}, \hat{p}_0^*\hat{E} \otimes I_p^{\otimes\mathbb{Z}})$, and morphisms given by maps $\hat{\sigma} : (\hat{E}, \hat{\alpha}) \to (\hat{F}, \hat{\beta})$in $\mathcal{HVBdl}^{\nabla}_{\mathbb{Z}_p}(S^3 \times \mathbb{R})(\hat{E}, \hat{F})$such that $\hat{\beta} \circ \hat{\sigma} = (\hat{\sigma} \otimes 1) \circ \hat{\alpha}$. The line bundle $I_p^{\otimes n}$ is non-trivial as a $\mathbb{Z}_p$-equivariant bundle without connection: the morphism $m : \hat{p}_0^*\hat{E} \otimes I_p^{\otimes\mathbb{Z}} \to \hat{p}_0^*\hat{E}$ defined by

$$m_{|(\hat{x},r)} : \hat{E}_{|(\hat{x},r)} \otimes I^{\otimes n}_{p|\hat{x}} \longrightarrow \hat{E}_{|(\hat{x},r)}, \quad \hat{e} \otimes (\hat{x}, z) \longmapsto z\,\hat{e} \, ,$$

is an isomorphism in $\mathcal{HVBdl}(S^3 \times \mathbb{R})$, but it is not $\mathbb{Z}_p$-equivariant as

$$m(\phi^{\hat{E}}_{\zeta}(\hat{e}) \otimes (\hat{x} \cdot \zeta, \zeta^{-n} \cdot z)) = \zeta^{-n} \cdot z\, \phi^{\hat{E}}_{\zeta}(\hat{e}) = \zeta^{-n}\, \phi^{\hat{E}}_{\zeta}(m(\hat{e} \otimes (\hat{x}, z))) .$$

We may use it to eliminate the additional line bundle with connection $(I_p^{\otimes n}, \mathrm{d}+\mathrm{i}\,n\,\kappa)$ in the bundle gerbe action by replacing the section

$(\hat{E}, \nabla^{\hat{E}}, \hat{\alpha})$ with an ordinary hermitean vector bundle with connection $(\hat{E}, \nabla^{\hat{E}} - \mathrm{i}\, a)$, where $a \in \Omega^1(S^3 \times \mathbb{R})$ is defined by $a_{|(\hat{x},r)} = r\, \kappa_{|\hat{x}}$; it is both $\mathbb{Z}_p$-equivariant, via the isomorphism $\phi^{\hat{E}}$ as before, and $\mathbb{Z}$-equivariant, via the isomorphism $\psi^{\hat{E}} = m \circ \hat{\alpha}$ which is parallel with respect to the modified connection $\nabla^{\hat{E}} - \mathrm{i}\, a$. However, the bundle $(\hat{E}, \phi^{\hat{E}}, \psi^{\hat{E}})$ is not $\mathbb{Z}_p \times \mathbb{Z}$-equivariant, as the fibrewise-linear actions of the groups obey non-trivial commutation relations

$$\phi_\zeta^{\hat{E}} \circ \psi_n^{\hat{E}} = \zeta^n\, \psi_n^{\hat{E}} \circ \phi_\zeta^{\hat{E}}$$

for all $\zeta \in \mathbb{Z}_p$ and $n \in \mathbb{Z}$. With $\zeta \in S^1$ a primitive pth root of unity, we call this a *ζ-twisted action* of the group $\mathbb{Z}_p \times \mathbb{Z}$, and the triple $(\hat{E}, \phi^{\hat{E}}, \psi^{\hat{E}})$ a $\mathbb{Z}_p \times_\zeta \mathbb{Z}$*-equivariant vector bundle* (with connection). Morphisms of such bundles still commute with the group actions, $\hat{\sigma} \circ \phi^{\hat{E}} = \phi^{\hat{E}} \circ \hat{\sigma}$ and $\hat{\sigma} \circ \psi^{\hat{E}} = \psi^{\hat{E}} \circ \hat{\sigma}$, and we denote by $\mathcal{HVB}\mathrm{dl}^{\nabla}_{\mathbb{Z} \times_\zeta \mathbb{Z}_p}(S^3 \times \mathbb{R})$ the category whose objects are triples $(\hat{E}, \phi^{\hat{E}}, \psi^{\hat{E}})$ and morphisms $\hat{\sigma}$ as above. Altogether, we have found an equivalence of rig categories

$$\Gamma(\mathbb{L}_p \times S^1, \mathcal{L}_p) \cong \mathcal{HVB}\mathrm{dl}^{\nabla}_{\mathbb{Z}_p \times_\zeta \mathbb{Z}}(S^3 \times \mathbb{R}).$$

For the category of higher functions, by descent along the surjective submersion $q_p \times \pi : S^3 \times \mathbb{R} \to \mathbb{L}_p \times S^1$ we also obtain an equivalence of rig categories

$$\mathcal{HVB}\mathrm{dl}^{\nabla}(\mathbb{L}_p \times S^1) \cong \mathcal{HVB}\mathrm{dl}^{\nabla}_{\mathbb{Z}_p \times \mathbb{Z}}(S^3 \times \mathbb{R})$$

Then $\mathcal{HVB}\mathrm{dl}^{\nabla}_{\mathbb{Z}_p \times_\zeta \mathbb{Z}}(S^3 \times \mathbb{R})$ is naturally a rig-module category over $\mathcal{HVB}\mathrm{dl}^{\nabla}_{\mathbb{Z}_p \times \mathbb{Z}}(S^3 \times \mathbb{R})$ under the usual tensor product of hermitean vector bundles with connection on S3×RS3×R. In particular, one sees that the obstruction for a section of L_p to descend to a higher function is precisely the ζ-twisted action of $\mathbb{Z}_p \times \mathbb{Z}$.

Finally, let us examine the dimensional reduction of the M2-brane system back to fundamental strings in the asymmetric orbifold of the SU(2) WZW model. The reduction described in Sect. 9.2 by integrating over the S^1-fibres can be generalised to any trivial S^1-bundle, and even to general circle bundles, by using transgression techniques (see [13] and references therein). The form of the class (9.2) suggests that the bundle gerbe $\mathcal{L}_p$ on $\mathbb{L}_p \times S^1$ should reduce precisely to the line bundle $J = \Pi_p^* K$ on $\mathbb{L}_p$. At the level of integer cohomology, this is implemented by the Gysin pushforward $\pi_{p!} : \mathrm{H}^3(\mathbb{L}_p \times S^1, \mathbb{Z}) \to \mathrm{H}^2(\mathbb{L}_p, \mathbb{Z})$ which maps the Dixmier–Douady class of L_p back to the corresponding Chern class: $\pi_{p!}(\delta_p) = c_1(J)$. Likewise, the H-flux (9.4) nicely reduces to the magic flux $F^J = \iota_{\partial_r} H$ on $\mathbb{L}_p$.

For the reduction of sections, previously we took Wilson loops (9.1) around the S^1-fibre of sections of a trivial bundle gerbe. In the present case our sections are no longer proper vector bundles, but we can still define Wilson loops along the S^1-fibre by regarding them as Wilson lines along unit length intervals in $\mathbb{R}$. If $\mathbb{R}$. If (E, ∇^E, α) is a bundle gerbe module for L_p of rank n (divisible by p), its holonomy around the S^1-fibre then translates to parallel transport in E along $\mathbb{R}$. Denoting by $P^E_{(x,r,s)}$ the parallel transport in E along $\mathbb{R}$ from (x, r) to $(x,r+s)$, we define an isomorphism by the composition

$$\mathrm{h}_{|(x,r)} = \alpha_{|(x,r,1)} \circ P^E_{(x,r,1)} : E_{|(x,r)} \xrightarrow{\cong} E_{|(x,r)} \otimes J_{|(x,\pi(r))}.$$

We then obtain a section $\mathrm{W}(E, \nabla^E, \alpha) : \mathbb{L}_p \to J$ by the analogue of the Wilson loop given by

$$x \longmapsto \mathrm{W}(E, \nabla^E, \alpha)(x) := \mathrm{tr}_{J_{|(x,\pi(r))}}\big(\mathrm{h}_{|(x,r)}\big) \in J_{|x} \qquad (9.6)$$

for any $r \in \mathbb{R}$, where the vector-valued trace is defined by

$$\mathrm{tr}_{J_{|(x,\pi(r))}}\big(\mathrm{h}_{|(x,r)}\big) := \sum_{i=1}^{n} \big\langle e_i \otimes k,\ \mathrm{h}_{|(x,r)}(e_i)\big\rangle_{E_{|(x,r)} \otimes J_{|x}}\, k$$

for any orthonormal basis $\{e_i\}$ of $E_{|(x,r)}$ and unit vector k in $J_{|x}$. To see that $\mathrm{W}(E,\nabla^E,\alpha)(x)$ is well defined, we use the fact that α is bicovariantly constant to compute

$$\mathrm{tr}_{J_{|(x,\pi(r+s))}}\left(\mathrm{h}_{|(x,r+s)}\right) = \sum_{i=1}^{n} \left\langle e_i \otimes k,\ \mathrm{h}_{|(x,r+s)}(e_i)\right\rangle_{E_{|(x,r+s)}\otimes J_{|x}} k$$

$$= \sum_{i=1}^{n} \left\langle e_i \otimes k,\ \left((P^E_{(x,r,s)} \otimes 1_{J_{|(x,\pi(r))}}) \circ \mathrm{h}_{|(x,r)} \circ P^E_{(x,r+s,-s)}\right)(e_i)\right\rangle_{E_{|(x,r+s)}\otimes J_{|x}} k$$

$$= \sum_{i=1}^{n} \left\langle P^{E*}_{(x,r,s)}(e_i) \otimes k,\ \mathrm{h}_{|(x,r)}\left(P^E_{(x,r+s,-s)}(e_i)\right)\right\rangle_{E_{|(x,r)}\otimes J_{|x}} k.$$

Since the parallel transport operator is unitary, we have $P^{E*}_{(x,r,s)} = P^{E-1}_{(x,r,s)} = P^E_{(x,r+s,-s)}$, and since the connection ∇^E is hermitean it follows that the set of vectors $\{P^E_{(x,r+s,-s)}(e_i)\}$ is an orthonormal basis of $E_{/(x,r)}$ As the vector-valued trace is independent of the choice of orthonormal basis, we thus find

$$\mathrm{tr}_{J_{|(x,\pi(r+s))}}\left(\mathrm{h}_{|(x,r+s)}\right) = \mathrm{tr}_{J_{|(x,\pi(r))}}\left(\mathrm{h}_{|(x,r)}\right)$$

and hence the definition of the Wilson loop (9.6) is independent of $r \in \mathbb{R}$. Thus, we obtain a well-defined map

$$\mathrm{W} : \Gamma(\mathbb{L}_p \times S^1, \mathcal{L}_p) \longrightarrow \Gamma(\mathbb{L}_p, J),$$

which as previously descends to 2-isomorphism classes of sections of L_p, and is compatible with direct sum, tensor product and the module actions.

10 OPEN PROBLEMS

In this paper we have shown how a bundle gerbe on a manifold *M* naturally gives rise to a 2-Hilbert space. The main ingredients in its construction are the notions of sections of a bundle gerbe, a hermitean 2-bundle metric, and integration of vector bundles with connection over the manifold *M*. Although this construction is already very well behaved, it can be improved in several ways. For instance, it would be desirable to have additive inverses for sections of a bundle gerbe, or more generally for morphisms of bundle gerbes. A general formalism for ring completions of rig categories has been developed in [4], but it remains to write down explicit models for ring completions of $\mathcal{HVB}\mathrm{dl}^{\nabla}(M)$ and $\mathcal{H}\mathrm{ilb}$. From the perspective of quantisation, additive inverses would be necessary to describe interesting interference phenomena.

It is still unclear whether a more general notion of morphisms of bundle gerbes can be formulated so that even bundle gerbes with non-torsion Dixmier–Douady class admit sections. For a detailed discussion of this problem, see [13]. Referring to our comments at the end of Sect. 7, it is possible that the topology on the objects of $\mathcal{H}$ilb has to be weakened in order to allow for such morphisms.

There are also other models for geometric representations of elements of $H^3(M, \mathbb{Z})$, for example, Azumaya algebra bundles [25, 32, 47].[32] It might be interesting to relate our constructions to this formalism similarly to [47]. So far, the endomorphism algebra bundles arising from our sections of bundle gerbes as their 2-endomorphisms are all finite-dimensional, but only since we have restricted our considerations to 2-morphisms which are bicovariantly constant.

The question of finding an analogue on $\mathcal{HVB}\mathrm{dl}^{\nabla}(M)$ of the Poisson algebra structure on $C^\infty(M, \mathbb{C})$ from geometric quantisation remains open. This structure might only exist on a suitable ring completion of $\mathcal{HVB}\mathrm{dl}^{\nabla}(M)$, but on the full subcategory of trivial hermitean line bundles with connection a Lie 2-algebra structure has been found in [42] (see [13]). Higher prequantisation would then require finding a representation of that higher Poisson algebra on the category of sections $\Gamma(M,L)$ of a prequantum bundle gerbe L on M. Finally, a notion of polarisation would be necessary to obtain a physically sensible 2-Hilbert space; such a notion has been proposed in [42]. It would be desirable to have a good physical example of complete higher quantisation, most probably from string theory or M-theory, against which a ematical formalism of 2-plectic quantisation could be confronted.

FOOTNOTES

1.If M is non-compact, we can instead consider the pre-Hilbert space to consist of compactly supported smooth sections.

2.We take an n-connected space M to have homotopy groups $\pi_i(M)=0$ for all $i=0,1,\ldots,n$.

3.Technically, we should require based paths with sitting instants here for concatenation of paths to give smooth paths. In the following all mapping spaces are assumed to consist of maps having appropriate sitting instants to make all the necessary gluing of maps smooth.

4.The base-point of $D^2 \subset \mathbb{R}^2$ is the point (1, 0), such that $S^1 \hookrightarrow D^2$ is a map of pointed manifolds.

5.This follows from the homotopy long exact sequence together with contractibility of the fibres of T^*Q.

6.The vector fields that forms of this type on mapping spaces act on are inserted into the first slots of the form under the integral.

7.The submersion property ensures that the fibres are smooth manifolds.

8.This is the convention that comes from viewing $Y^{[2]} \rightrightarrows Y$ as part of a simplicial set (the nerve of the Čech groupoid of $Y \to M$) and taking p_i to be the respective face maps.

9.An open covering $\{U_i\}_{i \in \Lambda}$ is *good* if all U_i and all possible finite intersections are contractible.

10.There are two versions of a categorified group, namely a category in groups, or a group in categories. The category $(\mathcal{H}\mathrm{ilb}^\times, \otimes)$ is a symmetric monoidal groupoid, i.e. it belongs to the latter class. We will encounter a similar ambiguity in Sect. 7.

11.It would be more precise to say that the sheaf of U(1) -valued transition functions gets replaced by the stack of transition line bundles with connection, and to consider local sections of this stack instead of generalised transition functions. We refrain from adapting this point of view for pedagogical reasons.

12.For example, the trivial bundle gerbe has only a single infinite-rank 1-endomorphism up to 2-isomorphism due to Kuiper's theorem. This trivialises the category of endomorphisms, analogously to how adding infinite-rank Hilbert bundles to the category of hermitean vector bundles on M trivialises $\mathcal{HVB}\mathrm{dl}(M)$.

13.In particular, $\mathcal{BG}\mathrm{rb}^\nabla$ is a 2-stack [38].

14.The set of maps from a set into a (commutative) monoid is a (commutative) monoid itself.

15.Functors from a category into a (symmetric) monoidal category form a (symmetric) monoidal category themselves.

16.Here we identify $\mathbb{C}^* = \mathbb{C}$ with pairing $\mathbb{C}^* \otimes \mathbb{C} \to \mathbb{C}, w \otimes z \mapsto w \cdot z$.

17.Because of the hermitean structure, the dual E^* is canonically isomorphic to the complex conjugate of E.

18.It is also compatible with pullbacks; whence, it is a transformation of 2-stacks.

19.These endomorphisms can analogously serve as higher functions, but $\mathcal{HVB}\mathrm{dl}^{\nabla}(M)$ is equivalent as a rig category and more convenient to work with.

20.This is true with our convention of taking 2-morphisms to be parallel. Without this feature the isomorphism (6.3) still holds if one omits the subscript 'par'.

21.Recall that we have already encountered a higher categorical version of this construction in Sect. 5.2.

22.In our language, an isomorphism is *stable* if it uses the coarsest common refinement of the coverings that the bundle gerbes are defined over.

23.One might also call ($\mathcal{H}$ilb$^{\times}$,$\otimes$) a different model for $\mathcal{B}\mathrm{U}(1)$.

24.We could probably drop the symmetric condition, at least at this stage, but we keep it here for convenience.

25.Recall from Sect. 5.5 that this means each set of morphisms $\mathcal{C}(a,b)$ is a finite-dimensional vector space, and composition is compatible with this structure.

26.If we were to extend this theory to encompass also infinite-dimensional Hilbert spaces, then the ground rig $\mathcal{H}$ilb$^{\infty}$ of possibly infinite-dimensional Hilbert spaces with continuous operators would be enriched in $\mathcal{V}$ect$^{\infty}$, the possibly infinite-dimensional vector spaces, rather than in $\mathcal{H}$ilb$^{\infty}$. There is still, however, additional structure on the morphism spaces: they are Banach spaces.

27.In a rig (and even generally in a ring) there is no notion of positive-definiteness.

28.This is true at least for the definition of morphisms of bundle gerbes employed here. In other places, for example, [51], an additional condition is used, relating the trace of the curvature of the 1-morphism to the curvings of the source and target gerbes. However, ematically, as well as in view of the DBI-action of D-branes, this condition appears to be unnatural.

29.Since $M=\mathbb{R}^2\times S^1$ is path-connected we can choose an arbitrary base-point. Here, for any topological group G, the classifying space $\mathsf{BG}\cong|\mathcal{B}\mathsf{G}|$ is the coarse moduli space of the classifying stack $\mathcal{B}\mathsf{G}=(\mathsf{G}\rightrightarrows *)$ of principal G -bundles.

30.Generally, on any manifold *M*, a bundle gerbe L admits a section $(E,\alpha):\mathcal{I}_0\rightarrow\mathcal{L}$ of finite-rank if and only if the Dixmier–Douady class DD(L) lies in the torsion subgroup Tor $(\mathrm{H}^3(M,\mathbb{Z}))$ of the degree 3 cohomology of *M*

[11]. In that case, the order of DD(L) is a divisor of the rank of every section (E,α).

31.Here and in the following we suppress pullbacks to $\mathbb{L}_p \times S^1$ for ease of notation.

32.Alternatively, the group $\mathrm{H}^3(M,\mathbb{Z})$ classifies principal $\mathsf{PU}(\mathcal{H})$ -bundles on M, where $\mathcal{H}$ is an infinite-dimensional separable Hilbert space, $\mathsf{U}(\mathcal{H})$ is the group of unitary endomorphisms of $\mathcal{H}$, and $\mathsf{PU}(\mathcal{H}) = \mathsf{U}(\mathcal{H})/\mathsf{U}(1)$ is an Eilenberg–MacLane space $K(2,\mathbb{Z})$, i.e. $\mathrm{H}^1(M,\mathsf{PU}(\mathcal{H})) \cong \mathrm{H}^3(M,\mathbb{Z})$, see, for example, [3]. By the Serre–Grothendieck theorem, every torsion class in $\mathrm{H}^3(M,\mathbb{Z})$ can be represented by a principal PU(n)-bundle on M.

ACKNOWLEDGEMENTS

This work was supported in part by the Action MP1405 QSPACE from the European Cooperation in Science and Technology (COST). The work of S.B. was supported by a James Watt Scholarship. The work of R.J.S. is supported in part by the Consolidated Grant ST/L000334/1 from the UK Science and Technology Facilities Council (STFC).

REFERENCES

1. Acharya, B.S.: On realizing N=1N=1 super Yang-Mills in M-theory (2000). arXiv:hep-th/0011089
2. Aschieri, P., Szabo, R.J.: Triproducts, nonassociative star products and geometry of RR-flux string compactifications. J. Phys Conf. Ser. **634**, 012004 (2015). arXiv:1504.03915
3. Atiyah, M.F., Segal, G.B.: Twisted K-theory. Ukr. Mat. Visn. **1**, 287–330 (2004). arXiv:.KT/0407054
4. Baas, N.A., Dundas, B.I., Richter, B., Rognes, J.: Ring completion of rig categories. J. Reine Angew.. **674**, 43–80 (2013). arXiv:0706.0531
5. Baez, J.C.: Higher-dimensional algebra II: 2-Hilbert spaces. Adv. . **127**, 125–189 (1997). arXiv:q-alg/9609018
6. Baez, J.C., Baratin, A., Freidel, L., Wise, D.K.: Infinite-dimensional representations of 2-groups. Mem. Am.. Soc. **vi+219**, 120 (2012). arXiv:0812.4969
7. Baez, J.C., Hoffnung, A.E., Rogers, C.L.: Categorified symplectic geometry and the classical string. Commun.. Phys. **293**, 701–725 (2010). arXiv:0808.0246
8. Blumenhagen, R., Deser, A., Lüst, D., Plauschinn, E., Rennecke, F.: Non-geometric fluxes, asymmetric strings and nonassociative geometry. J. Phys. A **44**, 385401 (2011). arXiv:1106.0316
9. Blumenhagen, R., Plauschinn, E.: Nonassociative gravity in string theory? J. Phys. A **44**, 015401 (2011). arXiv:1010.1263
10. Bongers, S.: Geometric quantization of symplectic and Poisson manifolds. M.Sc. thesis, Utrecht (2014). dspace.library.uu.nl/handle/1874/290019
11. Bouwknegt, P., Carey, A.L., ai, V., Murray, M.K., Stevenson, D.: Twisted K-theory and K-theory of bundle gerbes. Commun.. Phys. **228**, 17–49 (2002). arXiv:hep-th/0106194

12. Brylinski, J.L.: Loop Spaces, Characteristic Classes and Geometric Quantization. Birkhäuser, Basel (2008)
13. Bunk, S., Saemann, C., Szabo, R.J.: The 2-Hilbert space of a prequantum bundle gerbe (2016). arXiv:1608.08455
14. Chu, C.-S., Smith, D.J.: Towards the quantum geometry of the M5-brane in a constant CC-field from multiple membranes. JHEP **04**, 097 (2009). arXiv:0901.1847
15. Condeescu, C., Florakis, I., Lüst, D.: Asymmetric orbifolds, non-geometric fluxes and noncommutativity in closed string theory. JHEP **04**, 121 (2012). arXiv:1202.6366
16. Crane, L., Sheppeard, M.D.: 2-Categorical Poincaré representations and state sum applications. arXiv:/0306440
17. Crane, L., Yetter, D.N.: Measurable categories and 2-groups. Appl. Categ. Struct. **13**, 501–516 (2005). arXiv:/0305176
18. Davis, J.F., Kirk, P.: Lecture Notes in Algebraic Topology. American ematical Society, Providence (2001)
19. Fiorenza, D., Rogers, C., Schreiber, U.: Higher U(1)-gerbe connections in geometric prequantization. Rev. . Phys. **28**, 1650012 (2016). arXiv:1304.0236
20. Fiorenza, D., Sati, H., Schreiber, U.: A higher stacky perspective on Chern–Simons theory. In: Calaque, Damien, et al. (eds.) ematical Aspects of Quantum Field Theories. Springer, New York (2014). arXiv:1301.2580
21. Folland, G.B.: A Course in Abstract Harmonic Analysis. CRC Press, Boca Raton (2016)
22. Gukov, S., Rangamani, M., Witten, E.: Dibaryons, strings and branes in orbifold models. JHEP **12**, 025 (1998). arXiv:hep-th/9811048
23. Johnson, S.: Constructions with bundle gerbes, Ph.D. thesis (2002). arXiv:.DG/0312175
24. Kapranov, M.M., Voevodsky, V.A.: 2-Categories and Zamolodchikov tetrahedra equations. Proc. Symp. Pure . **56**, 177–260 (1994)
25. Kapustin, A.: D-branes in a topologically non-trivial BB-field. Adv. Theor. . Phys. **4**, 127–154 (2000). arXiv:hep-th/9909089
26. Karoubi, M.: Twisted bundles and twisted K-theory. Commun. . Phys. **228**, 17–49 (2002). arXiv:1012.2512
27. Karoubi, M.: K-Theory: An Introduction. Springer, New York (2009)
28. Kirillov, A.A.: Geometric quantization. Encycl. . Sci. **4**, 137–172 (1990)
29. Kostant, B.: Quantization and unitary representations. Lect. Notes . **170**, 87–208 (2006)
30. Lüst, D.: T-duality and closed string noncommutative (doubled) geometry. JHEP **12**, 084 (2010). arXiv:1010.1361
31. Maldacena, J.M., Moore, G.W., Seiberg, N.: Geometrical interpretation of D-branes in gauged WZW models. JHEP **07**, 046 (2001). arXiv:hep-th/0105038
32. ai, V., Melrose, R.B., Singer, I.M.: The index of projective families of elliptic operators. Geom. Topol. **9**, 341–373 (2005). arXiv:.DG/0206002
33. Murray, M.K.: Bundle gerbes. J. Lond. . Soc. **54**, 403–416 (1996). arXiv:dg-ga/9407015
34. Murray, M.K., Stevenson, D.: Bundle gerbes: stable isomorphism and local theory. J. Lond. . Soc. (2) **62**(3), 925–937 (2000). arXiv:/9908135
35. Mylonas, D., Schupp, P., Szabo, R.J.: Membrane sigma-models and quantization of non-geometric flux backgrounds. JHEP **09**, 012 (2012). arXiv:1207.0926
36. Nikolaus, T., Schreiber, U., Stevenson, D.: Principal ∞∞-bundles: general theory. J. Homotopy Relat. Struct. **10**, 749–801 (2015). arXiv:1207.0248
37. Nikolaus, T., Schreiber, U., Stevenson, D.: Principal ∞∞-bundles: presentations. J. Homotopy Relat. Struct. **10**, 565–622 (2015). arXiv:1207.0249
38. Nikolaus, T., Schweigert, C.: Equivariance in higher geometry. Adv. . **226**, 3367–3408 (2011). arXiv:1004.4558
39. Nuiten, J.: Cohomological quantization of local prequantum boundary field theory. M.Sc. thesis, Utrecht (2013). dspace.library.uu.nl/handle/1874/282756
40. Pressley, A., Segal, G.B.: Loop Groups. Oxford University Press, Oxford (1986)
41. Rogers, C.L.: Higher geometric quantization, talk at higher structures in Göttingen (2011), slides at ncatlab.org/nlab/files/RogersGottingen11.pdf
42. Rogers, C.L.: Higher symplectic geometry, Ph.D. thesis (2011). arXiv:1106.4068

43. Saemann, C., Szabo, R.J.: Groupoid quantization of loop spaces, PoS CORFU2011 (2011), 046. arXiv:1203.5921
44. Saemann, C., Szabo, R.J.: Quantization of 2-plectic manifolds. Progress in Operator Algebras, Noncommutative Geometry, and their Applications, The Theta Foundation, pp. 135–146 (2012). arXiv:1106.1890
45. Saemann, C., Szabo, R.J.: Groupoids, loop spaces and quantization of 2-plectic manifolds. Rev. . Phys. **25**, 1330005 (2013). arXiv:1211.0395
46. Schreiber, U.: Quantum 2-States: Sections of 2-vector bundles, talk at Fields Institute Toronto (2007). http://www..uni-hamburg.de/home/schreiber/atd.pdf
47. Schweigert, C., Tropp, C., Valentino, A.: A Serre–Swan theorem for gerbe modules on étale Lie groupoids. Theory Appl. Categ. **29**, 819–835 (2014). arXiv:1401.2824
48. Seiberg, N., Witten, E.: String theory and noncommutative geometry. JHEP **09**, 032 (1999). arXiv:hep-th/9908142
49. Szabo, R.J.: Quantum field theory on noncommutative spaces. Phys. Rep. **378**, 207–299 (2003). arXiv:hep-th/0109162
50. Waldorf, K.: overflow, comment (2011). overflow./a/72756/381
51. Waldorf, K.: More morphisms between bundle gerbes. Theory Appl. Categ. **18**, 240–273 (2007). arXiv:.CT/0702652

CHAPTER 9

On the Kinematics for the Closed Planar Homothetic Inverse Motions in Complex Plane

Onder Sener[1], Ayhan Tutar[1,2*]

[1] *Faculty of Art and Science, Ondokuz Mayis University, Samsun, Turkey*

[2] *Department of Mathematics, Faculty of Science, Kyrgyz-Türk Manas University, Bishkek, Kyrgyzstan*

ABSTRACT

In this paper, the Steiner area formula and the polar moment of inertia were expressed during one-parameter closed planar homothetic inverse motions in complex plane. The Steiner point was defined when the rotation number was different zero and it was called the Steiner normal when the rotation number was equal to zero. The fixed pole point was given with its components and its relation between Steiner point or Steiner normal was explained. The sagittal motion of a telescopic crane was considered as an example. This motion was described by a double hinge consisting of the fixed control panel of the telescopic crane and the moving arm of the telescopic crane. The theoretical concepts and results were applied for this motion.

Keywords: Steiner Formula, Polar Moment of İnertia, Planar Kinematics, Homothetic İnverse Motions, Complex Plane

1. INTRODUCTION

For a geometrical object rolling on a line and making a complete turn, some properties of the area of a path of a point were given by Steiner [1]. The Steiner area formula and the Holditch theorem during one-parameter closed planar homothetic motions were expressed by Tutar [2]. We want to explain the relation between the Steiner formula and the polar moment of inertia for the closed planar homothetic inverse motions in complex plane and find experimental data equivalents of the theoretical results in this study. We calculated the expression of the Steiner formula relative to the fixed coordinate system under one parameter closed planar motions in complex plane. If the points of the fixed plane which enclose the same area lie on a circle, then the centre of this circle is called the Steiner point (h=1) [3] [4]. If these points lie on a line, we use Steiner normal instead of Steiner point. Then we obtained the fixed pole point for the closed planar homothetic inverse motions. We dealt with the polar moment of inertia of a path generated by a closed planar homothetic inverse motion. Moreover, we expressed the relation between the area enclosed by a path and the polar moment of inertia. As an example, the sagittal motion of a telescopic crane which is described by a double hinge being fixed and moving was considered. The Steiner area formula, the fixed pole point and the polar moment of inertia were calculated for this motion. Furthermore, the relation between the Steiner formula and the polar moment of inertia was expressed.

2. CLOSED HOMOTHETIC INVERSE MOTIONS IN COMPLEX PLANE

We consider one parameter closed planar homothetic motion between two reference systems: the fixed E' and the moving E, with their origins (O, O') and orientations. Then, we take into account motion relative to the moving coordinate system (inverse motion). We know the motion defined by the transformation $X'(t) = h(t)\,Xe^{i\alpha} + U'(t)$ is called one-parameter closed planar homothetic direct motion in complex plane.

By taking displacement vector $OO' = U$ and $O'O = U'$, the total angle of rotation $\alpha(t)$, the motion defined by the transformation

$$X(t) = \frac{1}{h(t)}\left(X' - U'(t)\right)e^{-i\alpha(t)} \tag{1}$$

is called one-parameter closed planar homothetic inverse motion in complex plane and denoted by E'/E, where h is a homothetic scale of the motion E'/E, X and X' are the position vectors with respect to the moving and fixed rectangular coordinate systems of a point $X \in E$, respectively. The homothetic scale h and the vector X and U, U' are continuously differentiable functions of a real parameter t.

In Equation (1), $X(t)$ is the trajectory with the respect to the moving system of a point X' belonging to the fixed system.

If we consider the coordinates of the equation above,

$$X(t) = x_1(t) + ix_2(t), \quad X' = x_1' + ix_2',$$
$$U'(t) = u_1'(t) + iu_2'(t), \quad U(t) = u_1(t) + iu_2(t)$$

we can write

$$x_1(t) + ix_2(t) = \frac{1}{h(t)}\left[\left(x_1' - u_1'(t)\right) + i\left(x_2 - u_2'(t)\right)\right]\left(\cos\alpha(t) - i\sin\alpha(t)\right) \tag{2}$$

From Equation (2), the components of $X(t)$ may be given as

$$\left.\begin{aligned} x_1(t) &= \frac{1}{h(t)}\left[\cos(\alpha(t))\left(x_1' - u_1'(t)\right) + \sin(\alpha(t))\left(x_2' - u_2'(t)\right)\right] \\ x_2(t) &= \frac{1}{h(t)}\left[-\sin(\alpha(t))\left(x_1' - u_1'(t)\right) + \cos(\alpha(t))\left(x_2' - u_2'(t)\right)\right] \end{aligned}\right\} \tag{3}$$

If we show the coordinates of the Equation (1) as

$$X(t) = \begin{pmatrix} x_1(t) \\ x_2(t) \end{pmatrix}, \quad X' = \begin{pmatrix} x_1' \\ x_2' \end{pmatrix}, \quad U'(t) = \begin{pmatrix} u_1'(t) \\ u_2'(t) \end{pmatrix}, \quad U(t) = \begin{pmatrix} u_1(t) \\ u_2(t) \end{pmatrix}$$

and rotation matrix

$$R(t) = \begin{pmatrix} \cos(\alpha(t)) & -\sin(\alpha(t)) \\ \sin(\alpha(t)) & \cos(\alpha(t)) \end{pmatrix}, \tag{4}$$

we can obtain

$$X(t)=\frac{1}{h(t)}\left(R(t)\right)^{\mathrm{T}}\left(X'-U'(t)\right) \tag{5}$$

If Equation (3) differentiated, we have

$$\left.\begin{aligned}
\mathrm{d}x_1 &= -\frac{\mathrm{d}h}{h^2}\left[\cos\alpha\left(x_1'-u_1'\right)+\sin\alpha\left(x_2'-u_2'\right)\right]\\
&\quad+\frac{1}{h}\left[-\left(\cos\alpha\right)\mathrm{d}u_1'-\left(x_1'-u_1'\right)\left(\sin\alpha\right)\mathrm{d}\alpha-\left(\sin\alpha\right)\mathrm{d}u_2'+\left(x_2'-u_2'\right)\left(\cos\alpha\right)\mathrm{d}\alpha\right]\\
\mathrm{d}x_2 &= -\frac{\mathrm{d}h}{h^2}\left[-\sin\alpha\left(x_1'-u_1'\right)+\cos\alpha\left(x_2'-u_2'\right)\right]\\
&\quad+\frac{1}{h}\left[\left(\sin\alpha\right)\mathrm{d}u_1'-\left(x_1'-u_1'\right)\left(\cos\alpha\right)\mathrm{d}\alpha-\left(\cos\alpha\right)\mathrm{d}u_2'-\left(x_2'-u_2'\right)\left(\sin\alpha\right)\mathrm{d}\alpha\right]
\end{aligned}\right\} \tag{6}$$

2.1. The Steiner Formula for the Homothetic Inverse Motions

The formula for the area F' of a closed planar curve of the point X is given by

$$F'=\frac{1}{2}\oint\left(x_1\mathrm{d}x_2-x_2\mathrm{d}x_1\right) \tag{7}$$

If Equations (3) and (6) are replaced in Equation (7),

$$\begin{aligned}
2F' &= -\left(x_1'^2+x_2'^2\right)\oint\frac{1}{h^2}\mathrm{d}\alpha+x_1'\oint\left(2\frac{1}{h^2}u_1'\mathrm{d}\alpha-\frac{1}{h^2}\mathrm{d}u_2'\right)+x_2'\oint\left(2\frac{1}{h^2}u_2'\mathrm{d}\alpha+\frac{1}{h^2}\mathrm{d}u_1'\right)\\
&\quad-\oint\left\{\frac{1}{h^2}\left(u_1'^2+u_2'^2\right)\mathrm{d}\alpha-\frac{1}{h^2}\left(u_1'\mathrm{d}u_2'-u_2'\mathrm{d}u_1'\right)\right\}
\end{aligned} \tag{8}$$

is obtained. The integral coefficients in Equation (8) can be shown as

$$\left.\begin{aligned}
&\oint\left(2\frac{1}{h^2}u_1'\mathrm{d}\alpha-\frac{1}{h^2}\mathrm{d}u_2'\right)=a'^*\\
&\oint\left(2\frac{1}{h^2}u_2'\mathrm{d}\alpha+\frac{1}{h^2}\mathrm{d}u_1'\right)=b'^*\\
&-\oint\left\{\frac{1}{h^2}\left(u_1'^2+u_2'^2\right)\mathrm{d}\alpha-\frac{1}{h^2}\left(u_1'\mathrm{d}u_2'-u_2'\mathrm{d}u_1'\right)\right\}=c'
\end{aligned}\right\} \tag{9}$$

If we show the trajectory of the orjin of the fixed system by $F_{o'}'=F'\left(x_1'=0,x_2'=0\right)$, we can say

$$2F_{o'}'=c' \tag{10}$$

The coefficient m' is defined by

$$m'=\oint\frac{1}{h^2}\mathrm{d}\alpha=\frac{1}{h^2\left(t_0\right)}\oint\mathrm{d}\alpha=\frac{1}{h^2\left(t_0\right)}2\pi\nu \tag{11}$$

with the rotation number ν establishes whether the lines with F' = *const.* describing circles or straight lines. If $\nu\neq 0$, then we have circles. If $\nu=0$, the circles reduce to straight lines. If Equations (9), (10) and (11) are substituted in Equation (8), then

$$2\left(F'-F_{o'}'\right)=-\left(x_1'^2+x_2'^2\right)m'+a'^*x_1'+b'^*x_2' \tag{12}$$

can be written.

A Different Parametrization for the Integral Coefficients

Equation (5) by differentiation with respect to *t* yields

$$\mathrm{d}X=\frac{1}{h}\mathrm{d}\left(R^{\mathrm{T}}\right)\left(X'-U'\right)-\frac{1}{h}R^{\mathrm{T}}\mathrm{d}U'-\frac{\mathrm{d}h}{h^2}R^{\mathrm{T}}\left(X'-U'\right)$$

and if we use $X'=P'=\begin{pmatrix}p_1'\\p_2'\end{pmatrix}$ for the pole point, we can write

$$0=\mathrm{d}X=\frac{1}{h}\mathrm{d}\left(R^{\mathrm{T}}\right)\left(P'-U'\right)-\frac{1}{h}R^{\mathrm{T}}\mathrm{d}U'-\frac{\mathrm{d}h}{h^2}R^{\mathrm{T}}\left(P'-U'\right) \tag{13}$$

Then if $U' = \begin{pmatrix} u_1' \\ u_2' \end{pmatrix}$ is solved from Equation (13), we find

$$\left.\begin{aligned} u_1' &= p_1' + \frac{h}{h^2(\mathrm{d}\alpha)^2 + (\mathrm{d}h)^2}\left(\mathrm{d}h\mathrm{d}u_1' + h\mathrm{d}\alpha\mathrm{d}u_2'\right) \\ u_2' &= p_2' + \frac{h}{h^2(\mathrm{d}\alpha)^2 + (\mathrm{d}h)^2}\left(\mathrm{d}h\mathrm{d}u_2' - h\mathrm{d}\alpha\mathrm{d}u_1'\right) \end{aligned}\right\} \tag{14}$$

If Equation (14) is replaced in Equation (9),

$$\left.\begin{aligned} a'' &= 2\oint\frac{1}{h^2}p_1'\mathrm{d}\alpha + 2\oint\frac{\mathrm{d}\alpha}{h^3(\mathrm{d}\alpha)^2 + h(\mathrm{d}h)^2}\left(\mathrm{d}h\mathrm{d}u_1' + h\mathrm{d}\alpha\mathrm{d}u_2'\right) - \oint\frac{1}{h^2}\mathrm{d}u_2' \\ b'' &= 2\oint\frac{1}{h^2}p_2'\mathrm{d}\alpha + 2\oint\frac{\mathrm{d}\alpha}{h^3(\mathrm{d}\alpha)^2 + h(\mathrm{d}h)^2}\left(\mathrm{d}h\mathrm{d}u_2' - h\mathrm{d}\alpha\mathrm{d}u_1'\right) + \oint\frac{1}{h^2}\mathrm{d}u_1' \end{aligned}\right\} \tag{15}$$

can be rewritten. Also Equation (15) can be expressed as separately

$$a' := 2\oint\frac{1}{h^2}p_1'\mathrm{d}\alpha, \quad b' := 2\oint\frac{1}{h^2}p_2'\mathrm{d}\alpha \tag{16}$$

$$\begin{aligned} u_1' &:= 2\oint\frac{\mathrm{d}\alpha}{h^3(\mathrm{d}\alpha)^2 + h(\mathrm{d}h)^2}\left(\mathrm{d}h\mathrm{d}u_1' + h\mathrm{d}\alpha\mathrm{d}u_2'\right) - \oint\frac{1}{h^2}\mathrm{d}u_2', \\ u_2' &:= 2\oint\frac{\mathrm{d}\alpha}{h^3(\mathrm{d}\alpha)^2 + h(\mathrm{d}h)^2}\left(\mathrm{d}h\mathrm{d}u_2' - h\mathrm{d}\alpha\mathrm{d}u_1'\right) + \oint\frac{1}{h^2}\mathrm{d}u_1', \quad \mu' = \begin{pmatrix} u_1' \\ u_2' \end{pmatrix} \end{aligned} \tag{17}$$

Using Equations (16) and (17), we have

$$2\left(F' - F_{o'}'\right) = -\left(x_1'^2 + x_2'^2\right)m' + a'x_1' + b'x_2' + \mu_1'x_1' + \mu_2'x_2' \tag{18}$$

2.2. Steiner Point or Steiner Normal for the Homothetic Inverse Motions

By taking $m' \neq 0$, the Steiner point $S' = (s_1', s_2')$ for the closed planar homothetic inverse motion can be written as

$$s'_j = \frac{\oint \frac{1}{h^2} p'_j d\alpha}{\oint \frac{1}{h^2} d\alpha}, \quad j = 1,2 \tag{19}$$

Then

$$\oint \frac{1}{h^2} p'_1 d\alpha = s'_1 m', \quad \oint \frac{1}{h^2} p'_2 d\alpha = s'_2 m' \tag{20}$$

is found. If we consider Equation (19) and Equation (20) is replaced in Equation (17),

$$2\left(F' - F'_o\right) = -m'\left(x_1'^2 + x_2'^2 - 2s'_1 x'_1 - 2s'_2 x'_2\right) + \mu'_1 x'_1 + \mu'_2 x'_2 \tag{21}$$

is found. Equation (21) is called the Steiner area formula for the closed planar

homothetic inverse motion.

By dividing Equation (21) by m' and by completing the squares, one obtains

the equation of a circle,

$$\left(x'_1 - \left(s'_1 - \frac{\mu'_1}{2m'}\right)\right)^2 + \left(x'_2 - \left(s'_2 - \frac{\mu'_2}{2m'}\right)\right)^2 - \left(s'_1 - \frac{\mu'_1}{2m'}\right)^2 - \left(s'_2 - \frac{\mu'_2}{2m'}\right)^2 = -\frac{2\left(F' - F'_o\right)}{m'} \tag{22}$$

All the moving points of the fixed plane which pass around equal orbit areas under the motion E/E' lie on the same circle with the center

$$M' = \left(s'_1 - \frac{\mu'_1}{2m'}, s'_2 - \frac{\mu'_2}{2m'}\right) \tag{23}$$

in the fixed plane.

In the case of $h(t) = 1$, since $\mu'_1 = \mu'_2 = 0$, the point M' and the Steiner point S' coincide [3]. Also by taking $m' = 0$, if it is replaced in Equation (18), then we have

$$\left(a' + \mu_1'\right)x_1' + \left(b' + \mu_2'\right)x_2' - 2\left(F' - F_{o'}'\right) = 0 \tag{24}$$

Equation (24) is a straight line. If no complete loop occurs, then $\nu = 0$ and the circles are reduced to straight lines, in other words, to a circle whose center lies at infinity. The normal to the lines of equal areas in Equation (24) is given by

$$n' = \begin{pmatrix} a' + \mu_1' \\ b' + \mu_2' \end{pmatrix} \tag{25}$$

which is called the Steiner normal [5].

2.3. The Fixed Pole Point for the Homothetic Inverse Motions

Using Equation (14), if $P' = \begin{pmatrix} p_1' \\ p_2' \end{pmatrix}$ is solved,

$$\left.\begin{aligned} p_1' &= \frac{h}{h^2\left(\mathrm{d}\alpha\right)^2 + \left(\mathrm{d}h\right)^2}\left(-\mathrm{d}h\mathrm{d}u_1' - h\mathrm{d}\alpha\mathrm{d}u_2'\right) + u_1' \\ p_2' &= \frac{h}{h^2\left(\mathrm{d}\alpha\right)^2 + \left(\mathrm{d}h\right)^2}\left(-\mathrm{d}h\mathrm{d}u_2' + h\mathrm{d}\alpha\mathrm{d}u_1'\right) + u_2' \end{aligned}\right\} \tag{26}$$

is found for the fixed pole point P'.

For $m' \neq 0$, using Equations (11) and (19), we arrive at the relation in Equation (20) between the Steiner point and the pole point.

For $m' = 0$, using Equations (16), (17) and (25), we arrive at the relation between the Steiner normal and the pole point below:

$$\begin{pmatrix} a' \\ b' \end{pmatrix} = \begin{pmatrix} 2\oint \frac{1}{h^2} p_1' \mathrm{d}\alpha \\ 2\oint \frac{1}{h^2} p_2' \mathrm{d}\alpha \end{pmatrix} = n' - \mu' \tag{27}$$

2.4. The Polar Moments of Inertia for the Homothetic Inverse Motions

Because of their mathematical structure, these averages may be interpreted as polar moments of inertia T' for a path with closed homothetic motion. In this section we want to derive a formula combining T',m', n' and the area F' together.

A relation between the Steiner formula and the polar moment of inertia around the instantaneous pole can be found in [6]. Müller [3] proved a relation to the polar moment of inertia around the origin. Tölke [7] investigated the same for closed equiaffine mappings. Düldül and Kuruoğlu [8] generalized Müller's results for homothetic motion. If we use α as parametrization, we need to calculate

$$T' = \oint \left(x_1^2 + x_2^2\right) \mathrm{d}\alpha \tag{28}$$

along the path of X. Then, using Equation (3)

$$T' = \left(x_1'^2 + x_2'^2\right) m' + x_1' \oint \left(-2\frac{1}{h^2} u_1' \mathrm{d}\alpha\right) + x_2' \oint \left(-2\frac{1}{h^2} u_2' \mathrm{d}\alpha\right) + \oint \frac{1}{h^2}\left(u_1'^2 + u_2'^2\right) \mathrm{d}\alpha \tag{29}$$

is obtained.

We need calculate to the polar moments of inertia of the origin of the fixed system, therefore $T_{o'}' = T'\left(x_1' = 0, x_2' = 0\right)$, one obtains

$$T_{o'}' = \oint \frac{1}{h^2}\left(u_1'^2 + u_2'^2\right) \mathrm{d}\alpha \tag{30}$$

If Equation (30) is replaced in Equation (29),

$$T' - T_o' = \left(x_1'^2 + x_2'^2\right) m' + x_1' \oint \left(-2\frac{1}{h^2} u_1' \mathrm{d}\alpha\right) + x_2' \oint \left(-2\frac{1}{h^2} u_2' \mathrm{d}\alpha\right) \tag{31}$$

can be written. Also if Equation (14) is replaced in Equation (31),

$$T' - T'_o = \left(x_1'^2 + x_2'^2\right) m' + x_1' \left(-2\oint \frac{1}{h^2} p_1' \mathrm{d}\alpha - 2\oint \frac{\mathrm{d}\alpha}{h^3 (\mathrm{d}\alpha)^2 + h (\mathrm{d}h)^2} \left(\mathrm{d}h \mathrm{d}u_1' + h \mathrm{d}\alpha \mathrm{d}u_2' \right) \right)$$
$$+ x_2' \left(-2\oint \frac{1}{h^2} p_2' \mathrm{d}\alpha - 2\oint \frac{\mathrm{d}\alpha}{h^3 (\mathrm{d}\alpha)^2 + h (\mathrm{d}h)^2} \left(\mathrm{d}h \mathrm{d}u_2' - h \mathrm{d}\alpha \mathrm{d}u_1' \right) \right) \tag{32}$$

is found and by considering Equations (16) and (30) together, we arrive at the relation between the polar moments of inertia and the formula for the area **below:**

$$T' - T'_o = -2\left(F' - F'_o\right) + x_1' \left(-\oint \frac{1}{h^2} \mathrm{d}u_2' \right) + x_2' \left(\oint \frac{1}{h^2} \mathrm{d}u_1' \right) \tag{33}$$

3. APPLICATION: THE INVERSE MOTION OF THE TELESCOPIC CRANE

In the previous sections we emphasized three concepts: geometrical objects as the Steiner point or the Steiner normal, the pole point, and the polar moments of inertia for closed homothetic inverse motions in complex plane. In this section, we want to visualize the experimentally measured motion with these objects. Accordingly, we consider these characteristic directions for this motion.

We will show that how the kinematical objects which used in the previous sections can be applied. Dathe and Gezzi considered human gait in planar motions [5]. As an example, we have chosen the sagittal part of the movement of the telescopic crane at motion. We have chosen the telescopic crane, because the arm of telescopic crane can extend or retract during one parameter closed planar homothetic motion. The motion of telescopic crane has a double hinge and "a double hinge" means that it has two systems, a fixed arm and a moving arm of telescopic crane (Figure 1). There is a control panel of telescopic crane at the origin of fixed system. "L" arm can extend or retract by h parameter.

3.1. The Mathematical Model by Inverse Motion

We start by writing the equations of the double hinge in cartesian coordinates. Then we define using the condition $m = 0$, the Steiner normal and the total angle in relation to the double hinge.

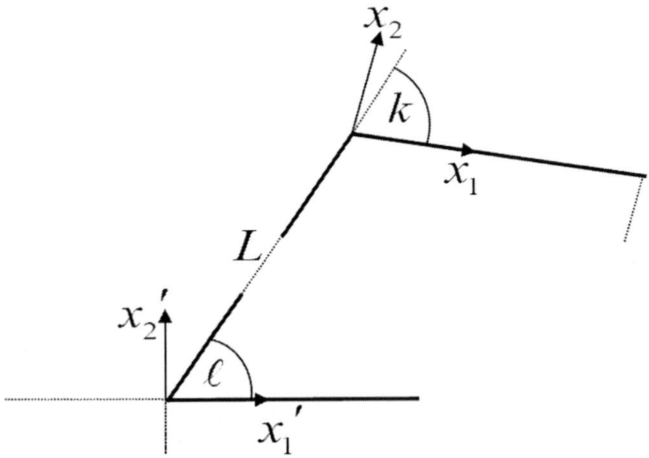

Figure 1. The arms of telescopic crane as a double hinge.

By taking displacement vector $OO' = U$ and $O'O = U'$, the total angle of rotation $l - k = \alpha$, the motion can be defined by the transformation

$$X(t) = \frac{1}{h(t)}\left(X' - U'(t)\right)e^{-i(\ell(t)-k(t))} \tag{34}$$

By taking

$$R(t) = \begin{pmatrix} \cos(\ell(t)-k(t)) & -\sin(\ell(t)-k(t)) \\ \sin(\ell(t)-k(t)) & \cos(\ell(t)-k(t)) \end{pmatrix},$$

$$U(t) = \begin{pmatrix} u_1(t) \\ u_2(t) \end{pmatrix} = \begin{pmatrix} -L\cos(k(t)) \\ -L\sin(k(t)) \end{pmatrix} \tag{35}$$

we have Equation (5) namely,

$$X(t) = \frac{1}{h(t)}\left(R(t)\right)^{\mathrm{T}}\left(X' - U'(t)\right)$$

Also we know that $U' = -RU$ and

$$U'(t) = \begin{pmatrix} u_1'(t) \\ u_2'(t) \end{pmatrix} = \begin{pmatrix} L\cos(\ell(t)) \\ L\sin(\ell(t)) \end{pmatrix} \tag{36}$$

can be written. So the double hinge may be written as

$$\left.\begin{aligned} x_1(t) &= \frac{1}{h(t)}\left[\cos(\ell(t)-k(t))(x_1'-L\cos(\ell))+\sin(\ell(t)-k(t))(x_2'-L\sin(\ell))\right] \\ x_2(t) &= \frac{1}{h(t)}\left[-\sin(\ell(t)-k(t))(x_1'-L\cos(\ell))+\cos(\ell(t)-k(t))(x_2'-L\sin(\ell))\right] \end{aligned}\right\} \tag{37}$$

where $\alpha = l - k$ is the resulting total angle.

We begin by calculating the time derivative of Equation (37). In this way, we obtain the velocities $\dot{x}_1(t), \dot{x}_2(t)$ which have to be inserted into Equation (7).

Then we have

$$\begin{aligned} x_1\dot{x}_2 - x_2\dot{x}_1 &= -\left(x_1'^2 + x_2'^2\right)\frac{1}{h^2}\left(\dot{\ell}-\dot{k}\right) + x_1'\left(2\frac{1}{h^2}L\cos\ell\left(\dot{\ell}-\dot{k}\right)-\frac{1}{h^2}L\dot{\ell}\cos\ell\right) \\ &\quad + x_2'\left(2\frac{1}{h^2}L\sin\ell\left(\dot{\ell}-\dot{k}\right)-\frac{1}{h^2}L\dot{\ell}\sin\ell\right)+\frac{1}{h^2}L^2\dot{k} \end{aligned} \tag{38}$$

We now integrate the previous equation using periodic boundary conditions by assuming that the integrands as periodic functions. The periodicity of f implies that integrals of the following types vanish $\oint \mathrm{d}f = \int_1^F \dot{f}\,\mathrm{d}t = f\big|_1^F = 0$. As a result of this, some of integrals of Equation (38) are not equal to zero. If we use Equation (38) in the area formula

$$F' = \frac{1}{2}\int_{t_1}^{t_2}\left(x_1\dot{x}_2 - x_2\dot{x}_1\right)\mathrm{d}t$$

we finally obtain

$$2F' = x_1' \int_{t_1}^{t_2} \left(2\frac{1}{h^2} L \cos\ell\left(\dot{\ell} - \dot{k}\right) - \frac{1}{h^2} L\dot{\ell}\cos\ell \right) dt$$
$$+ x_2' \int_{t_1}^{t_2} \left(2\frac{1}{h^2} L \sin\ell\left(\dot{\ell} - \dot{k}\right) - \frac{1}{h^2} L\dot{\ell}\sin\ell \right) dt \tag{39}$$

We can write the integral coefficients in Equation (39) as

$$\left. \begin{aligned} \int_{t_1}^{t_2} \left(2\frac{1}{h^2} L \cos\ell\left(\dot{\ell} - \dot{k}\right) - \frac{1}{h^2} L\dot{\ell}\cos\ell \right) dt = a'^* \\ \int_{t_1}^{t_2} \left(2\frac{1}{h^2} L \sin\ell\left(\dot{\ell} - \dot{k}\right) - \frac{1}{h^2} L\dot{\ell}\sin\ell \right) dt = b'^* \end{aligned} \right\} \tag{40}$$

Differentiating Equation (36) with respect to t and then replacing both of them in Equation (40), Equation (9) is found for application. In Section (2.1.1), using Equation (14),

$$a'^* = \underbrace{\int_{t_1}^{t_2} \left(2\frac{1}{h^2} p_1' d\alpha \right)}_{a'} + \underbrace{\int_{t_1}^{t_2} \left(\frac{2d\alpha}{h^3 (d\alpha)^2 + h(dh)^2} \left(dh du_1' + h d\alpha du_2' \right) - \frac{1}{h^2} du_2' \right)}_{\mu_1'}$$
$$b'^* = \underbrace{\int_{t_1}^{t_2} \left(2\frac{1}{h^2} p_2' d\alpha \right)}_{b'} + \underbrace{\int_{t_1}^{t_2} \left(\frac{2d\alpha}{h^3 (d\alpha)^2 + h(dh)^2} \left(dh du_2' - h d\alpha du_1' \right) - \frac{1}{h^2} du_1' \right)}_{\mu_2'} \tag{41}$$

is found and we have a straight line below:

$$2F' = \left(a' + \mu_1'\right) x_1' + \left(b' + \mu_2'\right) x_2' \tag{42}$$

In this case, we have the Steiner normal

$$n' = \begin{pmatrix} a' + \mu_1' \\ b' + \mu_2' \end{pmatrix} = L \begin{pmatrix} \int_{t_1}^{t_2} \left(2\frac{1}{h^2} \cos\ell\left(\dot{\ell} - \dot{k}\right) - \frac{1}{h^2} \dot{\ell}\cos\ell \right) dt \\ \int_{t_1}^{t_2} \left(2\frac{1}{h^2} \sin\ell\left(\dot{\ell} - \dot{k}\right) - \frac{1}{h^2} \dot{\ell}\sin\ell \right) dt \end{pmatrix} \tag{43}$$

3.2. The Fixed Pole Point of the Inverse Telescopic Crane Motion

If Equation (36) is replaced in Equation (26), the pole point $P' = \begin{pmatrix} p'_1 \\ p'_2 \end{pmatrix}$ with the components

$$\left.\begin{aligned} p'_1 &= \frac{h}{(\mathrm{d}h)^2 + h^2\left(\dot{\ell} - \dot{k}\right)^2}\left(\mathrm{d}hL\dot{\ell}\sin\ell - h\left(\dot{\ell} - \dot{k}\right)L\dot{\ell}\cos\ell\right) + L\cos\ell \\ p'_2 &= -\frac{h}{(\mathrm{d}h)^2 + h^2\left(\dot{\ell} - \dot{k}\right)^2}\left(\mathrm{d}hL\dot{\ell}\cos\ell + h\left(\dot{\ell} - \dot{k}\right)L\dot{\ell}\sin\ell\right) + L\sin\ell \end{aligned}\right\} \tag{44}$$

is found and also by using Equations (43) and (44), we reach at the relation between the Steiner normal and the pole point (Equation (27)).

3.3. The Polar Moments of Inertia of the Inverse Telescopic Crane Motion

Using Equations (28) and (37), if Equation (36) is replaced in Equation (29)

$$T' = x'_1 \oint \left(-2\frac{1}{h^2}L(\cos\ell)\left(\dot{\ell} - \dot{k}\right)\right)\mathrm{d}t + x'_2 \oint \left(-2\frac{1}{h^2}L(\sin\ell)\left(\dot{\ell} - \dot{k}\right)\right)\mathrm{d}t \tag{45}$$

is found. By considering Equations (41), (42) and (45) together, we arrive at the relation between the polar moments of inertia and the formula for the area below:

$$T' = -2F' - x'_1 L \oint \frac{1}{h^2}(\cos\ell)\dot{\ell}\,\mathrm{d}t - x'_2 L \oint \frac{1}{h^2}(\sin\ell)\dot{\ell}\,\mathrm{d}t \tag{46}$$

ACKNOWLEDGEMENTS

This study is supported by Ondokuz Mayıs University (Project No. PYO. FEN. 1904. 14.019).

REFERENCES

1. Steiner, J. (1840) Von dem Krümmungs-Schwerpuncte ebener Curven. Journal für die reine und angewandte Mathematik, 21, 33-63. https://doi.org/10.1515/crll.1840.21.33
2. Tutar, A. and Kuruoglu, N. (1999) The Steiner formula and the Holditch theorem for the Homothetic Motions on the Planar Kinematics. Mechanism and Machine Theory, 34, 1-6. https://doi.org/10.1016/S0094-114X(98)00028-7
3. Müller, H.R. (1978) Verallgemeinerung einer Formel von Steiner. Abhandlungen der Braunschweigischen Wissenschaftlichen Gesellschaft Band, 29, 107-113.
4. Müller, H.R. (1978) über Tr&aumlgheitsmomente bei Steinerscher Massenbelegung. Abhandlungen der Braunschweigischen Wissenschaftlichen Gesellschaft Band, 29, 115-119.
5. Dathe, H. and Gezzi, R. (2012) Characteristic Directions of Closed Planar Motions. Zeitschrift für Angewandte Mathematik und Mechanik, 92, 2-13. https://doi.org/10.1002/zamm.201100178
6. Blaschke, W. and Müller, H.R. (1956) Ebene Kinematik. R. Oldenbourg, München.
7. T&oumllke, J. (1978) Steiner-Formeln für die Bahnfl&aumlchen geschlossener &Aumlquiaffinbewegungen, Sitzungsberi. &Oumlsterreichische Akademie der Wissenschaften Mathematisch-Naturwissenschaftliche Klasse, 187, 325-337.
8. Kuruoglu, N., Düldül, M. and Tutar, A. (2003) Generalization of Steiner Formula for the Homothetic Motions on the Planar Kinematics. Applied Mathematics and Mechanics (English Edition), 24, 945-949. https://doi.org/10.1007/BF02446500

CHAPTER 10

A New Narrowband Phase Modulation Mathematical Identity

Shihadeh M. Saadeh

Department of Physics, Jordan University of Science and Technology, Irbid, Jordan

ABSTRACT

A new mathematical identity is suggested to describe narrow band phase modulation and other similar physical problems instead of using the Bessel function. Bessel functions are extensively used in mathematical physics [1,2], electromagnetic wave propagation and scattering [3,4], and communication system theory [3,5,6]. Such phenomena must often be approximated by appropriate formulas since there is no closed form solution or expression, which usually leads to complex mathematical solutions [5,7]. Comparisons are made between the exact solution numerically calculated and graphed with the new mathematical identities' prediction of phase modulation behavior. The proposed mathematical identity matches the results very well, leading to simpler analysis of such physical behavior.

Keywords: Phase Modulation; Bessel Functions; Optical Communication

1. INTRODUCTION

Frequency or phase modulation is an efficient form of communication, where information is transmitted over a carrier signal by changing the instantaneous frequency [4]. Phase modulation has been heavily employed over the last century, and is the bases of all modern digital communication systems. Bessel functions are used to better describe the mathematical equations that govern phase modulation.

The expressions $\varphi(t) = \delta \sin(\omega_m t)$ and $\cos[\varphi(t)] = \cos[\delta \sin(\omega_m t)]$ where δ and ω_m are the index and frequency of modulation, appear frequently in physics research and literature. They play a key role in the mathematical treatment of narrowband frequency modulation in communication and in antennas radiation analysis, design and studies [3,4].

Usually $\cos[\delta \sin(\omega_m t)]$ and $\sin[\delta \sin(\omega_m t)]$ are replaced by Bessel function series, which adds complexity to the mathematical treatment. In many cases this can lead to using mathematical approximations that reduce the accuracy of the analysis.

Previous published work done in collaboration with Prof. Salamo from the University of Arkansas dealt with laser beam propagation through two level samples [8]. While trying to better understanding the evolution of an electric field envelope as the previously studied laser beam propagated through the two level sample, and the effects of phase modulation by using custom in house developed software for a numerical based computer simulation study, it was observed that at a certain beer's absorption length inside the sample, the electric field envelope looks like an analytical function. During fitting trials of the observed electric field in several trigonometric form functions, an expression for the $\cos[\delta \sin(\omega_m t)]$ function in two terms was developed, and the idea of a alterative to the complex Bessel Functions approach was conceived [9-11]. This new expression was compared to the exact $\cos[\delta \sin(\omega_m t)]$, the agreement and fit are excellent for $\delta < 1$ i.e. for small index of modulation, which is the case in optical frequency modulation and in communications.

2. MATHEMATICAL DEVELOPMENT

Our first postulate is:

$$\cos[\delta \sin(\omega_m t)] = \cos\delta + (1 - \cos\delta)\cos^2(\omega_m t) \tag{1}$$

To verify the accuracy of our assumption, the end result undergoes vigorous numerical and analytical testing.

If the proposed postulate is true then it follows:

$$\cos\left[\delta\sin\left(\omega_m t\right)\right] = \cos\delta + \cos^2\left(\omega_m t\right) - \cos\delta\cos^2\left(\omega_m t\right) \quad (2)$$

$$\cos\left[\delta\sin\left(\omega_m t\right)\right] = \cos^2\left(\omega_m t\right) + \left(\cos\delta\right)\left[1 - \cos^2\left(\omega_m t\right)\right] \quad (3)$$

$$\cos\left[\delta\sin\left(\omega_m t\right)\right] = \cos^2\left(\omega_m t\right) + \left(\cos\delta\right)\sin^2\left(\omega_m t\right) \quad (4)$$

$$\sin^2\left[\delta\sin\left(\omega_m t\right)\right] = 1 - \cos^2\left[\delta\sin\left(\omega_m t\right)\right] \quad (5)$$

by substituting equation (4) in equation (5),

$$\sin^2\left[\delta\sin\left(\omega_m t\right)\right] = 1 - \left[\cos^2\left(\omega_m t\right) + \left(\cos\delta\right)\sin^2\left(\omega_m t\right)\right]^2 \quad (6)$$

$$\sin^2\left[\delta\sin\left(\omega_m t\right)\right] = \left[1 - \cos^2\left(\omega_m t\right) - \left(\cos\delta\right)\sin^2\left(\omega_m t\right)\right] \left[1 + \cos^2\left(\omega_m t\right) + \left(\cos\delta\right)\sin^2\left(\omega_m t\right)\right] \quad (7)$$

$$\sin^2\left[\delta\sin\left(\omega_m t\right)\right] = \left[\sin^2\left(\omega_m t\right) - \left(\cos\delta\right)\sin^2\left(\omega_m t\right)\right] \left[2 - \sin^2\left(\omega_m t\right) + \left(\cos\delta\right)\sin^2\left(\omega_m t\right)\right] \quad (8)$$

$$\sin^2\left[\delta\sin\left(\omega_m t\right)\right] = 2\left[\sin^2\left(\omega_m t\right) - \left(\cos\delta\right)\sin^2\left(\omega_m t\right)\right] - \left[\sin^2\left(\omega_m t\right) - \left(\cos\delta\right)\sin^2\left(\omega_m t\right)\right]^2 \quad (9)$$

$$\sin^2\left[\delta\sin\left(\omega_m t\right)\right] = 2\sin^2\left(\omega_m t\right)\left(1 - \cos\delta\right) - \sin^4\left(\omega_m t\right)\left(1 - \cos\delta\right)^2 \quad (10)$$

$$\sin^2\left[\delta\sin(\omega_m t)\right] = 4\sin^2\left(\frac{\delta}{2}\right)\sin^2(\omega_m t)\left[1-\sin^2(\omega_m t)\sin^2\left(\frac{\delta}{2}\right)\right] \tag{11}$$

$$\sin\left[\delta\sin(\omega_m t)\right] = 2\sin\left(\frac{\delta}{2}\right)\sin(\omega_m t)\sqrt{\left[1-sin^2(\omega_m t)sin^2\left(\frac{\delta}{2}\right)\right]} \tag{12}$$

The second postulate is:

$$\cos\left[\delta\cos(\omega_m t)\right] = \cos\delta + (1-\cos\delta)\sin^2(\omega_m t) \tag{13}$$

If this is a valid assumption, then:

$$\cos\left[\delta\cos(\omega_m t)\right] = \cos\delta + \sin^2(\omega_m t) - \cos\delta\sin^2(\omega_m t) \tag{14}$$

$$\cos\left[\delta\cos(\omega_m t)\right] = \sin^2(\omega_m t) + (\cos\delta)\left[1-\sin^2(\omega_m t)\right] \tag{15}$$

$$\cos\left[\delta\cos(\omega_m t)\right] = \sin^2(\omega_m t) + (\cos\delta)\cos^2(\omega_m t) \tag{16}$$

If that is right then:

$$\sin^2\left[\delta\cos(\omega_m t)\right] = 1-\cos^2\left[\delta\cos(\omega_m t)\right] \tag{17}$$

by substituting equation (16) in equation (17),

$$\sin^2\left[\delta\cos(\omega_m t)\right] = 1-\left[\sin^2(\omega_m t) + (\cos\delta)\cos^2(\omega_m t)\right]^2 \tag{18}$$

$$\sin^2\left[\delta\cos(\omega_m t)\right]=\left[1-\sin^2(\omega_m t)-(\cos\delta)\cos^2(\omega_m t)\right]$$
$$\left[1+\sin^2(\omega_m t)+(\cos\delta)\cos^2(\omega_m t)\right] \quad (19)$$

$$\sin^2\left[\delta\cos(\omega_m t)\right]=\left[\cos^2(\omega_m t)-(\cos\delta)\cos^2(\omega_m t)\right]$$
$$\left[2-\cos^2(\omega_m t)+(\cos\delta)\cos^2(\omega_m t)\right] \quad (20)$$

$$\sin^2\left[\delta\cos(\omega_m t)\right]=2\left[\cos^2(\omega_m t)-(\cos\delta)\cos^2(\omega_m t)\right]$$
$$-\left[\cos^2(\omega_m t)-(\cos\delta)\cos^2(\omega_m t)\right]^2 \quad (21)$$

$$\sin^2\left[\delta\cos(\omega_m t)\right]$$
$$=4\sin^2\left(\frac{\delta}{2}\right)\cos^2(\omega_m t)\left[1-\cos^2(\omega_m t)\sin^2\left(\frac{\delta}{2}\right)\right] \quad (22)$$

$$\sin\left[\delta\cos(\omega_m t)\right]$$
$$=2\sin\left(\frac{\delta}{2}\right)\cos(\omega_m t)\sqrt{\left[1-\cos^2(\omega_m t)\sin^2\left(\frac{\delta}{2}\right)\right]}. \quad (23)$$

From the above derivation, we get our new identities in equations (4), (12), (16) and (23).

3. TESTING THE SUGGESTED EXPRESSION

The new expression is tested in two approaches. The first approach is to analytically compare the result to the established phase modulation of light [12]. The second approach to compare the results with computer simulations.

3.1. Phase Modulation of Light

Amnon Yariv discuses phase modulation of light in chapter 9 of his book Introduction to Optical Electronics [12]. The electric field for the modulated light is given by:

$$E_{out} = A\cos\left[\omega t + \delta \sin(\omega_m t)\right] \tag{24}$$

where d is the phase modulation index and ω_m is the phase modulation frequency.

$$\begin{aligned} E_{out} &= A\cos(\omega_m t)\cos\left[\delta \sin(\omega_m t)\right] \\ &\quad - A\sin(\omega_m t)\sin\left[\delta \sin(\omega_m t)\right] \end{aligned} \tag{25}$$

The above book uses the Bessel function identities

$$\begin{aligned} &\cos\left[\delta \sin(\omega_m t)\right] \\ &= J_o(\delta) + 2J_o(\delta)\cos(2\omega_m t) + J_4(\delta)\cos(4\omega_m t) + \cdots \end{aligned} \tag{26}$$

$$\begin{aligned} &\sin\left[\delta \sin(\omega_m t)\right] \\ &= J_1(\delta)\sin(\omega_m t) + 2J_3(\delta)\sin(3\omega_m t) + \cdots \end{aligned} \tag{27}$$

$$\begin{aligned} E_{out} &= A\big[J_o(\delta)\cos(\omega t) \\ &+J_1(\delta)\cos(\omega + \omega_m)t - J_1(\delta)\cos(\omega - \omega_m)t \\ &+J_2(\delta)\cos(\omega + 2\omega_m)t + J_2(\delta)\cos(\omega - 2\omega_m)t \\ &+J_3(\delta)\cos(\omega + 3\omega_m)t - J_3(\delta)\cos(\omega - 3\omega_m)t \\ &+J_4(\delta)\cos(\omega + 4\omega_m)t - J_4(\delta)\cos(\omega - 4\omega_m)t + \cdots\big] \end{aligned} \tag{28}$$

However, it is generally assumed that only the first three terms are significant. This agrees with our experiments in which only the first three terms in the series are detected.

For small modulation, i.e. d < 1, $J_0(\delta) = 1$ and $J_1(\delta) = \sin(\delta/2)$ which leads to:

$$\begin{aligned} E_{out} = A\bigg[&\cos(\omega t) + \sin\left(\frac{\delta}{2}\right)\cos(\omega + \omega_m)t \\ &- \sin\left(\frac{\delta}{2}\right)\cos(\omega - \omega_m)t\bigg] \end{aligned} \tag{29}$$

If we use our suggested approximation instead of the Bessel function identities in E_{out}, then:

$$E_{out} = A\cos(\omega t)\cos\left[\delta\sin(\omega_m t)\right] - A\sin(\omega_m t)\sin\left[\delta\sin(\omega_{mt})\right] \quad (30)$$

By using the new identities of equations (4) and (12) in equation (30), we get:

$$E_{out} = A\left[\cos(\omega t)\cos^2(\omega_m t) + \cos(\omega t)(\cos\delta)\sin^2(\omega_m t)\right] - A\left[2\sin\left(\frac{\delta}{2}\right)\sin(\omega t)\sin\left(\frac{\delta}{2}\right)\sin(\omega_m t)\sqrt{\left(1-\sin^2\left(\frac{\delta}{2}\right)\sin^2(\omega_m t)\right)}\right] \quad (31)$$

$$E_{out} = A\left[\cos(\omega t)\left\{\cos^2(\omega_m t) + (\cos\delta)\sin^2(\omega_m t)\right\} - 2\sin\left(\frac{\delta}{2}\right)\sin(\omega t)\sin(\omega_m t)\sqrt{\left(1-\sin^2\left(\frac{\delta}{2}\right)\sin^2(\omega_m t)\right)}\right] \quad (32)$$

To check this formula against the previous formula of E_{out} we apply it for small δ where, $\cos\delta = 1$ and $\sin^2(\delta/2) = 0$. In this case,

$$E_{out} = A\left[\cos(\omega t) - 2\sin\left(\frac{\delta}{2}\right)\sin(\omega t)\sin(\omega_m t)\right] \quad (33)$$

If we use the identity:

$$-2\sin(\omega t)\sin(\omega_m t) = \cos(\omega + \omega_m)t - \cos(\omega - \omega_m)t \quad (34)$$

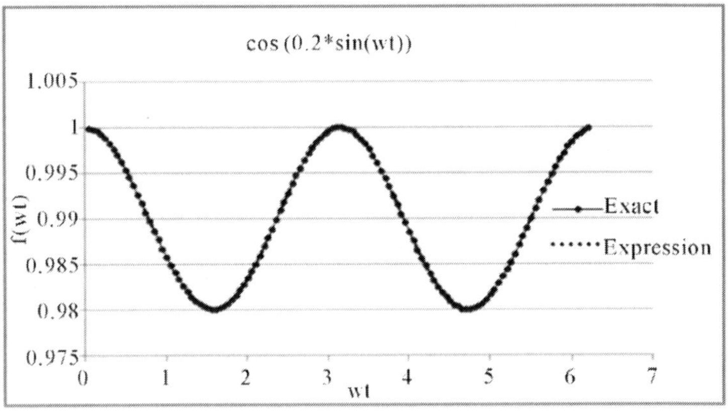
cos (0.2*sin(wt))
1.005
1
0.995
0.99
0.985
0.98
0.975
f(wt)
0
1
2
3
4
5
6
7
wt
Exact
Expression

(a)

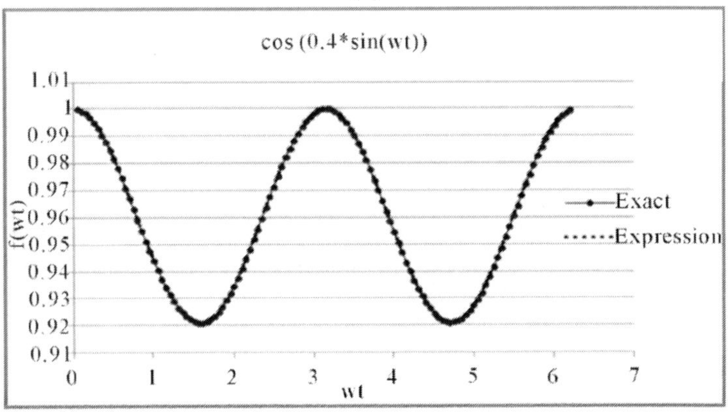
cos (0.4*sin(wt))
1.01
1
0.99
0.98
0.97
0.96
0.95
0.94
0.93
0.92
0.91
f(wt)
0
1
2
3
4
5
6
7
wt
Exact
Expression

(b)

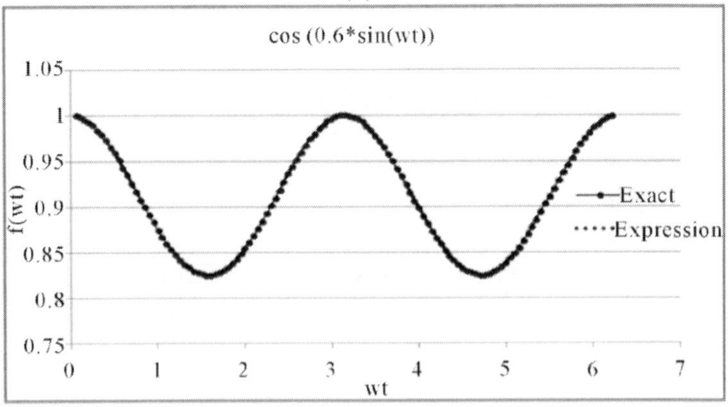
cos (0.6*sin(wt))
1.05
1
0.95
0.9
0.85
0.8
0.75
f(wt)
0
1
2
3
4
5
6
7
wt
Exact
Expression

(c)

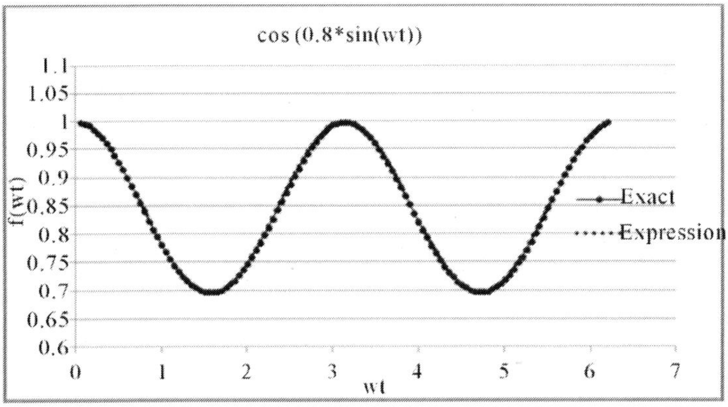

(d)

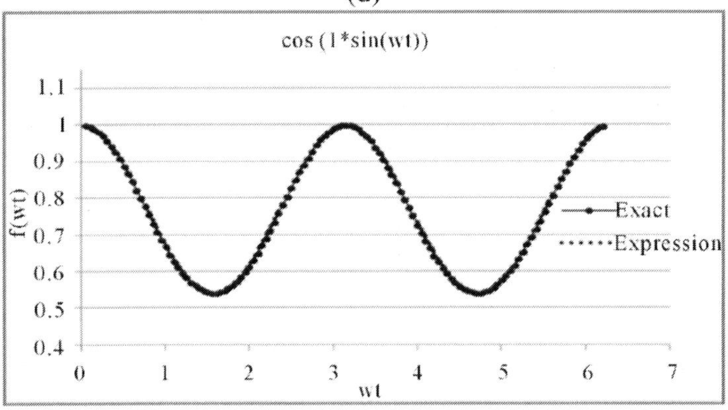

(e)

Figure 1. Comparison between exact results and proposed expression for $\cos\left[\delta\sin(\omega_m t)\right]$.

we get

$$E_{out} = A\left[\cos(\omega t) + \sin\left(\frac{\delta}{2}\right)\cos(\omega + \omega_m)t - \sin\left(\frac{\delta}{2}\right)\cos(\omega - \omega_m)t\right] \tag{35}$$

Which exactly matches the proposed solution?

3.2. Computer Generated Numerical Comparison

The graphs of Figure 1, compare the computer generated numerical results for the exact equation: $\cos\left[\delta \sin\left(\omega_m t\right)\right]$ at different values of d with that of the new proposed mathematical identity. Figure 2 shows the comparison for $\cos\left[\delta \cos\left(\omega_m t\right)\right]$.

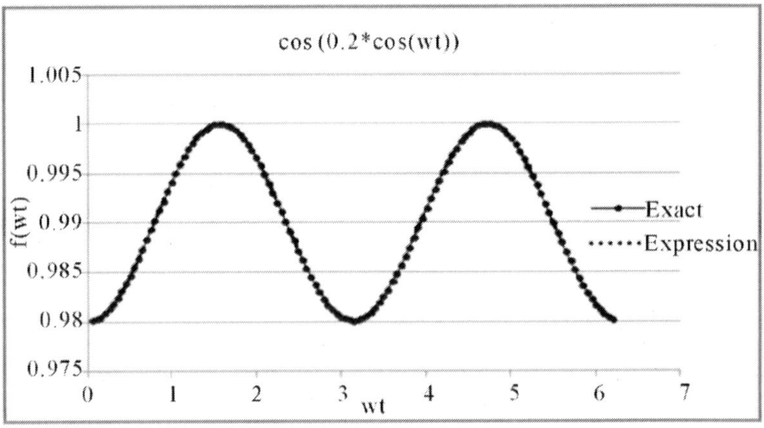

(a)

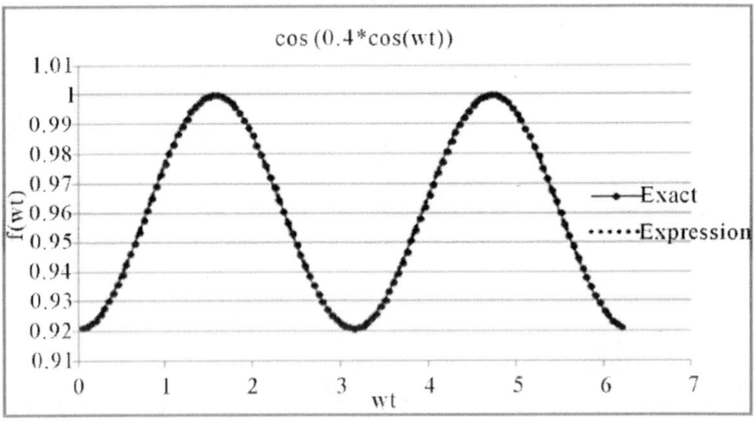

(b)

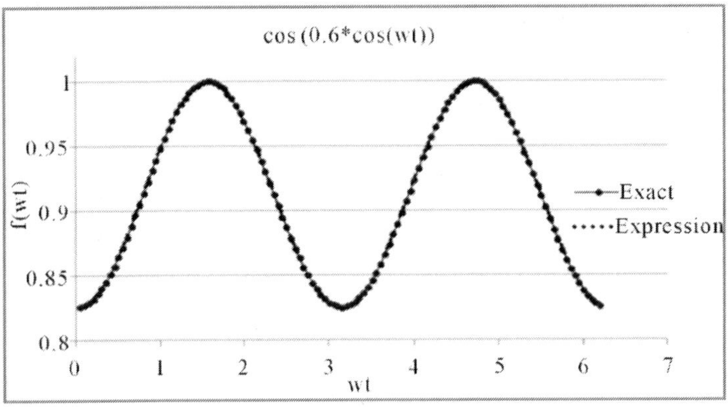

(c)

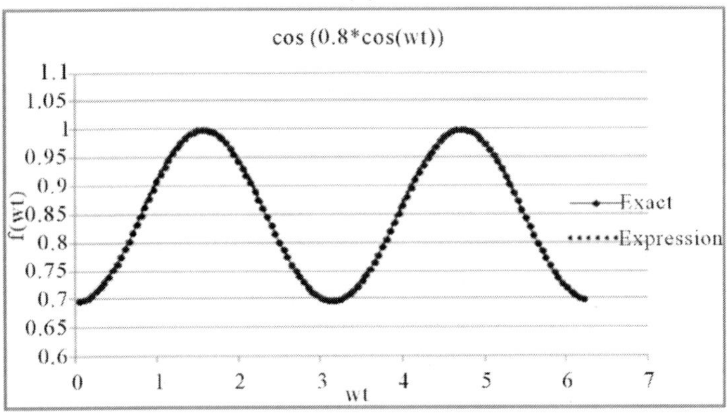

(d)

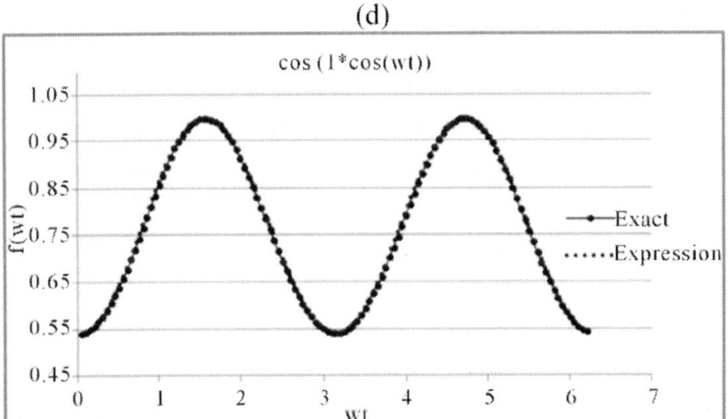

(e)

Figure 2. Comparison between exact results and proposed expression for $\cos\left[\delta\cos(\omega_m t)\right]$.

4. CONCLUSION

Shihadeh M. Saadeh (4), (12), (16) and (23) that describe narrow band phase modulation is presented. The proposed identity can also be used for similar physical phenomena's that use the Bessel function. The mathematical identity was shown to match analytical and numerical results very well. The new identities greatly reduce computation time and complexity of analytical treatment of such physical behavior.

REFERENCES

1. V. Alekseev and N. V. Sushilov, "Analytic Solutions of Bloch and Maxwell-Bloch Equations in the Case of Arbitrary Field Amplitude and Phase Modulation" Physical Review A, Vol. 46, No. 1, 1992, pp. 351-355. doi:10.1103/PhysRevA.46.351
2. N. Nayak and G. S. Agarwal, "Absorption and Fluorescence in Frequency-Modulated Fields under Conditions of Strong Modulation and Saturation" Physical Review A, Vol. 31, No. 5, 1985, pp. 3175-3182. doi:10.1103/PhysRevA.31.3175
3. Hund, "Frequency Modulation," McGraw-Hill, New York, 1942
4. J. D. Jackson, "Classical Electrodynamics," 3rd edition, Wiley, New York, 1998
5. J. G. Proakis and M. Salehi, "Communication Systems Engineering," Prentice Hall, Upper Saddle River, 2001
6. N. M. Blachman, "Noise and Its Effect on Communication," 2nd Edition, Krieger Publishing Co., Malabar, 1982
7. G. N. Watson, "A Treatise on the Theory of Bessel Functions," 2nd edition, Cambridge University Press, Cambridge, 1995
8. S. Saadeh, J. Shultz and G. Salamo, "Experimental Observation of Chirped Continuous Pulse-train Soliton Solutions to the Maxwell-Bloch Equations," Optics Express, Vol. 8, No. 2, 2001, pp. 153-158. doi:10.1364/OE.8.000153
9. L. A. Pipes, "Applied Mathematics for Engineers and Physicists," 2nd Edition, McGraw-Hill, New York, 1958
10. M. Abramowitz and I. A. Stegun, "Handbook of Mathematical Functions," National Bureau of Standards, Washington DC, 1964
11. N. M. Blachman and S. H. Mousavinezhad, "Trigonometric Approximation for Bessel Functions," IEEE Transactions on Aerospace and Electronic Systems, Vol. 22, No. 1, 1986, pp. 2-7. doi:10.1109/TAES.1986.310686
12. Yarvis, "Introduction to Optical Electronics," 2nd Edition, Holt McDougal, Geneva, 1977

CHAPTER 11

Poisson growth

Robb McDonald[1] and Mark Mineev-Weinstein[2,3]

1 Department of ematics, University College London, London, UK

2 International Institute of Physics, UFRN, Natal, Brazil

3 New Mexico Consortium, Los Alamos, NM 87545, USA

ABSTRACT

The two-dimensional free boundary problem in which the field is governed by Poisson's equation and for which the velocity of the free boundary is given by the gradient of the field—Poisson growth—is considered. The problem is a generalisation of classic Hele-Shaw free boundary flow or Laplacian growth problem and has many applications. In the case when the right hand side of Poisson's equation is constant, a formulation is obtained in terms of the Schwarz function of the free boundary. From this it is deduced that solutions of the Laplacian growth problem also satisfy the Poisson growth problem, the only difference being in their time evolution. The corresponding moment evolution equations, a Polubarinova–Galin type equation and a Baiocchi-type transformation for Poisson growth are also presented. Some explicit examples are given, one in which cusp formation is inhibited by the addition of the Poisson term, and another for a growing finger in which the Poisson term selects the width of the finger to be half that of the channel. For the more complicated case when the right hand side is linear in one space direction, the Schwarz function method is used to derive an exact solution describing a translating circular blob with changing radius.

Keywords: Laplacian growth, Hele-Shaw flows

1 INTRODUCTION

In the standard Hele-Shaw free boundary problem for fluid with zero surface tension, the pressure, or velocity potential, satisfies Laplace's equation in the fluid region, is constant on its boundary and the normal velocity of the interface is given by the gradient of the velocity potential in the normal direction. This simply formulated, but nonlinear, two-dimensional free boundary problem has a remarkable variety of applications occurring over a wide range of lengthscales including oil recovery, flow in porous media and injection moulding. Because of this, and with its rich ematical structure becomingly increasingly evident, Laplacian growth, as it is also known, has attracted much attention in the literature e.g. [9]. As a consequence, powerful ematical techniques, naturally formulated in terms of complex variables, have been developed leading to the derivation of a variety of exact solutions for Laplacian growth. These formulations include the Polubarinova–Galin equation [8, 20], Richardson's demonstration of an infinity of conservation laws for the moments of the fluid [21], and a formulation in terms of the Schwarz function for the free boundary e.g. [4].

The ematical problem in which Laplace's equation is replaced by Poisson's equation has received less attention, despite having important applications. When the right hand side of Poisson's equation is constant the model is equivalent to the evaporation of thin liquid films [2]. Moment-based methods were used by [2] to derive exact solutions, some of which displayed typical Laplacian growth type cusp formation on the boundary.

When the upper plate of a Hele-Shaw cell is raised or lowered, the thickness of the thin gap between the plates is time-dependent and this leads to the pressure inside the blob satisfying Poisson's equation with a time-dependent right-hand side. This time-dependence can be scaled out, so that the right-hand side becomes constant. Shelley et al. [22] studied this variable-gap Hele-Shaw problem and showed when the upper plate is raised instability arises, this being analogous to the evaporation case of [2]. Shelley et al. [22] examined the existence and regularity of solutions and found exact and numerical solutions with and without surface tension, including examples illustrating bubble fission and cusp formation. The multiply connected problem has been tackled by Crowdy and Kang [3]. They used associated quadrature domain identities to find explicit solutions which demonstrate that the usual well-posed 'squeeze' case for singly connected domains can break down in finite time in the multiply connected case. Another physical manifestation of Poisson growth is the debonding of adhesively joined surfaces e.g. [12]. The thin gap between the surfaces acts like a Hele-Shaw cell and air fingers penetrate the adhesive as the surfaces are separated.

In addition to Laplacian growth, several authors, e.g. [10, 13, 14], have considered other elliptic PDEs governing the interior of the fluid, as occurs, for example, in flow of fluid in inhomogeneous porous media. Few explicit solutions are known for these more general elliptic operators, one difficulty being their lack of conformal invariance.

The Poisson growth problem is formulated in Sect. 2 and its associated Schwarz function equation derived. This equation is then used to derive the moment evolution equations, a Polubarinova–Galin type equation and an associated Baiocchi-type transformation. Examples of explicit solutions are given in Sect. 3. The case when the right hand side of Poisson's equation is linear in one of the space variables is considered in Sect. 4. The Schwarz function equation is derived and used to find an exact solution consisting of a propagating circular blob with time-dependent radius.

2 SCHWARZ FUNCTION FORMULATION OF THE FREE BOUNDARY PROBLEM

2.1 Derivation of Schwarz function equation

Consider the free boundary problem for a domain $\Omega(t)$ in the (x,y)-plane

$$\begin{aligned} \nabla^2\phi &= -\beta + Q(x, y), \quad \text{in} \quad \Omega(t), \\ \phi &= 0 \quad \text{and} \quad v_n = \partial\phi/\partial n \quad \text{on} \quad \partial\Omega(t) \end{aligned} \tag{1}$$

Here β is a constant, Q represents any hydrodynamic singularities that are present (e.g. point sources or sinks), ϕ is a real velocity potential such that $\mathbf{u} = \nabla\phi$ is the velocity field, and v_n is the normal velocity of the boundary $\partial\Omega(t)$. The task is to find the evolution of the free boundary $\partial\Omega(t)$, which may be finite or infinite and singly or multiply-connected, given some initial boundary shape $\partial\Omega(0)$. The problem is illustrated in Fig. 1.

Equation (1) implies that $\nabla.\mathbf{u} = -\beta$; that is, depending on the sign of β there is a source ($\beta < 0$) or sink ($\beta > 0$) distributed uniformly over the extent of the fluid blob. The free boundary problem (1) is a generalisation of the standard Laplacian growth or Hele-Shaw problem ($\beta = 0$) and is referred to here as the Poisson growth problem. Introducing $z = x + iy$ and defining $\phi = \psi - \beta z\bar{z}/4$, it follows that $\nabla^2\psi = 0$ and so ψ can be considered the real part of an analytic function w. Let $g(z, t)$ be the Schwarz function of the free boundary $\partial\Omega(t)$ i.e. $g(z, t)$ is an analytic function in the neighbourhood of $\partial\Omega$ such that $g(z, t) = \bar{z}$ on $\partial\Omega$ [5]. Since $\phi = \psi - \beta z\bar{z}/4$ it follows that on $\partial\Omega(t)$

$$\bar{v} = \partial_z w - \beta \bar{z}/2 = \partial_z w - \beta g/2 \tag{2}$$

where *v* is the complex velocity field. It is known e.g. [1] that the boundary conditions on *∂Ω(t)* imply $2\bar{v} = \partial_t g$. Combining this with (2) gives

$$2\partial_z w = \partial_t g + \beta g = e^{-\beta t} \partial_t \left(g e^{\beta t} \right), \tag{3}$$

Equation (3) is referred to as the Schwarz function equation for Poisson growth. Lacey [11] discusses the behaviour of Schwarz function singularities in the case of Hele-Shaw flow with a time-dependent gap for the case when there is no hydrodynamic forcing. In particular he derives an explicit solution for the behaviour of an elliptical blob being squeezed between the plates of a Hele-Shaw cell.

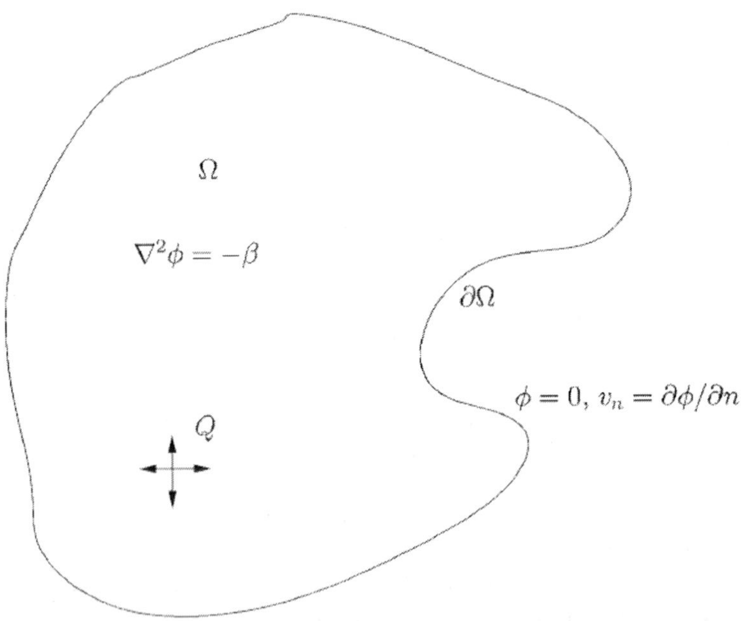

Figure 1 The Poisson growth problem with point source QQ

Strictly, the derivation of (3) applies on the free boundary only, but by analytic continuation it applies everywhere in the fluid blob $\Omega(t)$ where the terms of (3) are defined. If β=0 then (3) reduces to the well-known Schwarz function relation governing Laplacian growth e.g. [4]. Since the nature of the singularities of g and w in (3) are independent of β, an immediate implication is that solutions of the Poisson growth free boundary problem share the same geometries as the standard Hele-Shaw problem although they will in general have different evolution.

As noted in [7] there is a connection between the Poisson growth problem and the behaviour of a blob of fluid in a rotating Hele-Shaw cell of constant width. This is immediately evident by comparing the Schwarz function equations in each case. In the latter case the flow is subject to a centrifugal potential and has Schwarz function equation [15]

$$2\partial_z w = \partial_t g + \omega\partial_z(zg), \tag{4}$$

whereω is the rotation parameter.Rescaling z by the factor $\exp(\beta t)$ in (3) and choosing $\omega = \beta$ gives (4).

2.2 Moments

Suppose $Q(x, y) = Q\delta(x, y)$ where Q is constant, so that there is a point source of constant strength Q at the origin. Define the kth moment as

$$M_k = \iint_{\Omega(t)} z^k dA, \quad k = 0, 1, 2\cdots \tag{5}$$

Using the complex form of Green's theorem and (3) gives

$$\begin{aligned}\partial_t M_k &= \frac{1}{2i}\oint_{\partial\Omega(t)} z^k \partial_t g\, dz \\ &= -\frac{\beta}{2i}\oint_{\partial\Omega(t)} z^k g\, dz + \frac{1}{i}\oint_{\partial\Omega(t)} z^k \partial_z w\, dz \\ &= -\beta M_k + Q\delta_{k0}.\end{aligned} \tag{6}$$

[2,22] obtained the same evolution equations for the moments in the absence of point sources by direct integration of (1). More generally [7], write down an expression for the evolution of the integral of a harmonic function $u(z)$ integrated over the fluid domain in a variable-gap Hele-Shaw cell.

2.3 Polubarinova-Galian type equation

Denoting the moving boundary $z(s, t)$, where s is an arc-length parameter, the normal velocity of the boundary is

$$v_n = \mathrm{Im}(\bar{z}_t z_s) = \partial_n \phi = \partial_n(\psi - \beta z\bar{z}/4) \tag{7}$$

The complexified normal vector on $\partial\Omega$ is $-i_{zs}$ and $\nabla = 2\partial_{\bar{z}}$ in conjugate coordinates. Hence $\partial_n z\bar{z} = 2\mathrm{Im}(\bar{z}z_s)$, and (7) becomes

$$\mathrm{Im}\left[(\bar{z}_t + (\beta/2)\bar{z})z_s\right] = \partial_n \psi = \partial_s \psi^*, \tag{8}$$

where ψ^* is the harmonic conjugate of ψ. If there is a point source of strength Q inside the fluid blob at z=0 then on $\partial\Omega$, $\partial_s\psi^*>0$ and $Q\theta/2\pi=\psi^*=s$ may be used to parameterise the boundary giving

$$\mathrm{Im}\left[(\bar{z}_t + (\beta/2)\bar{z})z_\theta\right] = Q/2\pi \tag{9}$$

Equation (9) is of the type derived by [8, 20], and reduces to their equations when β=0.

2.4 A related Baiocchi-type transform

Define the real function Ψ, the Schwarz potential, by

$$\Psi = \frac{z\bar{z}}{4} - \frac{1}{4}\int^{z} g dz - \frac{1}{4}\overline{\int^{z} g dz} \tag{10}$$

Note that $\nabla^2\Psi$=1 in Ω and $\Psi_z = \Psi_{\bar{z}} = 0$ on $\partial\Omega$ which, in turn, implies $\Psi=\partial\Psi/\partial n=0$ on $\partial\Omega$. Applying the operator $e^{-\beta t}(\partial_t e^{\beta t}.)$e−βt(∂teβt.) to (10) and using (3) gives

$$\begin{aligned} \partial_t \Psi + \beta\Psi &= \frac{\beta}{4}z\bar{z} - \Re w, \\ &= \frac{\beta}{4}z\bar{z} - \psi, \\ &= -\phi. \end{aligned} \tag{11}$$

When β=0 this reduces to the standard Baiocchi transform for Hele-Shaw free boundary flow e.g. [4].

3 EXAMPLES

3.1 Circular blob

The boundary of a circle of radius $a(t)$ and centre z=0 has Schwarz function $g=a^2/z$. If the fluid region, which lies inside the circle, has no hydrodynamic singularities (i.e. Q=0) then w is analytic everywhere inside $\Omega(t)$. Comparing the behaviour of the singular terms in (3) gives

$$\frac{d}{dt}a^2 + \beta a^2 = 0, \tag{12}$$

or, $A(t)=A(0)\exp(-\beta t)$ where $A(t)=\pi a^2$ is the area of the circular blob. That is, the blob remains circular and centred at the origin with its area either increasing or decreasing exponentially according to the sign of β. Alternatively this result follows directly from (6) with the realization $M_0 \equiv A$ or from (9) using $z=a(t)e^{i\theta}$.

3.2 Stability of a circular blob and bubble

Let $z = a(\zeta + \epsilon\zeta^n)$, where $a = a(t)$ and $\epsilon=\epsilon(t)$ are real time-varying positive parameters with $0 < \epsilon \ll 1$, be a map from the interior of the unit ζ-disk to the interior of a perturbed circular blob of radius a. In polar coordinates the boundary of the blob is given by $r = a(1+\epsilon\cos(n-1)\theta)+O(\epsilon^2)$, $0 \leq \theta < 2\pi$, and so ϵ measures the amplitude of a sinusoidal perturbation to the circular boundary. The interest here is on the growth or decay of $\epsilon(t)$ with time. Since $\bar{\zeta} = \zeta^{-1}$ on the boundary of the blob, the Schwarz function of the blob boundary has behaviour as $z \to 0$

$$\begin{aligned} g(z,t) &= \frac{\epsilon a^{n+1}}{z^n} + \frac{a^2(1+\epsilon^2 n)}{z} + \mathcal{O}(1), \\ &\approx \frac{\epsilon a^{n+1}}{z^n} + \frac{a^2}{z} + \mathcal{O}(1), \end{aligned} \tag{13}$$

where the approximation $\epsilon^2 \ll n^{-1}$ has been used. For a point source of strength Q at z=0, the Schwarz function equation (3) becomes $\partial_t g+\beta g=Q/\pi z$

as $z \to 0$. Considering coefficients of the singular terms z^{-1} and z^{-n} in (3) using (13) gives two coupled ODEs for $\epsilon(t)$ and $a(t)$

$$\frac{d}{dt}(a^2) + \beta a^2 = \frac{Q}{\pi},$$
$$\frac{d}{dt}(\epsilon a^{n+1}) + \beta \epsilon a^{n+1} = 0. \tag{14}$$

Combining (14) into a single equation for ϵ gives

$$\frac{d\epsilon}{dt} + \frac{\epsilon}{2}(n+1)\left[\frac{Q}{\pi a^2} - \frac{(n-1)}{(n+1)}\beta\right] = 0 \tag{15}$$

Thus the flow is stable when $Q > \beta\pi a^2$. This criteria is equivalent to saying that stability follows when the blob expands in area since Q is the area flux due to a point source while $\beta\pi a^2$ is the rate of area loss due to the Poisson term. When $Q=0$ the blob is unstable if $\beta>0$ as also found by [22].

A similar analysis for a perturbed circular bubble in an infinite fluid (as opposed to a finite blob of fluid) using the map $z=a(\zeta^{-1}+\epsilon\zeta^n)$ from the interior of the unit ζ-disk to the exterior of the perturbed bubble and considering singular behaviour as $z \to \infty$ shows that a bubble is stable if $Q < \beta\pi a^2$.

3.3 Limaçon

Consider the time-dependent map from the interior of the unit ζ-circle to the limaçon-shaped fluid blob given by

$$z = a\zeta + b\zeta^2, \tag{16}$$

where $a(t)$ and $b(t)$ are real, time-dependent parameters to be found. A point source of strength Q is located at the origin and hence $\partial_z w \to Q/2\pi z + O(1)$ as $z \to 0$. From (16) as $z \to 0$ [see also (13)]

$$g = \frac{a^2 b}{z^2} + \frac{a^2 + 2b^2}{z} + \mathcal{O}(1) \tag{17}$$

Substituting (17) into (3) and comparing the singular behaviour of terms $O(z^{-1})$ and $O(z^{-2})$ gives the following pair of coupled ODEs for $a(t)$ and $b(t)$

$$\frac{d}{dt}(a^2 b) + \beta a^2 b = 0,$$

$$\frac{d}{dt}(a^2 + 2b^2) + \beta(a^2 + 2b^2) = \frac{Q}{\pi} \tag{18}$$

Note that (18) implies the existence of a steady state with b=0 and $a^2=Q/\pi\beta$ provided sgn($Q\beta$)>0. This corresponds to a circular blob of radius $\sqrt{Q/\pi\beta}$ in which the point source Q balances that due to a uniform sink $\beta\pi a^2$. Figure 2 shows the evolution of an initial limaçon with β=1 (i.e. 'evaporation') with and without the presence of a point source at the origin. In both cases the mass flux owing to the distributed sink causes the limaçon to shrink. When Q>0 the shrinkage is halted as $t\to\infty$ when the sink balances the mass source Q and the blob approaches a circle. When Q=0 (Laplacian growth) the problem is ill-posed and the limaçon forms a cusp singularity (i.e. becoming a cardioid) in finite time.

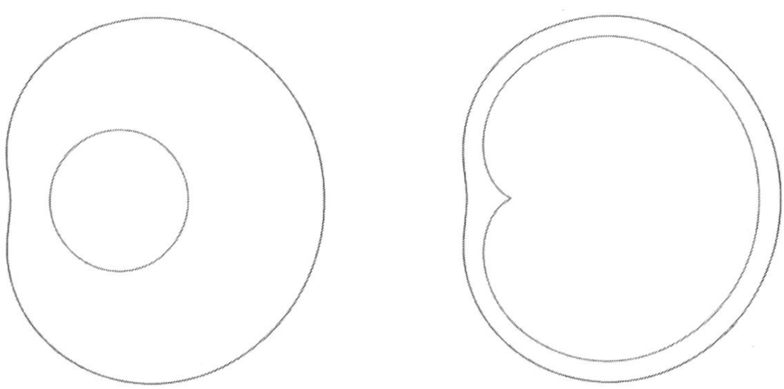

Figure 2 Evolution of a limaçon-shaped fluid blob with $a(0)$ = 1, $b(0)$ = .31 for β = 1. On the *left* the initial limaçon shrinks toward a circular blob shape as $t\to\infty$. This is the case when there is a point source of strength Q/π = .195 at the origin of the circle. On the *right* is the case when there is no point source present (Q = 0) and the blob begins to shrink forming a cusp singularity in finite time $t \approx .265$

3.4 A finger solution

Let

$$z = d(t) - \log \zeta + 2(1 - \lambda(t)) \log \left(\frac{1 + a(t)\zeta}{2} \right). \tag{19}$$

where a, d and λ are real time-dependent parameters, be a map from the interior of the unit ζ disk to a finger of length d occupying the fraction λ of a channel of width 2π. For the map to be well-defined it is required that $|a(t)| < 1$. The fluid speed as $\mathrm{Re}(z) \to \infty$ is unity.

The Schwarz function of the finger boundary is

$$g = d(t) + (2\lambda - 1) \log \zeta + 2(1 - \lambda) \log \left(\frac{\zeta + a}{2} \right). \tag{20}$$

which, as $z \to \infty$, has behavior

$$g \to (1 - 2\lambda)z + 2\lambda d - 4\lambda(1 - \lambda) \log 2 + 2(1 - \lambda) \log a + \mathcal{O}(z^{-1}) \tag{21}$$

The Schwarz function (20) has a logarithmic singularity at $\zeta = -a$ and because there are no point sources or sinks in the fluid, there are no singularities of the type

$(z - z(-a))^{-1}$ in $\partial_z w$, and hence (3) implies $z(-a) = const$ or

$$d - \log |a| + 2(1 - \lambda) \log \left(\frac{1 - a^2}{2} \right) = const \tag{22}$$

As $z \to \infty$, $\partial_z w = 1$ and (21) combined with (3) yields, considering terms of $O(z)$,

$$\frac{d\lambda}{dt} + \beta\lambda = \beta/2, \tag{23}$$

and considering $O(1)$ terms

$$\left(\frac{d}{dt}+\beta\right)\left[2\lambda d-4\lambda(1-\lambda)\log 2+2(1-\lambda)\log|a|\right]=2 \tag{24}$$

An immediate feature implied by (23) is that when$\beta > 0$ a steady state is reached with $\lambda = 1/2$. That is, the Poisson growth term selects the finger width to be precisely half the width of the channel. This is consistent with observations which show that $\lambda = 1/2$ is the preferred finger width in the Laplacian growth, or Hele-Shaw, problem where it is known as a Saffman–Taylor finger.

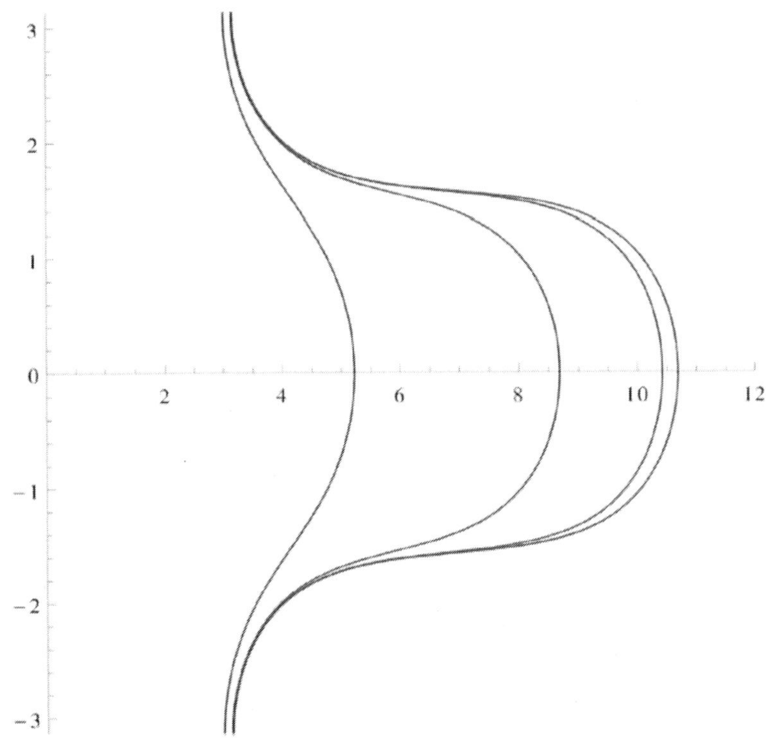

Figure 3 Evolution of a Poisson growth Saffman–Taylor finger, with $\beta = 0.2$, in a channel of width 2π. The finger propagates to the right with the interface shown at times 5, 10, 20 and 40, and the initial values of the parameters are $d(0) = 0$, $\lambda(0) = 0.45$ and $a(0) = 0.01$

Much theoretical effort has been made in seeking to explain selection of the half-width finger in Laplacian growth. This effort has either focussed on surface tension in providing the selection mechanism e.g. [17,23], or, more recently, through the nonlinear stability (without surface tension) of the time-evolving fingers [16]. The result here is consistent with that of [16] in that the Poisson term provides a perturbation to the evolving finger, permitting the half-width finger to be selected. Here, for large times, (24) implies the steady finger has length $d = 2/\beta - \log(a/2)$ where $a \to 1$ provided $\beta \ll 2$. A plot of an evolving finger given by (19) with $\beta = 0.2$ is shown in Fig. 3, where the time-dependent parameters have been found by solving ODEs (22), (23) and (24) numerically. If $\beta < 0$ the finger solution breaks down in finite time as it either becomes an infinitely thin finger when $\lambda(0) < 1/2$, or occupies the full width of the channel if $\lambda(0) > 1/2$.

3.5 Application to the geometry of valleys

The case when $\beta > 0$ is of possible relevance to the geometry of valleys cut by groundwater seepage [18]. The flow of groundwater in sandy soil is governed by

$$\nabla^2 h^2 = -\frac{2P}{K}, \tag{25}$$

where $h(x, y)$ be the elevation of the water table above some impermeable layer, $K > 0$ is the constant sand conductivity and $P > 0$ is the rate of precipitation (assumed constant over a timescale of many tens of years). Assuming the valley geometry is related to the local water table depth [18], its shape is governed by a Poisson growth problem (1) with $Q = 0$ and $\beta > 0$ proportional to P/K. As discussed in Sect. 3.4 this means that an individual valley in a periodic array tends to a steady shape with, for sufficiently small P, $a \approx 1$ and $\lambda = 1/2$, so that, from (19), the valley boundary is given by $x = d + \log(\cos y)$ i.e. a Saffman–Taylor finger. This is precisely the same boundary derived by [18] who used arguments based on field measurements which relate the flux of groundwater to the curvature of the valley finding excellent agreement between this theoretically predicted valley shape and field observations. While this close comparison is appealing, it is not clear that the assumption that the valley geometry is directly related to the local water table depth holds. In fact it seems likely that the general valley geometry problem is more complicated with the shape of the valley being determined by both a Poisson growth problem for the groundwater depth outside the valley itself, and a Laplacian growth problem inside the valley governing erosion of the valley walls [19]. This is an area of current investigation.

4 LINEAR VARIATION IN THE MASS FORCING

Consider the case in which the governing equation inside a finite fluid blob is

$$\nabla^2\phi = 8\gamma x, \tag{26}$$

where $\gamma > 0$ is a constant. This might occur, for example, when the 'evaporation' rate varies linearly in a particular direction and is such that when x changes sign the forcing goes from mass loss to mass injection. This may occur, for example, in a Hele-Shaw cell when the upper plate is tilted about the axis $x = 0$, although in this scenario the LHS is no longer the Laplacian operator–see [6].

Proceeding as in Sect. 2, let

$$\phi = \psi + \frac{\gamma}{2} z\bar{z}(z+\bar{z}), \tag{27}$$

where ψ is harmonic and is the real part of an associated complex analytic potential w. Thus, the complex velocity is

$$\bar{v} = \partial_z w + \gamma(2z\bar{z} + \bar{z}^2) \tag{28}$$

The boundary condition $2\bar{v} = \partial_t g$ and (28) combine to give the following Schwarz function relation valid on $\partial\Omega$

$$\partial_t g - 4\gamma z g - 2\gamma g^2 = 2\partial_z w, \tag{29}$$

which, again, by analytic continuation can be extended throughout Ω.

An exact solution is now described for (29): consider a circular blob of radius $a(t)$ centred on the Re(z)-axis at $z = b(t)$ with no hydrodynamic sources. The Schwarz function for the boundary of this blob is

$$g = b + \frac{a^2}{z-b} \tag{30}$$

Taking the time derivative of (30) and substituting into (29) and equating coefficients of the singular terms of $O(z-b)^{-1}$ and $O(z-b)^{-2}$ gives the following pair of coupled ODEs for $a(t)$ and $b(t)$

$$\frac{da}{dt} = 4\gamma ab,$$
$$\frac{db}{dt} = 2\gamma a^2. \tag{31}$$

By inspection (31) represents a circular blob translating in the positive direction. The radius of the blob decreases if it is centred in the region $x < 0$ and increases if $x > 0$. An exact solution to (31) is obtained by differentiating the equation for db/dt in (31) with respect to time and then substituting for ada/dt using the other equation of (31) to find an differential equation in b alone which is, after integrating once,

$$\frac{db}{dt} = 4\gamma b^2 + C, \tag{32}$$

where C is constant. Once db/dt is known the area of the circular blob πa^2 can be found from (31) i.e. $a^2 = db/dt/2\gamma$. At $t = 0$ let $a = a0 > 0$ and $b = b0$.

If $b_0^2 > a_0^2/2$ then (32) gives

$$b = -d_0 \coth\left[4d_0\gamma t - \coth^{-1}(b_0/d_0)\right], \tag{33}$$

where $d_0 = \sqrt{b_0^2 - a_0^2/2}$. The solution (33) behaves differently according to the sign of b_0: (i) For $b_0 > 0$ (33) implies $b(t) \to \infty$ in the finite time $t^* = \coth^{-1}(b_0/d_0)/4_\gamma d_0$. In this case the problem is ill-posed with the blob propagating infinite distance and becoming infinitely large at $t = t^*$. (ii) For $b_0 < 0$, as $t \to \infty$, $b \to -d_0$ monotonically (i.e. b approaches $-d_0$ smoothly from below) and $a(t)^2 = 2d_0^2 \operatorname{cosech}^2\left[4d_0\gamma t - \coth^{-1}(b_0/d_0)\right] \to 0$ as $t \to \infty$. This corresponds to a diminishing circular blob which 'evaporates' completely as it approaches $x = -d_0$ as $t \to \infty$.

If $b_0^2 > a_0^2/2$ then the solution to (32) is

$$b = e_0 \tan\left[4e_0 d_0\gamma t + \tan^{-1}(b_0/e_0)\right], \tag{34}$$

Where $e_0 = \sqrt{a_0^2/2 - b_0^2}$. In this case if $b0 < 0$ then the blob initially shrinks while propagating toward positive x, before crossing into the region $x > 0$ where it begins to grow eventually becoming unbounded in size in finite time $t^{**} = [\pi/2 - \tan^{-1}(b_0/e_0)]/(4_\gamma e_0 d_0)$. An example showing the behaviour of $a(t)$ and $b(t)$ of a circular blob first decreasing in size and then increasing once its centre crosses $x = 0$ is shown in Fig. 4.

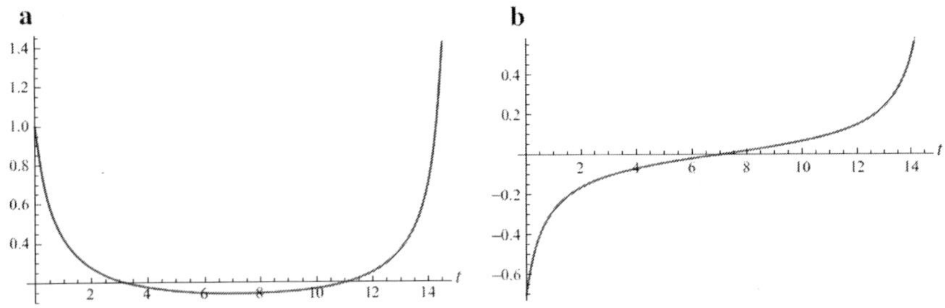

Figure 4 Behaviour of the blob radius $a(t)$ and centre $x = b(t)$ subject to a linearly varying mass forcing with initial conditions $a(0) = 1$ and $b(0) = -0.7$. Here $\gamma = 1/2$

5 CONCLUSIONS

The Poisson growth problem has been formulated in terms of the Schwarz function of the free boundary for the cases when the right hand side is either constant or varies linearly in one space direction. Simple examples of exact solutions have been derived. Although not presented here, it is straightforward to find the Schwarz function equation when the right hand side of Poisson's equation depends only on the radial coordinate $r^2 = z\bar{z} = zg$ on $\partial\Omega$, in which case a trivial solution is a non-translating circular blob with time-dependent area. It is of interest to explore the possibility of formulating Schwarz function equations and finding explicit solutions for other forms of functional dependence of the right hand side of Poisson's equation.

REFERENCES

1. Abanov, A., Mineev-Weinstein, M., Zabrodin, A.: Self-similarity in Laplacian growth. Physica D **235**, 62–71 (2007)
2. Agam, O.: Viscous fingering in volatile thin films. Phys. Rev. E **79**, 021603 (2009)
3. Crowdy, D., Kang, H.: Squeeze flow of multiply-connected fluid domains in a Hele-Shaw cell. Nonlinear Sci. **11**, 279–304 (2001)
4. Cummings, L.J., Howison, S., King, J.: Two-dimensional Stokes and Hele-Shaw flows with free surfaces. Eur. J. Appl. . **10**, 635–680 (1999)
5. Davis, P.J.: The Schwarz function and its applications. In: The Carus ematical Monographs No. 17. The ematical Association of America (1974)

6. Dias, E.O., Miranda, J.A.: Taper-induced control of viscous fingering in variable-gap Hele-Shaw flows. Phys. Rev. E **87**, 053015 (2013)
7. Entov, V.M., Etingof, P.I., Kleinbock, D.Y.: On nonlinear interface dynamics in Hele-Shaw flows. Eur. J. Appl. . **6**, 399–420 (1995)
8. Galin, L.A.: Unsteady filtration with a free surface. Dokl. Akad. Nauk SSSR **47**, 246–249 (1945)
9. Gustafsson, B., Vasil'ev, A.: Conformal and Potential Analysis in Hele-Shaw Cells. Birkhauser, Basel (2006)
10. Khavinson, D., Mineev-Weinstein, M., Putinar, M.: Planar elliptic growth. Complex. Anal. Operat. Theory **3**, 425–451 (2009)
11. Lacey, A.A.: Bounds on solutions of one-phase Stefan problems. Eur. J. Appl. . **6**, 509–516 (1995)
12. Lindner, A., Derks, D., Shelley, M.J.: Stretch flow of thin layers of Newtonian liquids: fingering patterns and lifting forces. Phys. Fluids **17**, 072107 (2005)
13. Loutsenko, I., Yermolayeva, O.: Non-Laplacian growth. Physica D **235**, 56–61 (2007)
14. Lundberg, E.: Laplacian growth, elliptic growth, and singularities of the Schwarz potential. J. Phys. A . Theor. **44**, 135202 (2011)
15. McDonald, N.R.: Generalized Hele-Shaw flow: a Schwarz function approach. Eur. J. Appl. **22**, 517–532 (2011)
16. Mineev-Weinstein, M.: Selection of the Saffman-Taylor finger width in the absence of surface tension: an exact result. Phys. Rev. Lett. **80**, 2113 (1998)
17. Pelce, P. (ed.): Dynamics of Curved Fronts. Academic Press, New York (1988)
18. Petroff, A.P., Devauchelle, O., Abrams, D.M., Lobkovsky, A.E., Kudrolli, A., Rothman, D.H.: Geometry of valley growth. J. Fluid Mech. **673**, 245–254 (2011)
19. Petroff, A.P., Devauchelle, O., Kudrolli, A., Rothman, D.H.: Four remarks on the growth of channel networks. Comptes Rendus Geoscience **344**, 33–40 (2012)
20. Polubarinova-Kochina, P.Y.: On a problem of the motion of the contour of a petroleum shell. Dokl. Akad. Nauk SSSR **47**, 254–257 (1945)
21. Richardson, S.: Hele-Shaw flows produced by injection of fluid into a narrow channel. J. Fluid Mech. **56**, 609–618 (1972)
22. Shelley, M.J., Tian, F.R., Wlodarski, K.: Hele-Shaw flow and pattern formation in a time-dependent gap. Nonlinearity **10**, 1471–1495 (1997)
23. Vanden-Broeck, J.M.: Fingers in a Hele-Shaw cell with surface tension. Phys. Fluids **26**, 2033 (1983)

CHAPTER 12

Exact Traveling Wave Solutions of Nonlinear PDEs in Mathematical Physics

Jameel F. Alzaidy

Mathematics Department, Faculty of Science, Taif University, Taif, KSA

ABSTRACT

In the present article, we construct the exact traveling wave solutions of nonlinear PDEs in mathematical physics via the variant Boussinesq equations and the coupled KdV equations by using the extended mapping method and auxiliary equation method. This method is more powerful and will be used in further works to establish more entirely new solu-tions for other kinds of nonlinear partial differential equations arising in mathematical physics.

Keywords: Extended Mapping Method; Auxiliary Equation Method; The Variant Boussinesq Equations; The Coupled KdV Equations; Traveling Wave Solutions

1. INTRODUCTION

The effort in finding exact solutions to nonlinear equations is important for the understanding of most nonlinear physical phenomena. For instance, the nonlinear wave phenomena observed in fluid dynamics, plasma and optical fibers are often modeled by the bell shaped sech solutions and the kink shaped tanh solutions. Many effective methods have been presented, such as inverse scattering transform method [1], Bäcklund transformation [2], Darboux transformation [3], Hirota bilinear method [4], variable separation approach [5], various tanh methods [6-9], homogeneous balance method [10], similarity reductions method [11,12], (G'/G)-expansion method [13], the reduction mKdV equation method [14], the tri-function method [15,16], the projective Riccati equation method [17], the Weierstrass elliptic function method [18], the Sine-Cosine method [19,20], the Jacobi elliptic function expansion [21,22], the complex hyperbolic function method [23], the truncated Painleve' expansion [24], the F-expansion method [25], the rank analysis method [26], the ansatz method [27,28], the exp-function expansion method [29], the sub-ODE method [30], and so on.

The main objective of this paper is used the extended mapping method and auxiliary equation method to construct the exact solutions for nonlinear evolution equations in the mathematical physics via the variant Boussinesq equations and the coupled KdV equations.

2. DESCRIPTION OF THE EXTENDED MAPPING METHOD

Suppose we have the following nonlinear PDE:

$$F\left(u,u_t,u_x,u_{tt},u_{xx},u_{xt},\cdots\right)=0, \tag{1}$$

where $u=u(x,t)$ is an unknown function, F is a polynomial in $u=u(x,t)$ and its various partial derivatives in which the highest order derivatives and nonlinear terms are involved. In the following we give the main steps of a deformation method:

Step 1. The traveling wave variable

$$u\left(x,t\right)=u\left(\xi\right),\ \xi=k\left(x-\omega t\right), \tag{2}$$

where k and ω are the wave number and the wave speed, respectively. Under the transformation (2), Equation (1) becomes an ordinary differential equation (ODE) as

$$P\left(u,-k\omega u',ku',k^2\omega^2u'',k^2u'',\cdots\right)=0, \tag{3}$$

Where $' = \frac{d}{d\xi}$

Step 2. Suppose that the solution Equation (3) has the following form:

$$u(\xi) = a_0 + \sum_{i=1}^{n} \left(a_i f^i(\xi) + b_i f^{-i}(\xi) \right)$$
$$+ \sum_{i=2}^{n} c_i f^{i-2}(\xi) f'(\xi)$$
$$+ \sum_{i=-1}^{n} d_i f^i(\xi) f'(\xi), \qquad (4)$$

where a_0, a_i, b_i, c_i and d_i are constants to be deter-mined later, while , $f(\xi)$ satisfies the auxiliary ordinary differential equation:

$$\left[f'(\xi) \right]^2 = p f^4(\xi) + q f^2(\xi) + r, \qquad (5)$$

where *p*, *q* and *r* are arbitrary constants.

Step 3. The positive integer "*n*" can be determined by considering the homogeneous balance between the high-est derivative term and the nonlinear terms appearing in Equation (3). Therefore, we can get the value of *n* in Equation (4).

Step 4. Substituting Equation (4) into Equation (3) with the condition (5), we obtain polynomial in $f^i(\xi)[f'(\xi)]^j$, $(i = \cdots, -2, -1, 0, 1, 2, \cdots; j = 0, 1)$. Set-ting each coefficient of this polynomial to be zero, yields a set of algebraic equations for a_0, a_i, b_i, c_i, d_i and k expressed by p, q, r.

Step 6. Since the general solutions of auxiliary Equa-tion (5) have been well known for us (see Appendix A), then substituting the obtained coefficients and the general solution of Equation (5) into Equation (4), we have the travelling wave solutions of the nonlinear PDE (1).

3. APPLICATIONS OF THE METHOD

In this section, we apply the extended mapping method to construct the exact solutions for the variant Boussi-nesq equations and the coupled KdV equations. Which are very important nonlinear evolution equations in mathematical physics and have been paid attention by many researchers.

3.1. Example 1. The Variant Boussinesq Equations

We start with the variant Boussinesq equations [31] in the following form:

$$v_t = \alpha u_{xxx} - u_x - (uv)_x, \tag{6}$$

$$u_t = 3\alpha u_{xxt} - v_x - uu_x, \tag{7}$$

where α is a constant. As models for water waves, u is the velocity and v is the total depth. Wang [32] obtained their solitary wave solutions by using homogeneous bal-ance method while Fan *et al.* [33] got a series of new traveling wave solutions of this system by using an alge-braic method. The traveling wave variables below

$$u(x,t) = u(\xi),\ v(x,t) = v(\xi),\ \xi = k(x - \omega t), \tag{8}$$

permit us converting the Equations (6) and (7) into ODEs for $u(x,t) = u(\xi)$ and $v(x,t) = v(\xi)$ as follows:

$$-\omega v' + u' + (uv)' - \alpha k^2 u''' = 0, \tag{9}$$

and

$$-\omega u' + uu' + v' + 3\alpha k^2 \omega u'' = 0, \tag{10}$$

On integrating Equations (9) and (10) with respect to ξ once, we get

$$L_1 - \omega v + u + uv - \alpha k^2 u'' = 0, \tag{11}$$

and

$$L_2 - \omega u + \frac{u^2}{2} + v + 3\alpha k^2 \omega u'' = 0, \tag{12}$$

where $L1$ and $L2$ are integration constants.

Suppose that the solutions of Equations (11) and (12) can be expressed by

$$u(\xi) = a_0 + \sum_{i=1}^{n}\left(a_i f^i(\xi) + b_i f^{-i}(\xi)\right)$$
$$+ \sum_{i=2}^{n} c_i f^{i-2}(\xi) f'(\xi)$$
$$+ \sum_{i=-1}^{-n} d_i f^i(\xi) f'(\xi), \tag{13}$$

$$v(\xi) = \alpha_0 + \sum_{i=1}^{m}\left(\alpha_i f^i(\xi) + \beta_i f^{-i}(\xi)\right)$$
$$+ \sum_{i=2}^{m} \gamma_i f^{i-2}(\xi) f'(\xi)$$
$$+ \sum_{i=-1}^{-m} e_i f^i(\xi) f'(\xi), \tag{14}$$

where a_0, a_i, b_i, c_i, d_i, α_0, α_i, β_i, γ_i, and e_i are constants to be determined later.

Considering the homogeneous balance between the highest order derivatives and the nonlinear terms in Equations (11) and (12), we get $n = m = 2$, hence the exact solutions of Equations (11) and (12) can be rewritten as:

$$u(\xi) = a_0 + a_1 f(\xi) + b_1 \frac{1}{f(\xi)} + a_2 f^2(\xi) + b_2 \frac{1}{f^2(\xi)}$$
$$+ c_2 f'(\xi) + d_1 \frac{f'(\xi)}{f(\xi)} + d_2 \frac{f'(\xi)}{f^2(\xi)}, \tag{15}$$

and

$$v(\xi) = \alpha_0 + \alpha_1 f(\xi) + \beta_1 \frac{1}{f(\xi)} + \alpha_2 f^2(\xi) + \beta_2 \frac{1}{f^2(\xi)}$$
$$+ \gamma_2 f'(\xi) + e_1 \frac{f'(\xi)}{f(\xi)} + e_2 \frac{f'(\xi)}{f^2(\xi)}. \tag{16}$$

where a_0, a_1, a_2, b_1, b_2, c_2, d_1, d_2, α_0, α_1, α_2, β_1, β_2, γ_2, e_1 and e_2 are constants to be determined later. Substitut-ing Equations (15) and (16) with the condition (5) into Equations (11) and (12) and collecting all terms with the same power of $f^i(\xi)[f'(\xi)]^j$,

(i=...,-2,-1,0,1,2,...;j=0,1). Setting each coefficients of this polynomial to be zero, we get a system of alge-braic equations which can be solved by Maple or Mathematica. Thus we obtain the two sets of solutions as:

Case 1:

$$a_0 = a_1 = b_1 = b_2 = d_1 = d_2 = \alpha_1 = \beta_1 = \beta_2 = e_1 = e_2 = 0,$$
$$a_2 = -\frac{p\left(1+6p\omega^2\right)}{q\omega},\ c_2 = -\frac{\sqrt{p}\left(1+6\omega^2\right)}{q\omega},$$
$$\alpha_0 = \frac{1}{18}\left(-15+\frac{1}{\omega^2}\right),\ \alpha_2 = \frac{p\left(1+6\omega^2\right)}{6q\omega^2},$$
$$\gamma_2 = \frac{\sqrt{p}\left(1+6\omega^2\right)}{6q\omega^2},\ L_1 = \frac{q^2\omega^2\left(1-15\omega^2\right)+pr\left(1+6\omega^2\right)^2}{18q^2\omega^3},$$
$$L_2 = \frac{-3pr\left(1+6\omega^2\right)^2+q^2\left(15\omega^2-1\right)}{18q^2\omega^2},\ k = \pm\frac{1}{3\omega}\sqrt{\frac{1+6\omega^2}{2q\alpha}}. \tag{17}$$

Case 2:

$$a_0 = a_1 = b_1 = a_2 = c_2 = d_1 = \alpha_1 = \beta_1 = \alpha_2 = e_1 = 0,$$
$$b_2 = -\frac{r\left(1+6r\omega^2\right)}{q\omega},\ d_2 = \frac{\sqrt{r}\left(1+6\omega^2\right)}{q\omega},$$
$$\alpha_0 = \frac{1}{18}\left(-15+\frac{1}{\omega^2}\right),\ \beta_2 = \frac{r\left(1+6\omega^2\right)}{6q\omega^2},$$
$$e_2 = -\frac{\sqrt{r}\left(1+6\omega^2\right)}{6q\omega^2},\ L_1 = \frac{q^2\omega^2\left(1-15\omega^2\right)+pr\left(1+6\omega^2\right)^2}{18q^2\omega^3},$$
$$L_2 = \frac{-3pr\left(1+6\omega^2\right)^2+q^2\left(15\omega^2-1\right)}{18q^2\omega^2},\ k = \pm\frac{1}{3\omega}\sqrt{\frac{1+6\omega^2}{2q\alpha}}. \tag{18}$$

Note that, there are other cases which are omitted here. Since the solutions obtained here are so many, we just list some of the exact solutions corresponding to Case 2 to illustrate the effectiveness of the extended mapping method.

Substituting (18) into Equation (15) and Equation (16) yields

$$u(\xi)=-\frac{r\left(1+6r\omega^{2}\right)}{q\omega}\left(\frac{1}{f^{2}(\xi)}\right)+\frac{\sqrt{r}\left(1+6\omega^{2}\right)}{q\omega}\left(\frac{f'(\xi)}{f^{2}(\xi)}\right), \tag{19}$$

and

$$v(\xi)=\frac{1}{18}\left(-15+\frac{1}{\omega^{2}}\right)+\frac{r\left(1+6\omega^{2}\right)}{6q\omega^{2}}\left(\frac{1}{f^{2}(\xi)}\right)-\frac{\sqrt{r}\left(1+6\omega^{2}\right)}{6q\omega^{2}}\left(\frac{f'(\xi)}{f^{2}(\xi)}\right), \tag{20}$$

where

$$\xi=\pm\frac{1}{3\omega}\sqrt{\frac{1+6\omega^{2}}{2q\alpha}}(x-\omega t) \tag{21}$$

According to the Appendix A, we have the following families of exact solutions:

Family 1. If $r=1$, $q=-(1+m^2)$, $p=m^2$, $f(\xi)=sn(\xi)$, then we get

$$u(\xi)=\frac{\left(1+6\omega^{2}\right)}{\left(1+m^{2}\right)\omega}\left[ns^{2}(\xi)-cs(\xi)ds(\xi)\right], \tag{22}$$

and

$$v(\xi)=\frac{1}{18}\left(-15+\frac{1}{\omega^2}\right)$$
$$-\frac{\left(1+6\omega^2\right)}{6\omega^2\left(1+m^2\right)}\left[ns^2(\xi)-cs(\xi)ds(\xi)\right], \tag{23}$$

where

$$\xi=\pm\frac{i}{3\omega}\sqrt{\frac{1+6\omega^2}{2\alpha\left(1+m^2\right)}}(x-\omega t) \tag{24}$$

Family 2. If $r=1-m^2$, $q=2m^2-1$, $p=-m^2$, $f(\xi)=cn(\xi)$, then we get

$$u(\xi)=\frac{\left(m^2-1\right)\left[1+6\omega^2\left(1-m^2\right)\right]}{\left(2m^2-1\right)\omega}nc^2(\xi)$$
$$-\frac{\sqrt{1-m^2}\left(1+6\omega^2\right)}{\left(2m^2-1\right)\omega}dc(\xi)sc(\xi), \tag{25}$$

and

$$v(\xi)=\frac{1}{18}\left(-15+\frac{1}{\omega^2}\right)+\frac{\left(1-m^2\right)\left(1+6\omega^2\right)}{6\omega^2\left(2m^2-1\right)}nc^2(\xi)$$
$$+\frac{\sqrt{1-m^2}\left(1+6\omega^2\right)}{6\omega^2\left(2m^2-1\right)}dc(\xi)sc(\xi), \tag{26}$$

where

$$\xi=\pm\frac{1}{3\omega}\sqrt{\frac{1+6\omega^2}{2\alpha\left(2m^2-1\right)}}(x-\omega t) \tag{27}$$

Family 3. If $r=m^2-1$, $q=2-m^2$, $p=-1$, $f(\xi)=dn(\xi)$, then we get

$$u(\xi)=\frac{(1-m^2)\left[1+6\omega^2(m^2-1)\right]}{(2-m^2)\omega}nd^2(\xi)-\frac{m^2\sqrt{m^2-1}(1+6\omega^2)}{(2-m^2)\omega}sd(\xi)cd(\xi), \tag{28}$$

and

$$v(\xi)=\frac{1}{18}\left(-15+\frac{1}{\omega^2}\right)+\frac{(m^2-1)(1+6\omega^2)}{6\omega^2(2-m^2)}nd^2(\xi)+\frac{m^2\sqrt{m^2-1}(1+6\omega^2)}{6\omega^2(2-m^2)}sd(\xi)cd(\xi), \tag{29}$$

where

$$\xi=\pm\frac{1}{3\omega}\sqrt{\frac{1+6\omega^2}{2\alpha(2-m^2)}}(x-\omega t) \tag{30}$$

Family 4. If $r=m^2$, $q=-(1+m^2)$, $p=-1$, $f(\xi)=dc(\xi)$, then we get

$$u(\xi)=\frac{m^2(1+6r\omega^2)}{\omega(1+m^2)}cd^2(\xi)-\frac{m(1-m^2)(1+6\omega^2)}{\omega(1+m^2)}sd(\xi)nd(\xi), \tag{31}$$

and

$$v(\xi)=\frac{1}{18}\left(-15+\frac{1}{\omega^2}\right)-\frac{m^2\left(1+6\omega^2\right)}{6\omega^2\left(1+m^2\right)}cd^2(\xi)$$
$$+\frac{m\left(1-m^2\right)\left(1+6\omega^2\right)}{6\omega^2\left(1+m^2\right)}sd(\xi)nd(\xi), \tag{32}$$

where

$$\xi=\pm\frac{i}{3\omega}\sqrt{\frac{1+6\omega^2}{2\alpha\left(1+m^2\right)}}(x-\omega t) \tag{33}$$

Family 5. If r=1, q=2-m^2, p=1-m^2, $f(\xi)$=$sc(\xi)$, then we get

$$u(\xi)=-\frac{\left(1+6\omega^2\right)}{\omega\left(2-m^2\right)}\left[cs^2(\xi)-ds(\xi)ns(\xi)\right], \tag{34}$$

and

$$v(\xi)=\frac{1}{18}\left(-15+\frac{1}{\omega^2}\right)$$
$$+\frac{\left(1+6\omega^2\right)}{6\omega^2\left(2-m^2\right)}\left[cs^2(\xi)-ds(\xi)ns(\xi)\right], \tag{35}$$

where

$$\xi=\pm\frac{1}{3\omega}\sqrt{\frac{1+6\omega^2}{2\alpha\left(2-m^2\right)}}(x-\omega t) \tag{36}$$

Family 6. If $r=\frac{1}{4},\ q=\frac{1}{2}\left(1-2m^2\right),\ p=\frac{1}{4}$, $f(\xi)$=$ns(\xi)$ ±$cs(\xi)$, then we get

$$u(\xi) = -\frac{(2+3\omega^2)}{4\omega(1-2m^2)[ns(\xi) \pm cs(\xi)]^2} - \frac{(1+6\omega^2)dn(\xi)}{\omega(1-2m^2)[cn(\xi) \pm 1]}, \tag{37}$$

and

$$v(\xi) = \frac{1}{18}\left(-15 + \frac{1}{\omega^2}\right) + \frac{(1+6\omega^2)}{12\omega^2(1-2m^2)[ns(\xi) \pm cs(\xi)]^2} + \frac{(1+6\omega^2)dn(\xi)}{6\omega^2(1-2m^2)[cn(\xi) \pm 1]}, \tag{38}$$

where

$$\xi = \pm\frac{1}{3\omega}\sqrt{\frac{1+6\omega^2}{\alpha(1-2m^2)}}(x-\omega t) \tag{39}$$

Similarly, we can write down the other families of exact solutions of Equations (6) and (7) which are omitted for convenience.

3.2. Example 2. The Coupled KdV Equations

In this subsection, we consider the coupled KdV Equations [34]:

$$u_t = 6vv_x - 6\alpha uu_x - \alpha u_{xxx}, \tag{40}$$

$$v_t = -3\alpha uv_x - \alpha v_{xxx}, \tag{41}$$

where α is a constant. The traveling wave variables (8) permit us converting the Equations (40) and (41) into ODEs in the forms:

$$\omega u' + 6vv' - 6\alpha uu' - \alpha k^2 u''' = 0, \tag{42}$$

$$\omega v' - 3\alpha uv' - \alpha k^2 v''' = 0 \tag{43}$$

Suppose that the solutions of Equations (42) and (43) can be expressed by

$$\begin{aligned} u(\xi) &= a_0 + \sum_{i=1}^{n}\left(a_i f^i(\xi) + b_i f^{-i}(\xi)\right) \\ &+ \sum_{i=2}^{n} c_i f^{i-2}(\xi) f'(\xi) \\ &+ \sum_{i=-1}^{-n} d_i f^i(\xi) f'(\xi), \end{aligned} \tag{44}$$

$$\begin{aligned} v(\xi) &= \alpha_0 + \sum_{i=1}^{m}\left(\alpha_i f^i(\xi) + \beta_i f^{-i}(\xi)\right) \\ &+ \sum_{i=2}^{m} \gamma_i f^{i-2}(\xi) f'(\xi) \\ &+ \sum_{i=-1}^{-m} e_i f^i(\xi) f'(\xi), \end{aligned} \tag{45}$$

where a_0, a_1, a_2, b_1, b_2, c_2, d_1, d_2, α_0, α_1, α_2, β_1, β_2, γ_2, e_1 and e_2 are constants to be determined later.

Considering the homogeneous balance between the highest order derivatives and the nonlinear terms in Equation (63) and Equation (64), we get n = m = 2, so Equations (44) and (45) can be rewritten as:

$$\begin{aligned} u(\xi) &= a_0 + a_1 f(\xi) + b_1 \frac{1}{f(\xi)} + a_2 f^2(\xi) + b_2 \frac{1}{f^2(\xi)} \\ &+ c_2 f'(\xi) + d_1 \frac{f'(\xi)}{f(\xi)} + d_2 \frac{f'(\xi)}{f^2(\xi)}, \end{aligned} \tag{46}$$

$$v(\xi)=\alpha_0+\alpha_1 f(\xi)+\beta_1\frac{1}{f(\xi)}+\alpha_2 f^2(\xi)+\beta_2\frac{1}{f^2(\xi)}$$
$$+\gamma_2 f'(\xi)+e_1\frac{f'(\xi)}{f(\xi)}+e_2\frac{f'(\xi)}{f^2(\xi)}, \tag{47}$$

where a_0, a_1, a_2, b_1, b_2, c_2, d_1, d_2, α_0, α_1, α_2, β_1, β_2, γ_2, e_1 and e_2 are constants to be determined later. Substituting Equation (46) and Equation (47) with the condition (5) into Equation (42) and Equation (43) and collecting all terms with the same power of, $f^i(\xi)[f'(\xi)]^j$, ($i=\ldots,-2,-1,0,1,2,\ldots;j=0,1$). Setting each coefficients of this polynomial to be zero, we get a system of algebraic equations which can be solved by Maple or Mathematica to get the following sets of solutions:

Case 1:

$$a_0=a_1=b_1=c_2=d_1=d_2=\alpha_0=\alpha_1$$
$$=\beta_1=\gamma_2=e_1=e_2=0,$$
$$a_2=-\frac{\omega p}{q\alpha},\ b_2=-\frac{\omega r}{q\alpha},\ \alpha_2=\frac{\omega p}{q\sqrt{2\alpha}},$$
$$\beta_2=\frac{\omega r}{q\sqrt{2\alpha}},\ k=\pm\frac{1}{2}\sqrt{\frac{\omega}{\alpha q}}. \tag{48}$$

Case 2:

$$a_0=a_1=b_1=b_2=d_1=d_2=\alpha_0=\alpha_1$$
$$=\beta_1=\beta_2=e_1=e_2=0,$$
$$a_2=-\frac{2\omega p}{q\alpha},\ c_2=\frac{2\omega\sqrt{p}}{q\alpha},\ \alpha_2=-\frac{\omega p}{q}\sqrt{\frac{2}{\alpha}},$$
$$\gamma_2=\frac{\omega}{q}\sqrt{\frac{2p}{\alpha}},\ k=\pm\sqrt{\frac{\omega}{q\alpha}}. \tag{49}$$

Note that, there are other cases which are omitted here. Since the solutions obtained here are so many, we just list some of the exact solutions corresponding to Case 2 to illustrate the effectiveness of the extended mapping method.

Substituting (49) into Equation (46) and Equation (47) yields

$$u(\xi)=-\frac{2\omega p}{q\alpha}f^2(\xi)+\frac{2\omega\sqrt{p}}{q\alpha}f'(\xi), \tag{50}$$

and

$$v(\xi)=-\frac{\omega p}{q}\sqrt{\frac{2}{\alpha}}f^2(\xi)+\frac{\omega}{q}\sqrt{\frac{2p}{\alpha}}f'(\xi), \tag{51}$$

where

$$\xi=\pm\sqrt{\frac{\omega}{q\alpha}}(x-\omega t) \tag{52}$$

According to the Appendix A, we have the following families of exact solutions:

Family 1. If $r=1$, $q=2m^2-1$, $p=m^2(m^2-1)$, $f(\xi)=sd(\xi)$, then we get

$$u(\xi)=\frac{2\omega m^2(1-m^2)}{\alpha(2m^2-1)}sd^2(\xi)+\frac{2\omega m\sqrt{m^2-1}}{\alpha(2m^2-1)}nd(\xi)cd(\xi), \tag{53}$$

and

$$v(\xi)=\frac{\omega m^2(1-m^2)}{(2m^2-1)}\sqrt{\frac{2}{\alpha}}sd^2(\xi)+\frac{\omega m}{(2m^2-1)}\sqrt{\frac{2(m^2-1)}{\alpha}}nd(\xi)cd(\xi), \tag{54}$$

where

$$\xi=\pm\sqrt{\frac{\omega}{\alpha\left(2m^2-1\right)}}(x-\omega t) \tag{55}$$

Family 2. If $r=m^2(m^2-1)$, $q=2m^2-1$, $p=1$, $f(\xi)=ds(\xi)$, then we get

$$u(\xi)=-\frac{2\omega}{\alpha\left(2m^2-1\right)}\left[ds^2(\xi)+cs(\xi)ns(\xi)\right], \tag{56}$$

and

$$v(\xi)=-\frac{\omega}{2m^2-1}\sqrt{\frac{2}{\alpha}}\left[ds^2(\xi)+cs(\xi)ns(\xi)\right], \tag{57}$$

where

$$\xi=\pm\sqrt{\frac{\omega}{\alpha\left(2m^2-1\right)}}(x-\omega t) \tag{58}$$

Family 3. If $r=\frac{m^2}{4},\ q=\frac{1}{2}\left(m^2-2\right),\ p=\frac{m^2}{4}$, $f(\xi)=sn(\xi)\pm icn(\xi)$, then we get

$$u(\xi)=-\frac{\omega m^2}{\alpha\left(m^2-2\right)}\left[sn(\xi)\pm icn(\xi)\right]^2+\frac{2\omega m}{\alpha\left(m^2-2\right)}\left[cn(\xi)dn(\xi)\mp isn(\xi)dn(\xi)\right], \tag{59}$$

and

$$v(\xi)=-\frac{\omega m^2}{m^2-2}\sqrt{\frac{2}{\alpha}}\left[sn(\xi)\pm icn(\xi)\right]^2 +\frac{\omega m}{m^2-2}\sqrt{\frac{2}{\alpha}}\left[cn(\xi)dn(\xi)\mp isn(\xi)dn(\xi)\right], \tag{60}$$

where

$$\xi=\pm\sqrt{\frac{2\omega}{\alpha\left(m^2-2\right)}}(x-\omega t) \tag{61}$$

Family 4. If $r=1$, $q=-(1+m^2)$, $p=m^2$, $f(\xi)=sn(\xi)$, then we get

$$u(\xi)=\frac{2\omega m^2}{\alpha\left(1+m^2\right)}sn^2(\xi) -\frac{2\omega m}{\alpha\left(1+m^2\right)}cn(\xi)dn(\xi), \tag{62}$$

and

$$v(\xi)=\frac{\omega m^2}{1+m^2}\sqrt{\frac{2}{\alpha}}sn^2(\xi) -\frac{\omega m}{1+m^2}\sqrt{\frac{2}{\alpha}}cn(\xi)dn(\xi), \tag{63}$$

where

$$\xi=\pm i\sqrt{\frac{\omega}{\alpha\left(1+m^2\right)}}(x-\omega t) \tag{64}$$

Family 5. If $r=1-m^2$, $q=2m^2-1$, $p=-m^2$, $f(\xi)=cn(\xi)$, then we get

$$u(\xi)=\frac{2\omega m^2}{\alpha(2m^2-1)}cn^2(\xi)-\frac{2im\omega}{\alpha(2m^2-1)}sn(\xi)dn(\xi), \tag{65}$$

and

$$v(\xi)=\frac{\omega m^2}{2m^2-1}\sqrt{\frac{2}{\alpha}}cn^2(\xi)-\frac{im\omega}{2m^2-1}\sqrt{\frac{2}{\alpha}}sn(\xi)dn(\xi), \tag{66}$$

where

$$\xi=\pm\sqrt{\frac{\omega}{\alpha(2m^2-1)}}(x-\omega t) \tag{67}$$

Family 6. If $r=1-m^2$, $q=2-m^2$, $p=1$, $f(\xi)=cs(\xi)$, then we get

$$u(\xi)=-\frac{2\omega}{\alpha(2-m^2)}\left[cs^2(\xi)+ns(\xi)ds(\xi)\right], \tag{68}$$

and

$$v(\xi)=-\frac{\omega}{2-m^2}\sqrt{\frac{2}{\alpha}}\left[cs^2(\xi)+ns(\xi)ds(\xi)\right], \tag{69}$$

where

$$\xi=\pm\sqrt{\frac{\omega}{\alpha(2-m^2)}}(x-\omega t) \tag{70}$$

4. CONCLUSION AND DISCUSSION

In this article, we have found the exact solutions of the variant Boussinesq equations and the coupled KdV equations by using the extended mapping method and the auxiliary equation method. Also, we conclude according to the Appendix B that our results in terms of Jacobi elliptic functions generate into hyperbolic functions when m ® 1 and generate into trigonometric functions when m ® 0. This method provides a powerful mathematical tool to obtain more general exact solutions of a great many nonlinear PDEs in mathematical physics.

REFERENCES

1. M. J. Ablowitz and P. A. Clarkson, "Soliton, Nonlinear Evolution Equations and Inverse Scattering," Cambridge University Press, Cambridge, 1991. doi:10.1017/ CBO9780 511623998
2. C. H. Gu, et al., "Soliton Theory and Its Application," Zhejiang Science and Technology Press, Zhejiang, 1990.
3. V. B. Matveev and M. A. Salle, "Darboux Transformation and Soliton," Springer, Berlin, 1991.
4. R. Hirota, "The Direct Method in Soliton Theory," Cambridge University Press, Cambridge, 2004. doi:10.1017/CBO9780511543043
5. S. Y. Lou and J. Z. Lu, "Special Solutions from Variable Separation Approach: Davey-Stewartson Equation," Journal of Physics A: Mathematical and General, Vol. 29, No. 14, 1996, pp. 4209-4215. doi:10.1088/0305-4470/29/14/038
6. E. J. Parkes and B. R. Duffy, "Travelling Solitary Wave Solutions to a Compound KdV-Burgers Equation," Physics Letters A, Vol. 229, No. 4, 1997, pp. 217-220. doi:10.1016/S0375-9601(97)00193-X
7. E. Fan, "Extended Tanh-Function Method and Its Applications to Nonlinear Equations," Physics Letters A, Vol. 277, No. 4-5, 2000, pp. 212-218. doi:10.1016/S0375-9601(00)00725-8
8. Z. Y. Yan, "New Explicit Travelling Wave Solutions for Two New Integrable Coupled Nonlinear Evolution Equations," Physics Letters A, Vol. 292, No. 1-2, 2001, pp. 100-106. doi:10.1016/S0375-9601(01)00772-1
9. Y. Chen and Y. Zheng, "Generalized Extended TanhFunction Method to Construct New Explicit Exact Solutions for the Approximate Equations for Long Water Waves," International Journal of Modern Physics C, Vol. 14, No. 5, 2003, pp. 601-611. doi:10.1142/S0129183103004760
10. M. L. Wang, "Application of a Homogeneous Balance Method to Exact Solutions of Nonlinear Equations in Mathematical Physics," Physics Letters A, Vol. 216, No. 1-5, 1996, pp. 67-75. doi:10.1016/0375-9601(96)00283-6
11. G. W. Bluman and S. Kumei, "Symmetries and Differential Equations," Springer-Verlag, New York, 1989.
12. P. J. Olver, "Applications of Lie Groups to Differential Equations," Springer-Verlag, New York, 1986. doi:10.1007/978-1-4684-0274-2
13. E. M. E. Zayed and K. A. Gepreel, "The $\left(\frac{G'}{G}\right)$-Expansion Method for Finding Traveling Wave Solutions of Nonlinear PDEs in Mathematical Physics," Journal of Mathematical Physics, Vol. 50, No. 1, 2009, Article ID: 013502. doi:10.1063/1.3033750

14. Z. Y. Yan, "A Reduction mKdV Method with Symbolic Computation to Constract New Doubly-Periodic Solutions for Nonlinear Wave Equations," International Journal of Modern Physics C, Vol. 14, No. 5, 2003, pp. 661-672. doi:10.1142/S0129183103004814
15. Z. Y. Yan, "The New Tri-Function Method to Multiple Exact Solutions of Nonlinear Wave Equations," Physica Scripta, Vol. 78, No. 3, 2008, Article ID: 035001. doi:10.1088/0031-8949/78/03/035001
16. Z. Y. Yan, "Periodic, Solitary and Rational Wave Solutions of the 3D Extended Quantum Zakharov-Kuznetsov Equation in Dense Quantum Plasmas," Physics Letters A, Vol. 373, No. 29, 2009, pp. 2432-2437. doi:10.1016/j.physleta.2009.04.018
17. D. C. Lu and B. J. Hong, "New Exact Solutions for the (2+1)-Dimensional Generalized Broer-Kaup System," Applied Mathematics and Computation, Vol. 199, No. 2, 2008, pp. 572-580. doi:10.1016/j.amc.2007.10.012

A. V. Porubov, "Periodical Solution to the Nonlinear Dissipative Equation for Surface Waves in a Convecting Liquid," Physics Letters A, Vol. 221, No. 6, 1996, pp. 391-394. doi:10.1016/0375-9601(96)00598-1

18. M. Wazwaz, "The Tanh and Sine-Cosine Method for Compact and Noncompact Solutions of Nonlinear Klein Gordon Equation," Applied Mathematics and Computation, Vol. 167, No. 2, 2005, pp. 1179-1195. doi:10.1016/j.amc.2004.08.006
19. Z. Y. Yan and H. Q. Zhang, "New Explicit Solitary Wave Solutions and Periodic Wave Solutions for WhithamBroer-Kaup Equation in Shallow Water," Physics Letters A, Vol. 285, No. 5-6, 2001, pp. 355-362. doi:10.1016/S0375-9601(01)00376-0
20. D. C. Lu, "Jacobi Elliptic Functions Solutions for Two Variant Boussinesq Equations," Chaos, Solitons and Fractals, Vol. 24, No. 5, 2005, pp. 1373-1385. doi:10.1016/j.chaos.2004.09.085
21. Z. Y. Yan, "Abundant Families of Jacobi Elliptic Functions of the (2+1) Dimensional Integrable Davey-Stawartson-Type Equation via a New Method," Chaos, Solitons and Fractals, Vol. 18, No. 2, 2003, pp. 299-309. doi:10.1016/S0960-0779(02)00653-7
22. C. L. Bai and H. Zhao, "Generalized Method to Construct the Solitonic Solutions to (3+1)-Dimensional Nonlinear Equation," Physics Letters A, Vol. 354, No. 5-6, 2006, pp. 428-436. doi:10.1016/j.physleta.2006.01.084
23. F. Cariello and M. Tabor, "Similarity Reductions from Extended Painleve' Expansions for Nonintegrable Evolution Equations," Physica D: Nonlinear Phenomena, Vol. 53, No. 1, 1991, pp. 59-70. doi:10.1016/0167-2789(91)90164-5
24. M. Wang and X. Li, "Extended F-Expansion and Periodic Wave Solutions for the Generalized Zakharov Equations," Physics Letters A, Vol. 343, No. 1-3, 2005, pp. 48- 54. doi:10.1016/j.physleta.2005.05.085
25. X. Feng, "Exploratory Approach to Explicit Solution of Nonlinear Evolution Equations," International Journal of Theoretical Physics, Vol. 39, 2000, p. 222. doi:10.1023/A:1003615705115
26. J. L. Hu, "Explicit Solutions to Three Nonlinear Physical Models," Physics Letters A, Vol. 287, No. 1-2, 2001, pp. 81-89. doi:10.1016/S0375-9601(01)00461-3
27. J. L. Hu, "A New Method for Finding Exact Traveling Wave Solutions to Nonlinear Partial Differential Equations," Physics Letters A, Vol. 286, No. 2-3, 2001, pp. 175-179. doi:10.1016/S0375-9601(01)00291-2
28. J. H. He and X. H. Wu, "Exp-Function Method for Nonlinear Wave Equations," Chaos, Solitons and Fractals, Vol. 30, No. 3, 2006, pp. 700-708. doi:10.1016/j.chaos.2006.03.020
29. X. Z. Li and M. L. Wang, "A Sub-ODE Method for Finding Exact Solutions of a Generalized KdV-mKdV Equation with Higher Order Nonlinear Terms," Physics Letters A, Vol. 361, No. 1-2, 2007, pp. 115-118. doi:10.1016/j.physleta.2006.09.022
30. E. M. E. Zayed and K. A. Gepreel, "Three Types of TravelingWave Solutions for Nonlinear Evolution Equations Using the $\left(\frac{G'}{G}\right)$-Expansion Method," International Journal of Nonlinear Science, Vol. 7, 2009, p. 501.
31. M. L. Wang, "Solitary Wave Solutions for Variant Boussinesq Equations," Physics Letters A, Vol. 199, No. 3-4, 1995, pp. 169-172. doi:10.1016/0375-9601(95)00092-H

32. E. G. Fan and Y. C. Hon, "A Series of Travelling Wave Solutions for Two Variant Boussinesq Equations in Shallow Water Waves," Chaos, Solitons and Fractals, Vol. 15, No. 3, 2003, pp. 559-566. doi:10.1016/S0960-0779(02)00144-3
33. R. Hirota and J. Satsuma, "Soliton Solutions of Coupled Korteweg-de Vries Equation," Physics Letters A, Vol. 85, No. 8-9, 1981, pp. 407-408. doi:10.1016/0375-9601(81)90423-0
34. K. A. Gepreel, "Exact Solutions for Nonlinear PDEs with the Variable Coefficients in Mathematical Physics," Journal of Information and Computing Science, Vol. 6, No. 1, 2011, pp. 3-14.

CHAPTER 13

Some Mathematical and Physical Remarks on Surreal Numbers

Juan Antonio Nieto

Facultad de Ciencias Fsico-Matemáticas de la Universidad Autónoma de Sinaloa, Culiacán, México

ABSTRACT

We make a number of observations on Conway surreal number theory which may be useful, for further developments, in both mathematics and theoretical physics. In particular, we argue that the concepts of surreal numbers and matroids can be linked. Moreover, we established a relation between the Gonshor approach on surreal numbers and tensors. We also comment about the possibility to connect surreal numbers with supersymmetry. In addition, we comment about possible relation between surreal numbers and fractal theory. Finally, we argue that the surreal structure may provide a different mathematical tool in the understanding of singularities in both high energy physics and gravitation.

Keywords: Surreal Numbers, Supersymmetry, Cosmology

1. INTRODUCTION

Surreal numbers are a fascinating subject in mathematics. Such numbers were invented, or discovered, by the mathematician John Horton Conway in the 70's [1] [2]. Roughly speaking, the key Conways idea is to consider a surreal number in terms of previously created dual sets X_L and X_R . Here, L stands for left and R for right. One of the interesting things is that such numbers contain many well known ordered fields, including integer numbers, the dyadic rationals, the real numbers and hyperreals, among other numerical structures. Moreover, the structure of surreal numbers leads to a system where we can consider the concept of infinite number as naturally and consistently as any "ordinary" numbers.

It turns out that in contrast to the inductive Conway definition of surreal numbers, Gonshor [3] proposed in 1986 another definition which is based on a sequence of dual pluses and minuses {+, -}. Gonshor itself proves that his definition of surreal numbers is equivalent to the Conway definition.

In this article, we shall make a number of remarks on surreal number theory which we believe can be useful in both scenarios: mathematics and physics. In particular, we shall established a connection between surreal numbers and tensors. Secondly, we shall show that surreal numbers can be linked to matroids. Moreover, we shall argue that surreal numbers may be connected with spin structures and therefore may provide an interesting development in supersymmetry. We also comment about the possibility that surreal numbers are connected with fractal theory. Finally, we also mention that concepts of infinitely small and infinitely large in surreal numbers may provide a possible solution for singularities in both high energy physics and gravitation.

Technically, this work is organized as follows. In Section 2, we briefly review the Conway definition of surreal numbers. In Section 3, we also briefly review the Gonshor definition of a surreal number. In Section 4, we established a connection between surreal numbers and tensors. In Section 5, we comment about the possibility that surreal numbers and matroids are related. Moreover, in Section 6 we mention number of possible applications of the surreal number theory, division algebras, supersymmetry, black holes and cosmology.

2. CONWAY FORMALISM

Let us write a surreal number by

$$x = \{X_L \mid X_R\} \tag{1}$$

and call X_L and X_R the left and right sets of x, respectively. Conway develops the surreal numbers structure S from two axioms:

Axiom 1. Every surreal number corresponds to two sets X_L and X_R of previously created numbers, such that no member of the left set $X_L \in X_L$ is greater or equal to any member X_R of the right set X_R.

Let us denote by the symbol $\ngeq$ the notion of no greater or equal to. So the axiom establishes that if x is a surreal number then for each $X_L \in X_L$ and $X_R \in X_R$ one has $X_L \ngeq X_R$. This is denoted by $X_L \ngeq X_R$.

Axiom 2. One number $x = \{X_L \mid X_R\}$ is less than or equal to another number $y = \{Y_L \mid Y_R\}$ if and only the two conditions $X_L \ngeq y$ and $x \ngeq Y_R$ are satisfied.

This can be simplified by saying that $x \leq y$ if and only if $X_L \ngeq y$ and $x \ngeq Y_R$.

Observe that Conway definition relies in an inductive method; before a surreal number x is introduced one needs to know the two sets X_L and X_R of surreal numbers. Thus, since each surreal number x corresponds to two sets X_L and X_R of previous numbers then one wonders what do one starts on the zeroth day or 0-day? If one denotes the empty set by $\varnothing$ then one defines the zero as

$$0 = \{\varnothing \mid \varnothing\}. \tag{2}$$

Using this, one finds that in the first day or 1-day one gets the numbers

$$-1 = \{\varnothing \mid 0\}, \qquad \{0 \mid \varnothing\} = +1 \tag{3}$$

In the 2-day one has

$$-2 = \{\varnothing \mid 1\}, \quad \frac{1}{2} = \{-1 \mid 0\}, \quad \{0 \mid 1\} = \frac{1}{2}, \quad \{1 \mid \varnothing\} = +2. \tag{4}$$

While in the 3-day one obtains

$$-3=\{\varnothing \mid 2\},\ -\frac{3}{2}=\{-1 \mid 0\},\ \frac{1}{2}=\{0 \mid 1\},\quad +2=\{2 \mid \varnothing\}. \tag{5}$$

The process continues as the following theorem establishes:

Theorem 1. Suppose that the different numbers at the end of n-day are

$$-3=\{\varnothing \mid 2\},\ -\frac{3}{2}=\{-1 \mid 0\},\ \frac{1}{2}=\{0 \mid 1\},\quad +2=\{2 \mid \varnothing\}. \tag{6}$$

Then the only new numbers that will be created on the $(n + 1)$-day are

$$\{\varnothing \mid x_1\},\{x_1 \mid x_2\},\cdots,\{x_{m-1} \mid x_m\},\{x_m \mid \varnothing\}. \tag{7}$$

Furthermore, for positive numbers one has

$$\{x_m \mid \varnothing\} = x_m + 1 \tag{8}$$

and

$$\{x_m \mid x_{m+1}\} = \frac{x_m + x_{m+1}}{2}. \tag{9}$$

While defining

$$-x = \{-X_R \mid -X_L\}, \tag{10}$$

for negative numbers one gets

$$-x = \{-X_R \mid -X_L\}, \tag{11}$$

and

$$\{-x_m \mid -x_{m+1}\} = -\frac{(x_m + x_{m+1})}{2}. \tag{12}$$

Thus, at the n-day one obtains $2^n + 1$ numbers all of which are of form

$$x = \frac{m}{2^n}, \tag{13}$$

where m is an integer and n is a natural number, $n > 0$. Of course, the numbers (13) are dyadic rationals which are dense in the reals R. Let us recall this theorem:

Theorem 2. The set of dyadic rationals is dense in the reals R.

Proof: Assume that $a > b$, with a and b elements of the reals R. By Archimedean property exist $n \in N$ such that

$$0 < \frac{1}{n} < b - a, \tag{14}$$

which implies

$$0 < \frac{1}{2^n} < \frac{1}{n} < b - a.$$

Thus, one has

$$1 < 2^n b - 2^n a. \tag{15}$$

As the distance between $2^n b$ and $2^n a$ is grater than 1, there is an integer m such that

$$2^n a < m < 2^n b \tag{16}$$

and therefore

$$a < \frac{m}{2^n} < b. \tag{17}$$

So, the set of dyadic rationals are dense in R.

The sum and product of surreal numbers are defined as

$$x+y=\{X_L+y,x+Y_L \mid X_R+y,x+Y_R\} \tag{18}$$

and

$$xy=\{X_Ly+xY_L-X_LY_L,X_Ry+xY_R-X_RY_R \mid X_Ly+xY_R-X_LY_R,X_Ry+xY_L-X_RY_L\}. \tag{19}$$

The importance of (18) and (19) is that allow us to prove that the surreal number structure is algebraically a closed field. Moreover, through (18) and (19) it is also possible to show that the real numbers R are contained in the surreals *S* (see Ref. [1] for details).

3. GONSHOR FORMALISM

In 1986, Gonshor [3] introduced a different but equivalent definition of surreal numbers.

Definition 1. A surreal number is a function μ from initial segment of the ordinals into the set{+, -}.

For instance, if μ is the function so that μ (1) = +, μ (2) = -, μ (3) = μ (4) = +then μ is the surreal number(++-+). In the Gonshor approach the expressions (3)-(5) becomes: 1-day

$$-1=(-),\quad (+)=+1, \tag{20}$$

in the 2-day

$$-2=(--),\ -\frac{1}{2}=(-+),\ (+-)=+\frac{1}{2},\quad (++)=+2, \tag{21}$$

and 3-day

$$-3=(---),\ -\frac{3}{2}=(--+),\ -\frac{3}{4}=(-+-),\ -\frac{1}{4}=(-++)$$

$$(+--)=+\frac{1}{4},\ (+-+)=+\frac{3}{4},\ (++-)=+\frac{3}{2},\ (+++)=+3, \tag{22}$$

respectively. Moreover, in Gonshor approach one finds the different numbers through the formula

$$n|a| + \frac{|b|}{2} + \sum_{i=i}^{q} \frac{|c_i|}{2^{i+1}}, \tag{23}$$

where a, b, c_1,..., $c_q \in \{+,-\}$ and $a \neq b$. Furthermore, one has $|+| = +$ and $|-| = -$. As in the case of Conway definition through (23) one gets the dyadic rationals. Observe that the values in (20), (21) and (22) are in agreement with (23). Just for clarity, let us consider the additional example:

$$(++-+-+) = 2 - \frac{1}{2} + \frac{1}{4} - \frac{1}{8} + \frac{1}{16} = \frac{27}{16}. \tag{24}$$

By the defining the order $x < y$ if $x(\alpha) < y(\alpha)$, where α is the first place where x and y differ and the convention - < 0 < +, it is possible to show that the Conway and Gonshor definitions of surreal numbers are equivalent (see Ref. [3] for details).

4. SURREAL NUMBERS AND TENSORS

Let us introduce a p-tensor [4] ,

$$t_{\mu_1\mu_2\cdots\mu_p}, \tag{25}$$

where the indices μ_1, μ_2,..., μ_p run from 1 to 2. Of course p indicates the rank of $t_{\mu_1\mu_2\cdots\mu_p}$. In tensorial analysis, (25) is a familiar object. One arrives to a link with surreal numbers by making the identification 1 → + and 2 → -. For instance, the tensor t_{1121} in the Gonshor notation becomes

$$t_{1121} \to t_{++-+} \to (++-+) \tag{26}$$

In terms of $t_{\mu_1\mu_2\cdots\mu_p}$, the expressions (18), (19) and (20) read

$$-1 = t_2, \qquad t_1 = +1, \tag{27}$$

in the 2-day

$$-2 = t_{22}, \quad -\frac{1}{2} = t_{21}, \quad t_{12} = \frac{1}{2}, \quad t_{11} = 2, \tag{28}$$

and 3-day

$$-3 = t_{111}, \quad -\frac{3}{2} = t_{112}, \quad -\frac{3}{4} = t_{121}, \quad -\frac{1}{4} = t_{122},$$
$$t_{211} = +\frac{1}{4}, \quad t_{212} = +\frac{3}{4}, \quad t_{221} = +\frac{3}{2}, \quad t_{222} = +3, \tag{29}$$

respectively.

Formally, one note that there is a duality between positive and negative labels in surreal numbers. In fact, one can prove that this is general for any n-day. This could be anticipated because according to Conway definition (1) a surreal number can be written in terms of the dual pair left and right sets X_L and X_R. Further, the concept of duality it is even clearer in the Gonshor definition of surreal numbers since in such a case one has a functions μ with codominio in the dual set {+, -}. In terms of the tensor $t_{\mu_1\mu_2\cdots\mu_p}$ in (25) such a duality can be written in the form

$$t_{\mu_1\mu_2\cdots\mu_p} + (-1)^q \varepsilon_{\mu_1\nu_1}\varepsilon_{\mu_2\nu_2}\cdots\varepsilon_{\mu_p\nu_p} t^{\nu_1\nu_2\cdots\nu_p} = 0, \tag{30}$$

where

$$\varepsilon_{\mu\nu} = \begin{pmatrix} 0 & 1 \\ -1 & 0 \end{pmatrix}. \tag{31}$$

It is interesting to observe that the 2-day corresponds to

$$t_{\mu_1\mu_2} = 2\begin{pmatrix} 1 & 0 \\ 0 & -1 \end{pmatrix} + \frac{1}{2}\begin{pmatrix} 0 & 1 \\ -1 & 0 \end{pmatrix}. \tag{32}$$

If one introduces the notation

$$\eta_{\mu\nu} = \begin{pmatrix} 1 & 0 \\ 0 & -1 \end{pmatrix}, \tag{33}$$

one discovers that (32) can be written as

$$t_{\mu\nu} = 2\eta_{\mu\nu} + \frac{1}{2}\varepsilon_{\mu\nu}. \tag{34}$$

It is worth mentioning that, in general any 2 × 2-matrix $\Omega_{\mu\nu}$ can be written as

$$\Omega_{\mu\nu} = x\delta_{\mu\nu} + y\varepsilon_{\mu\nu} + r\eta_{\mu\nu} + s\lambda_{\mu\nu}. \tag{35}$$

Here, one has

$$\delta_{\mu\nu} \equiv \begin{pmatrix} 1 & 0 \\ 0 & 1 \end{pmatrix} \tag{36}$$

and

$$\lambda_{\mu\nu} \equiv \begin{pmatrix} 0 & 1 \\ 1 & 0 \end{pmatrix}. \tag{37}$$

The set of matrices (31), (33), (36) and (37) determine a basis for any 2 × 2-matrix belonging to the set of 2 × 2-matrices which we denote by $M(2, R)$.

It is interesting that by setting $r = 0$ and $s = 0$ in (4) one gets the complex structure $\Omega_{\mu\nu} \rightarrow z_{\mu\nu}$, namely

$$z_{\mu\nu} = x\delta_{\mu\nu} + y\varepsilon_{\mu\nu}. \tag{38}$$

In fact, in the typical notation of a complex number (38) becomes $z = x + iy$. Observe also that when $\det \Omega \neq 0$ one obtains the group $GL(2, R)$ from $M(2, R)$. If one further requires that $\det \Omega \neq 1$, then one gets the elements of the subgroup $SL(2, R)$. It is worth mentioning that the fundamental matrices

$\delta_{\mu\varpi}$, $\eta_{\mu\varpi}$, $\lambda_{\mu\varpi}$ and $\varepsilon_{\mu\varpi}$ given in (31), (33), (36) and (37) not only form a basis for $M(2, R)$ but also determine a basis for the Clifford algebras $C(2, 0)$ and $C(1, 1)$. In fact, one has the isomorphisms $M(2, R) \sim C(2, 0) \sim C(1, 1)$. There exist a theorem that establishes that all the others higher dimensional algebras of any signature $C(a, b)$ can be constructed from the building blocks $C(2, 0)$, $C(1, 1)$and $C(0, 2)$ (see Ref. [5] and references therein). So a connection of these developments with surreal numbers seems to be a promising scenario.

5. SURREAL NUMBERS AND MATROIDS

For a definition of a non-oriented matroid see Ref. [6] and for oriented matroid see Ref. [7] (see also Refs. [8] [9] [10] [11] [12] and references therein). Here, we shall focus in some particular cases of oriented matroids. First, assume that $\chi^{\mu\nu}$ satisfies the Grassmann-Plücker relation

$$\chi^{\mu[\nu}\chi^{\alpha\beta]} = 0. \tag{39}$$

Here, the bracket $[\nu\alpha\beta]$ means completely antisymmetric. In this case, the ground set of a 2-rank oriented matroid $M = (E, \chi^{\mu\nu})$ is

$$E = \{1, 2, 3, 4\} \tag{40}$$

and the alternating map becomes

$$\chi^{\mu\nu} \to \{-1, 0, 1\}. \tag{41}$$

The $\chi^{\mu\nu}$ function can be identified with a 2-rank chirotope. The collection of bases for this oriented matroid is

$$B = \{\{1, 2\}, \{1, 3\}, \{1, 4\}, \{2, 3\}, \{2, 4\}, \{3, 4\}\}, \tag{42}$$

which can be obtained by just given values to the indices μ and ν in $\chi^{\mu\nu}$.Actually, the pair (E, B) determines a 2-rank uniform non-oriented ordinary matroid.

Let us consider the underlying ground bitset (from bit and set) [13] [14]

$$\mathcal{E} = \{1, 2\} \tag{43}$$

and the pre-ground set

$$E_0 = \{(1,1),(1,2),(2,1),(2,2)\} \tag{44}$$

One finds a relation between E_0 and E by comparing (40) and (44). In fact, one has

$$\begin{aligned}(1,1) &\leftrightarrow \mathbf{1}, \quad (1,2) \leftrightarrow \mathbf{2},\\ (2,1) &\leftrightarrow \mathbf{3}, \quad (2,2) \leftrightarrow \mathbf{4}.\end{aligned} \tag{45}$$

This can be understood considering that (45) is equivalence relation by making the identification of indices $\{a, b\} \leftrightarrow \mu$, ..., etc. Observe that considering this identifications the family of bases (42) becomes

$$\begin{aligned}\mathcal{B}_0 = \{&\{(1,1),(1,2)\},\{(1,1),(2,1)\},\{(1,1),(2,2)\},\\ &\{(1,2),(2,1)\},\{(1,2),(2,2)\},\{(2,1),(2,2)\}\}.\end{aligned} \tag{46}$$

It turns out that the chiritope χ^{μ_ν} can be associated with a 2-qubit system. So the pair (E_0, B_0) can be identified with a qubitoid (a combination of qubit and matroid).

The procedure can be generalized to higher dimensions. For instance, consider the pre-ground set

$$\begin{aligned}E_0 = \{&(1,1,1,1),(1,1,1,2),(1,1,2,1),(1,1,2,2),\\ &(1,2,1,1),(1,2,1,2),(1,2,2,1),(1,2,2,2)\\ &(2,1,1,1),(2,1,1,2),(2,1,2,1),(2,1,2,2)\\ &(2,2,1,1),(2,2,1,2),(2,2,2,1),(2,2,2,2)\}\end{aligned} \tag{47}$$

It is not difficult to see that by making the identifications

$$
\begin{array}{lll}
(1,1,1,1) \leftrightarrow \mathbf{1} & (1,1,1,2) \leftrightarrow \mathbf{2} & (1,1,2,1) \leftrightarrow \mathbf{3} \\
(1,1,2,2) \leftrightarrow \mathbf{4} & (1,2,1,1) \leftrightarrow \mathbf{5} & (1,2,1,2) \leftrightarrow \mathbf{6} \\
(1,2,2,1) \leftrightarrow \mathbf{7} & (1,2,2,2) \leftrightarrow \mathbf{8} & (2,1,1,1) \leftrightarrow \mathbf{9} \\
(2,1,1,2) \leftrightarrow \mathbf{10} & (2,1,2,1) \leftrightarrow \mathbf{11} & (2,1,2,2) \leftrightarrow \mathbf{12} \\
(2,2,1,1) \leftrightarrow \mathbf{13} & (2,2,1,2) \leftrightarrow \mathbf{14} & (2,2,2,1) \leftrightarrow \mathbf{15} \\
(2,2,2,2) \leftrightarrow \mathbf{16}, & &
\end{array} \tag{48}
$$

one obtains a relation between the pre-ground set E_0 given in (47) and the ground set

$$
E = \{\mathbf{1}, \mathbf{2}, \cdots, \mathbf{15}, \mathbf{16}\}. \tag{49}
$$

This can be again understood by considering that (49) is equivalent to make the identification of indices $(a, b, c, d) \leftrightarrow \mu, \ldots$ etc. It turns out that considering these relations one finds that the collection of bases $\mathcal{B}$ contains $\binom{16}{2} = 120$ two-element sub-set of the 16-element set E, given in (49). This 2-element subset can be obtained by considering a lexicographic order of all 120 two-subsets of {1,2,...,15,16}. One finds that the first terms of $\mathcal{B}_0$ look like

$$
\begin{aligned}
\mathcal{B}_0 = \{ & \{(1,1,1,1),(1,1,1,2)\},\{(1,1,1,1),(1,1,2,1)\},\{(1,1,1,1),(1,1,2,2)\}, \\
& \{(1,1,1,1),(1,2,1,1)\},\{(1,1,1,1),(1,2,1,2)\},\cdots\}.
\end{aligned} \tag{50}
$$

(See Refs. [13] and [14] for details.)

The method, of course, can be extended to 2^{2n+1}-dimensions, $n = 0,1,2,\ldots$ and can be connected to N-qubit system. However, it is worth mentioning that the complete classification of N-qubit systems is a difficult, or perhaps an impossible task. In reference [15] an interesting development for characterizing a subclass of N-qubit entanglement has been considered. An attractive aspect of this construction is that the N-qubit entanglement can be understood in geometric terms. The idea is based on the bipartite partitions of the Hilbert space in the form $C^{2^N} = C^L \otimes C^l$, with $L = 2^{N-n}$ and $l = 2^n$. Such a partition allows a geometric interpretation in terms of the complex Grassmannian variety $Gr(L,l)$ of l-planes in C^L via the Plücker embedding. In this case, the Plucker coordinates of the Grassmannians are natural invariants of the theory.

There are a number of ways in which one can connect matroids with surreal numbers. First, one may think in the bitset given in (43) in the Gonshor form

$$\mathcal{E} = \{1,2\} \rightarrow \{-,+\}. \tag{51}$$

Second, the numbers of any the ground set in matroid theory

$$E = \{\mathbf{1},\mathbf{2},\mathbf{3},\cdots\}, \tag{52}$$

can be written in terms of the surreal numbers as

$$\{\{0,\varnothing\},\{1,\varnothing\},\{2,\varnothing\},\cdots\}. \tag{53}$$

In this context the basis set $\mathcal{B}$ will be also written in terms of the surreal numbers. Third, another possibility is also to identify the chirotope map $\chi \rightarrow \{-1,0,1\}$ in terms of the surreal numbers $\chi \rightarrow \{\{\varnothing,0\},\{\varnothing,\varnothing\},\{0,\varnothing\}\}$.

Of course, it will interesting to fully develop these possible links between matroids and surreal numbers. But even at these stage one note that the key concept in both matroid theory and surreal numbers theory is duality. This is because in matroid theory it is known that in matroid theory there is a key theorem that every matroid $\mathcal{M}$ has a dual $\mathcal{M}^*$, while in surreal number theory duality is everywhere. In a sense this is because a surreal numbers $x = \{X_L, X_R\}$ is defined in terms of two dual sets X_L and X_R. So one wonders whether in surreal number theory exist a theorem establishing that for every surreal number set $\mathcal{S}$ there exist a dual surreal number set. $\mathcal{S}^*$

6. VARIOUS MATHEMATICAL AND PHYSICAL POSSIBLE APPLICATIONS

In this section we shall describe an additional number of possible applications of surreal numbers in mathematics and physics. Although such a description will be brief the main idea is to stimulate further research in the area. One may think that our proposals are in a sense for experts in the topic but in fact the main intention is to call the attention of mathematicians and physicist telling them look here are a number of subjects in which you have the opportunity to participate.

I. Applications in mathematics:

(a) Division algebras

There is a celebrated Hurwitz theorem:

Theorem (Hurwitz, 1898): Every normed algebra over the reals with an identity is isomorphic to one of following four algebras: the real numbers, the complex numbers, the quaternions, and the Cayley (octonion) numbers.

Moreover, the Hurwitz theorem is closely related with the parallelizable spheres S^1, S^3 and S^7 [16] and the remarkable theorem that only exist division algebras in 1, 2, 4 and 8 dimensions [17] [18]. So, one wonders what could be the corresponding Hurwitz theorem and these remarkable developments on division algebras if one extend the real numbers to surreal numbers. In this context, it has been proved in Refs. [19] [20] [21] that for normalized qubits the complex 1-qubit, 2-qubit and 3-qubit are deeply related to division algebras via the Hopf maps, $S^3 \xrightarrow{S^1} S^2$, $S^7 \xrightarrow{S^3} S^4$ and $S^{15} \xrightarrow{S^7} S^8$, respectively. It seems that there does not exist a Hopf map for higher N-qubit states. So, from the perspective of Hopf maps, and therefore of division algebras, one arrives to the conclusion that 1-qubit, 2-qubit and 3-qubit are more special than higher dimensional qubits (see Refs. [22] [23] for details). Again one wonders whether surreal numbers can contribute in this qubits theory framework.

II. Applications in physics:

(a) Supersymmetry: For finite sets X_L and X_R, one of the key tools in surreal numbers are integers n and dyadic rationals $d = \frac{m}{2^k}$. For n = 0,1 and 2 and $d = \frac{1}{2}, \frac{3}{2}$ one recalls the spin structure of supersymmetry. So one wonders **if** for instance spin $\frac{1}{4}$ may be a prediction of surreal number theory. Remarkable, this spin has been proposed in N = 1 supersymmetry in connection with anyons (see Refs. [24] [25] and references therein). Thus, for finite sets X_L and X_R, surreal numbers $x = \{ X_L, X_R \}$ and in the Gonshor approach, one finds that bosons can be identified with s integer spin and fermions with dyadic rational with spin

$$n|a| + \frac{|b|}{2} + \sum_{i=1}^{q} \frac{|c_i|}{2^{i+1}},$$

given in (23). One can even think in this expression as the eigenvalues of a ket $|n, s_1, s_2, s_3, \cdots\rangle$. Here we made the associations

$$s_1 \to \frac{|b|}{2}, s_2 \to \frac{|c_1|}{4}, s_3 \to \frac{|c_2|}{8}, \cdots \tag{54}$$

and so on. Thus, in this framework, it seems the whole structure of surreal numbers can be identified with a kind of supersymmetric approach.

(c) Black-holes

Consider the Schwarzschild metric [26]

$$ds^2 = -\left(1 - \frac{2GM}{c^2 r}\right) dt^2 + \frac{dr^2}{\left(1 - \frac{2GM}{c^2 r}\right)} + r^2\left(d\theta^2 + \sin^2\theta d\phi^2\right), \tag{55}$$

where M is the source mass, G is the Newton gravitational constant and c is the light velocity. There are a number of observations that one can make about (55). First, notice that in this expression all quantities are real numbers. Second there are two type of singularities, namely in $r = r_s = \frac{2GM}{c^2}$ and $r = 0$. It is known that using Kruskal coordinates it is possible to show that the singularity at $r = r$ is simply a coordinate singularity. However the singularity at $r = 0$ is a true physical singularity of spacetime. First of all, in this context, when one referes about singularity in terms of real numbers one means that in the limit $r \to 0$ one obtains the expression $\frac{2GM}{c^2 r} \to \infty$ (see Ref. [26] and references therein for details).

From the point of view of surreal numbers theory the singularity $\frac{2GM}{c^2 r} \to \infty$, when $r \to 0$, is not a real problem because in such a mathematical theory all kind of infinite large and infinite small are present. So by a assuming that all quantities in the line element given in (55) is written in terms of surreal numbers

$$d\mathcal{S}^2 = -\left(1 - \frac{2G\mathcal{M}}{c^2 \mathcal{R}}\right) d\mathcal{T}^2 + \frac{d\mathcal{R}^2}{\left(1 - \frac{2G\mathcal{M}}{c^2 \mathcal{R}}\right)} + \mathcal{R}^2\left(d\theta^2 + \sin^2\theta d\phi^2\right), \tag{56}$$

the problem of singularities in black-hole physics no longer exist!

(d) Cosmology

In the Friedmann cosmological equation [26]

$$\frac{1}{a^2}\frac{da(t)}{dt}+\frac{k}{a^2}-\frac{8\pi\rho_m}{3}-\frac{8\pi\rho_r}{3}=0, \quad (57)$$

one assumes that the matter density is given by

$$\rho_m=\frac{\rho_{0m}}{a^3}, \quad (58)$$

while the radiation energy density is

$$\rho_r=\frac{\rho_{0r}}{a^4}, \quad (59)$$

where k, ρ_{0m} and ρ_{0r} are constants. So, even if one does not consider the solution of (57) the expressions (58) and (59) tell us that there is a 'big-bang' singularity at $a \to 0$. In fact, when $a \to 0$ one has

$$\rho_m \to \infty \quad (60)$$

and

$$\rho_r \to \infty. \quad (61)$$

Just as in the case of black-holes these singularities are related to the fact that one is considering real numbers structure in the length scale a as well as in the time evolution parameter *t*. Again, one wonders what formalism one may obtain by replacing a by some kind of surreal length scale $\mathcal{A}$ and the time parameter *t* by a surreal time parameter $\mathcal{T}$. Of course, this in turn will imply that the whole gravitational theory must be modified with surreal numbers structure.

Another possibility is to identify the whole evolution of the surreal numbers structure with a cosmological model in the sense that in 0-day one has the scalar field particle of 0-spin (the Higgs field?), in the 1-day one has (-1) -spin and 1-spin (the photon?) and the 2-day one obtains the (-2) -spin, $\left(-\frac{1}{2}\right)$-spin, $\frac{1}{2}$-spin 2-spin (graviton and fermion?) and so on. Following this idea one may even identify the 0-day and 0-spin with the big bang and since $0=\{\varnothing,\varnothing\}$ one can say that everything in our universe started with vacuum state $\varnothing$.

(e) Fractals

It is known that fractals and dyadic fractions are deeply related. Much of this relationship can be explained by infinite binary tree which can be viewed as a certain subset of the modular group $PSL(2,Z)$ (the general linear group of 2 by 2 matrices over the integers). The subset is essentially the dyadic grupoid or dyadic monoid. This in turn provides the natural setting for the symmetry and self-similarity of many fractals. Moreover, it is also that these groups and the rational numbers can be connected with dyadic subsets [27].

7. FINAL REMARKS

Due to the fact that duality is the underlying concept in both surreal numbers and matroid theory, we believe that it is a matter of time that these two mathematical scenarios are considered as important tools in physics and in particular in high energy physics and gravity.

From the serious difficulties with infinities in black-hole physics and cosmology as well as in higher energy physics it seems to us that surreal numbers theory offers a new view for a solution, instead of thinking that the infinities are the enemies in quantum and classical physical theory incorporate them in a natural way as surreal numbers framework suggests.

It turns out that surreal numbers can be understood as a particular case of games [2] (see also Ref. [28]) which is a fascinating mathematical theory. In fact, games can be added and substracted forming an Abelian group and a sub-group of games is identified with surreal numbers which can also be multiplied and form a field. As we mentioned before, this field contains the real numbers among many other numbers structures. The key additional condition for reducing a game $x = \{x_L,..., x_R,...\}$ to a surreal number is that x_L and x_R are surreal numbers and satisfy $x_L < x_R$. So, one wonders whether game theory may lead to even more interesting applications that those presented in this work.

Finally, we believe that it is just a matter of time for the recognition of the surreal numbers structure as one of the key mathematical tools in superstring theory [29] [30] [31]. This is because although the problems of some infinities are solved there remain always additional problems with the emergency of new infinities. This phenomena may be traced back to the fact that the action in superstring theory is written in terms of real functions (target space-time coordinates) rather that surreal functions.

ACKNOWLEDGEMENTS

I would like to thank to P. A. Nieto, C. Garca-Quintero and A. Meza for helpful comments. This work was partially supported by PROFAPI/2013.

REFERENCES

1. Knuth, D.E. (1974) Surreal Numbers: How Two Ex-Students Turned on to Pure Mathematics and Found Total Happiness: A Mathematical Novelette. Addison-Wesley Publishing Co., Massachusetts.
2. Conway, J.H. (1976) On Number and Games. London Mathematical Society Monographs, Academic Press, London.
3. Gonshor, H. (1986) An Introduction to the Theory of Surreal Numbers. London Mathematical Society Lectures Notes Series: 110, Cambridge University Press, Cambridge.
4. Sokolnikoff, I.S. (1951) Tensor Analysis. Chapman & Hall, New York.
5. Atiyah, M.F., Bott, R. and Shapiro, A. (1964) Topology, 3, 3. https:/doi.org/10.1016/0040-9383(64)90003-5
6. Oxley, J.G. (1992) Matroid Theory. Oxford Science Publications. Oxford University Press, New York.
7. Björner, A., Las Vergnas, M., Sturmfels, B., White, N. and Ziegler, G.M. (1993) Oriented Matroids. Cambridge University Press, Cambridge.
8. Nieto, J.A. (2004) Advances in Theoretical and Mathematical Physics, 8, 177. arXiv: hep-th/0310071.
9. Nieto, J.A. (2006) Advances in Theoretical and Mathematical Physics, 10, 747. arXiv: hep-th/0506106.
10. Nieto, J.A. (2004) Journal of Mathematical Physics, 45, 285. arXiv: hep-th/0212100. https:/doi.org/10.1063/1.1625416
11. Nieto, J.A. (2014) Nuclear Physics B, 883, 350-372. arXiv:1402.6998 [hep-th]. https:/doi.org/10.1016/j.nuclphysb.2014.04.001
12. Nieto, J.A. and Marin, M.C. (2000) Journal of Mathematical Physics, 41, 7997-8005. hep-th/0005117. https:/doi.org/10.1063/1.1319518
13. Nieto, J.A. (2013) Physics Letters B, 718, 1543-1547. arXiv:1210.0928 [hep-th]. https:/doi.org/10.1016/j.physletb.2012.12.034
14. Nieto, J.A. (2010) Physics Letters B, 692, 43-46. e-Print: arXiv:1004.5372 [hep-th]. https:/doi.org/10.1016/j.physletb.2010.07.010
15. Levay, P. (2005) Journal of Physics A, 38, 9075-9085. https:/doi.org/10.1088/0305-4470/38/41/016
16. Adams, J.F. (1962) Annals of Mathematics, 75, 603-632. https:/doi.org/10.2307/1970213
17. Bott, R. and Milnor, J. (1958) Bulletin of the American Mathematical Society, 64, 87-89. https:/doi.org/10.1090/S0002-9904-1958-10166-4
18. Kerbvair, M. (1958) Proceedings of the National Academy of Sciences of the United States of America, 44, 286.
19. Mosseri, R. and Dandoloff, R. (2001) Journal of Physics A, 34, 10243-10252. https:/doi.org/10.1088/0305-4470/34/47/324
20. Mosseri, R. (2003) Two and Three Qubits Geometry and Hopf Fibrations. arXiv:quant-ph/0310053. https:/doi.org/10.1103/PhysRevD.75.024024
21. Bernevig, B.A. and Chen, H.D. (2003) Journal of Physics A, 36, 8325-8339. https:/doi.org/10.1088/0305-4470/36/30/309
22. Levay, P. (2007) Physical Review D, 75, Article ID: 024024. e-Print: hep-th/0610314. https:/doi.org/10.1103/PhysRevD.75.024024

23. Levay, P. and Szalay, S. (2011) Physical Review D, 83, Article ID: 045005. e-Print: arXiv:1011.4180 [hep-th]. https:/doi.org/10.1103/PhysRevD.83.045005
24. Mezincescu, L. and Townsend, P.K. (2010) Physical Review Letters, 105, Article ID: 191601. arXiv:1008.2334 [hep-th]. https:/doi.org/10.1103/PhysRevLett.105.191601
25. Soroking, D. (2002) Fortschritte der Physik, 50, 724-728. https:/doi.org/10.1002/1521-3978(200205)50:5/7<724::AID-PROP724>3.0.CO;2-J
26. Misner, C.W., Thorne, K.S. and Wheeler, J.A. (1973) Gravitation. W.H. Freeman, San Francisco.
27. Vepstas, L. (2004) The Minkowski Question Mark, GL(2,Z) and the Modular Group. http://www.linas.org/math/chap-minkowski.pdf
28. Schleicher, D. and Stoll, M. (2006) Moscow Mathematical Journal, 6, 359-388. arXiv:math/0410026 [math.CO].
29. Nambu, Y. (1970) Duality and Hydrodynamics. Lectures at the Copenhagen Conference.
30. Goto, T. (1971) Progress of Theoretical Physics, 46, 1560-1569. https:/doi.org/10.1143/PTP.46.1560
31. Green, M., Schwarz, J. and Witten, E. (1987) Superstring Theory. Cambridge University Press, Cambridge.

CHAPTER 14

Physics in Discrete Spaces: On Some Cosmological Issues

Pierre Peretto

Laboratory of Physics and Modelling of Condensed Matter, Grenoble, France

ABSTRACT

This article is an application of the theory of discrete spaces to cosmology. Its conclusions are necessarily speculative. An interesting aspect is that it gives possible solutions to many pending problems within a unique framework. Let us cite a scenario for the Big-Bang that avoids any initial mathematical singularity, an interpretation of dark matter that does not involve any hadronic matter, a description of the formation of stellar and galactic black holes and, for the later, a description of quasars, their characteristics and their source of energy. Finally dark energy is also given an interpretation through modifications of the laws of gravity.

Keywords: Big Bang, Dark Matter, Black Holes, Quasars, Bekenstein-Hawking Entropy, Cosmological Constant

1. INTRODUCTION

Cosmology is a branch of physics that aims at giving a comprehensive interpretation of the universe as a whole. Theoretical cosmology is based upon General Relativity whose central idea is the introduction of a site-depen- dent metric that is a specific organization of space-time. In preceding articles [1] -[4], we have introduced another specific type of space-time organization (obviously not at variance with General Relativity); according to which the universe would be made of a set of very simple interacting binary physical units called cosmic bits, a discrete model of space-time must have deep consequences in cosmology. In this contribution, we use the theory of discrete spaces to give possible interpretations of various cosmological pending problems. Some aspects of the present contribution have already been presented in former articles. They are more precisely discussed and gathered in the present paper.

2. ON THE BIG-BANG PARADIGM

2.1. The Classical Approach of the Big-Bang Theory

The observation that the galaxies are moving away from one another led to the conclusion that far in the past, some 13.7 billions years ago, the dimension of the universe was extremely small. This conclusion is strongly supported by the theory of General relativity which concludes that the universe is a dynamic system that either expands or contracts. The study of the equations of General relativity shows that they can admit singular solutions indeed: At its very beginning the universe as a whole would be reduced to a single point that is a mathematical singularity.

Most physicists, however, are not pleased with mathematical singularities. To avoid such a situation several remedies have been put forward all based upon the idea that there existed a pre Big-Bang state. Let us cite:

1) The metrics of our space-time is 4-dimensional and Lorentzian. In the pre Big-Bang state it would have been 4-dimensional and Euclidean [5].

2) A close idea is that time could have been imaginary in the pre Big-Bang state [6].

3) Finally it has also been suggested that the universe could be cyclic that is a universe with an expansion behaviour following a contraction behaviour [7].

Another difficulty was found in the observation that the young universe is homogeneous in spite of the fact that its parts could not be related by relativistic causality. The solution, put forward by Guth, is the phenomenon of inflation [8].

2.2. A Reminder of the Model of Discrete Spaces

According to previous contributions, more specifically in Ref. [1], the universe is comprised of a countable set of cosmic bits a = 1,2,... that are very simple physical systems whose states σ_a are completely determined by one binary variable $\sigma_a = \pm 1$. Cosmic bits are similar to (classical) Ising spins. Is the set of cosmic bits a finite set? Since there is no physical reason to limit its size, the set of cosmic bits is assumed to be infinite but, as we shall see, this does not mean that the observable universe is infinite. The, non measurable, size of a cosmic bit would be of the order of the Planck length, that is $l_{Cb} = l_{Pl} \cong 1.6 \times 10^{-33}$ cm lcb = lpl ≅ 1.6×10^{-33} cm.

The cosmic bits interact with each other through random two-body connexions such as $\pm J^{(2)}\sigma_a\sigma_b$ and random four-body connexions such as $\pm J^{(4)}\sigma_a\sigma_b\sigma_c\sigma_d$. The universe therefore may be seen as a sort of Ising spin-glass. The interplay between the interactions generates a granular structure whose grains are called world points. A world point is made of n ≅ 10^{46} cosmic bits all connected with each others through binary ferromagnetic interactions $-J^{(2)}\sigma_a\sigma_b$ (a,b = 1,.., n). The size l^* of a world point would be of the order of $l^* \cong 0.5 \times 10^{-21}$.

Finally the system is subject to a degree of disorder whose intensity b is similar to the inverse of some temperature although the meaning of this temperature has nothing to do with ordinary temperature. The role of b is extremely important because b is the driving principle that builds all physical objects, space-time itself [1], quantum states [2], fields [3], and particles [4]. Its role, we shall see, would also be central in cosmology.

To summarize we put forward that the universe may be considered as a ferromagnetic powder (a special sort of spin glass) whose properties can be found by using the classical tools of statistical physics.

2.3. A New Interpretation of the Big-Bang

The ferromagnetic interactions inside a world point generate a polarization $\varphi = \langle s \rangle$ defined by the thermal average of the order parameter s [1].

$$s = \frac{1}{n} \sum_{a=1,\cdots,n} \sigma_a .$$

The polarization φ is a solution of the mean field equation:

$$\phi = tanh\,(bJ\varphi) \tag{1}$$

where $J = nJ^{(2)}$. The polarization φ of a world point i may be seen as the length of a d-dimensional vector ϕ_i

$$\phi_i = \begin{pmatrix} \varphi_{i1} \\ \varphi_{i2} \\ \vdots \\ \varphi_{id} \end{pmatrix}.$$

If $bJ < 1$ the only solution of Equation (1) is $\varphi = 0$. The vacuum state ϕ^{vaccum} of the universe is then symmetrical. If $bJ > 1$ there are non-zero solutions of Equation (1). The vacuum state of the universe is asymmetrical. The dimension d of the internal space of a world point is given by the following equation (see Ref. [1]):

$$d = \text{Int}(bj). \tag{2}$$

When the polarization φ vanishes, that is when $bJ < 1$, everything disappears. There is no more space-time, no quantum states, no fields, and no particles. It is tempting to consider that this situation corresponds to a pre Big-Bang state. According to this point of view the Big-Bang must be considered as a (second order) phase transition between $bJ < 1$ and $bJ > 1$. How can we imagine, in the framework of discrete spaces, the nature of the pre Big Bang state?

We are compelled to define the pre Big-Bang as a set of parameters given a priori. This set of parameters has an ontological character that is they do not need further explanations. The parameters are the following:

1) A infinite countable set of binary units.

2) A random distribution of 2-body interactions and 4-body interactions.

3) A noise distribution.

The parameters are generated freely, as in cards deals for example, except for two constraints.

- There is a correlation between the signs of 2-body $\pm J^{(2)}$ and 4-body $\pm J^{(4)}$ interactions (for more details see Ref. [1]).
- The cosmic parameters b_i and b_j of two neighbouring world points i and j are close to one another.

Since no time has been generated so far we can see the pre Big-Bang state as a state frozen in an eternal present. We can consider either that the present has no beginning and no end, or alternatively that the beginning and the end of the present are one and the same thing. If we adopt the first point of view, that is the universe has no beginning, then there is no room for an act of creation. If we adopt the second point of view, the present, a notion then consubstantial with the pre Big Bang state, can be seen as an act of creation. Which point of view is adopted is a matter of personal philosophy.

2.4. The Very Beginning of the Universe

The states $\Psi = \{\phi_i\} i = 1,...,N$ of a universe made of N world points in states ϕ_i are approximately obtained by minimizing the Lagrangian $\Lambda(\Psi)$ (see Ref. [1]):

$$\Lambda(\psi) = \psi^{T}\left(\Delta \otimes \{G\}\right)\psi \tag{3}$$

Δ is a $n \times n$ random square matrix whose elements Δ_{ij} describe the random (Gaussian) interaction between world points i an j. $\{G\}$is a set of N $d \times d$ internal interaction matrices G_i that describe the interactions between the d components $\varphi_{i\mu}$ $\mu = 1,...,d$ of world point i.

In Ref. [1] we have shown that in spaces devoid of particles one has $G_i = G^{vaccum}$ and

$$G^{\text{vacuum}} = \begin{pmatrix} -1/c^2 & & & \\ & 1 & & \\ & & \ddots & \\ & & & 1 \end{pmatrix} = \begin{pmatrix} 1-bJ & & & \\ & 1 & & \\ & & \ddots & \\ & & & 1 \end{pmatrix} \tag{4}$$

with a metric tensor g given by

$$g = \begin{pmatrix} \operatorname{sgn}(1-bJ) & & & \\ & 1 & & \\ & & \ddots & \\ & & & 1 \end{pmatrix}$$

c is the, dimensionless, speed of light. It is given by $c = 1/\sqrt{bJ-1}$. From the formula, $\sin^2(\theta_W) = 1/bJ$, where is the Weinberg angle of electroweak interactions (see Ref. [3]), and the experimental value $sin^2(\theta_W) = 0.231$ one finds $bJ = 4.32$ that is $bJ > 1$. We conclude that the dimension d of our universe is d = 4, that the vacuum is asymmetrical and that its metrics is Minkowskian.

In the pre Big Bang state, for world points where bJ < 1, the speed of light c is an imaginary number and the metrics is Euclidean as proposed in Refs. [5] and [6].

Space-time, quantum states, fields and matter spontaneously appear in regions of the pre Big Bang state where $bJ > 1$. We define our universe as a region of the pre Big-Bang state where $4 < bJ < 5$. According to this point of view an infinite number of universes could emerge from the pre Big-Bang state but most of these universes are quite different from ours and entirely out of reach. It must be noted that the speed of light $c = 1/\sqrt{bJ - 1}$ diverges at the transition $bJ = 1$ so possibly solving the inflation problem [8].

3. DARK MATTER

3.1. Introducing Dark Matter

At the beginning of the 20th century the astronomical observation revealed an anomaly of the speed of stars rotating around the centre of their galaxies [9]. The observed speed is constant whereas, according to Newton dynamics, it would have to decay as $1/\sqrt{r}$. It has been put forward that the anomaly could be explained by the presence of some invisible matter, called dark matter that forms a halo around the galaxies. Dark matter, although invisible, reveals its presence, as ordinary matter does, by the bending of space-time. This effect is observed indeed and allows, for example, the drawing of dark matter maps to be carried out.

Obviously the question is to determine the nature of dark matter that is the sort of particles it is made of. Various hypotheses have been proposed such as cold wandering bodies or neutrinos etc. but all these possibilities have been ruled out and most astrophysicists agree today that dark matter cannot be made of hadrons. In other respects Milgrom has proposed another explanation for the rotation anomaly and dark matter seems to be no longer necessary. He suggests that the Newton law of gravitation strength must be modified along the following formula

$$f(r) = -GMm\mu(r)/r^2$$

with $\mu(r) = 1$ for $r < r_M$ and $\mu(r) \propto$ for $r > r_M$.

G is the gravitation constant and r_M is the Milgrom parameter [3]. This is the MOND theory (for Modified Newton Dynamics). The agreement with observation is striking and rather convincing and it seemed that dark matter could be ignored. That is not the case.

The gravitation forces are, in the MOND theory, exactly known. They are central and enable an exact calculation of the motions of galaxies in a cluster of galaxies to be carried out. The observation of the galaxy cluster 1E0657-56 (the Bullet) does not support the calculations. Dark matter is still necessary to explain the observation and also, for example, to account for the formation of galaxies.

Moreover the Milgrom interactions have no observable effect upon the geometry of space-time whatsoever and cannot explain the observed maps of dark matter. Besides its unknown nature dark matter gives rise to other problems. For example it seems that there is no dark matter in our solar system whereas the solar system is placed right inside the disk of the Milky Way where dark matter must be present.

3.2. A New Interpretation of the Dark Matter

The model of discrete space-time that we put forward may provide an alternative interpretation.

Let us consider the metric matrix of vacuum (Equation (4)). This matrix is sensitive to variations of cosmic noise b and since the speed of light is given by $c = 1/\sqrt{bJ-1}$ a variation of cosmic noise b leads to a variation of the speed of light c. Space then behaves as a refractory medium and a non uniform repartition of cosmic noise b reveals itself by the bending of light rays exactly as gravitational lenses would do.

Since the trajectories of massive bodies follow the geodesics of space-time a non uniform cosmic noise repartition has the same effects as those induced by massive bodies. Given a repartition of cosmic noise throughout the whole (infinite) universe, our universe, we have said, can be defined as a (probably finite) region where $4 \leq bJ < 5$. Dark matter would be present in parts of our universe where the cosmic noise value b would be in between these two limits.

This is the case in our solar system (where $bJ \cong 4.32$) but the noise distribution would be uniform enough to make unobservable in the solar system the bending of light and therefore the presence of dark matter.

4. BLACK HOLES

4.1. Two Sorts of Black Holes

Black holes are regions of our universe where the Lorenz metrics is strongly modified or even disappears altogether. These objects have been experimentally observed, not directly because they are invisible, but through their gravitational effects. Two types of black holes have been identified.

- The stellar black holes with masses of the order of 15 Sun masses. Their presence is revealed by the motion of binary stars where a star orbits around a point associated with no visible object.
- The galactic (or super-massive) black holes that are placed in the vicinity of the centres of galaxies. Their presence is revealed by the simultaneous motions of numerous stars orbiting around a point still associated with no visible object. The masses of galactic black holes are enormous. The galactic black hole of the Milky Way, our own galaxy, is called Sgr A*. Its mass exceeds several millions Sun masses but galactic black holes with masses higher than several billions Sun masses are observed in the distant universe.

Black holes give rise to many questions such as:

1) Are stellar black holes and galactic black holes two aspects of the same physical object?
2) What is the mechanism of their creations?
3) What is the nature of the matter they are made of? etc.

The model of discrete spaces that we propose may give possible answers to these questions.

4.2. Stellar Black Holes

The solution, by Schwarzschild, of the equation of General Relativity in spherically symmetrical systems shows that the space-time metrics is singular at a distance given by $r_s = 2Gm/c^2$ (the Schwarzschild radius). An object, smaller than r_S, with mass m seems to display the characteristics of a stellar black hole. The first question is to devise a mechanism that creates such objects. The idea is that strong enough gravitation forces cause, through an enormous increase of densities, a series of modifications of the very nature of usual matter.

- A first step that we find in ordinary stars, such as in the Sun itself, is the dissociation of electrons from their nuclei. This state of matter is called plasma.
- In stronger gravitational fields the components of nuclei, protons and neutrons dissociate in turn and give rise to a new state of matter, a quantum gas with about one electron for one neutron and one proton. This is the material the white dwarfs are made of.
- The next, and up to now the last, step is the absorption of electrons by the protons. Matter is then entirely made of neutrons. This is the material the neutron stars (or pulsars) are made of.

The existence of stellar black holes seems to require another step in the transformations of matter that is the transformation of particles of the Standard Model themselves. In Ref. [4] we have proposed a structure of the elementary particles that accounts for most of the properties of the Standard Model. In this model the elementary particles are built around specific world points, called the seeds of the particles whose internal spaces display their symmetry. For large enough densities the effect of 4-bodies interactions $J^{(4)}$ is overcome and the world point structure is modified so that different world points can merge and give rise to larger world points. In our interpretation, a stellar black hole is a macroscopic version of the internal space of elementary particles. In a black hole, as in internal spaces of elementary particles, space and time are meaningless concepts and there is no specific orientation of the internal space. Seen from the outside a stellar black hole seems to be a no-universe region that stretches the fabric of surrounding ordinary space-time so giving rise to the observed gravitational effects. Moreover the transformation of ordinary matter into black holes amounts to a simple reorganization, not a vanishing, of its degrees of freedom. Therefore, in our interpretation, the entropy of a black hole remains finite and the physics of a stellar black hole is no more in contradiction with the second principle of thermodynamics.

To make this discussion more complete let's have a look at the "no-hair theorem".

If we merge Q world points together, the polarization φ_{bh} of the black hole is given, as in Equation (1), by

$$\varphi_{bh} = \tanh(QbJ\varphi_{bh}).$$

In stellar black holes Q is a large number that is $\varphi_{bh} \cong 1$.

In other respects the polarization is a solution of the mean field equation

$$\varphi_{bh,\mu} = \tanh\left(\left(QbJ/d_{bh}\right)\varphi_{bh,\mu}\right).$$

For the components to have a physical meaning it is necessary that

$$QbJ/d_{bh} > 1.$$

This determines the internal dimensionality of a stellar black hole: $d_{bh} = QbJ$.

Since Q is generally a very great number, the internal dimensionality of a stellar black hole, if not infinite, is extremely large which means that the notion of dimensionality loses, in actual fact, its meaning.

The polarization component $\varphi_{bh,\mu}$ of φ_{bh} $(\mu = 1,\cdots,d_{bh})$, is a solution of the mean field equation

$$\varphi_{bh,\mu} \cong \tanh\left(bJ\varphi_{bh,\mu}\right).$$

The cosmic noise b into a stellar black hole tends towards the cosmic noise of its surrounding, that is $bJ \cong 4.324.32$. Then we have $\varphi bh, \mu \cong \pm 1$ where the sign is chosen at random.

According to our interpretation a stellar black hole would be similar to a "hairy", hollow, d_{bh}-dimensional sphere.

Seen from the outside a stellar black hole seems to be a no-universe region that stretches the fabric of surrounding ordinary space-time so giving rise to the observed gravitational effects. Moreover the transformation of ordinary matter into black holes amounts to a simple reorganization, not a vanishing, of its degrees of freedom. Therefore, in our interpretation, the entropy of a black hole remains finite and the physics of a stellar black hole is no more in contradiction with the second principle of thermodynamics. This approach also suggests a new derivation of the Bekenstein Hawking entropy formula:

Let A be the surface of a stellar black hole. Since the size of a cosmic bit is of the order of the Planck length, the number n_{cbh} of cosmic bits that form the surface of the stellar black hole is $n_{cbh} = A/l_{pl})^2$. Two of these cosmic bits define a specific dimension and therefore the maximal number n_{dbh} of dimensions of a black hole is $n_{dbh} = n_{cbh}/2$. The polarization φ_{bh} along one dimension is $\varphi_{bh} \cong \pm 1$ and the number W of configurations of a black hole is $W = 2^{ndbh}$.

Finally the black hole entropy is given by

$$S = \mathrm{Ln}(W) = n_{dbh}\mathrm{Ln}(2) = (\mathrm{Ln}(2)/2)\,A/(l_{Pl})^2 = 0.347\,A/(l_{Pl})^2$$

a formula close to but not identical to the Bekenstein formula [10].

$$S = 1/4\,A/(l_{Pl})^2 = 0.250\,A/(l_{Pl})^2$$

4.3. Galactic (Super-Massive) Black Holes

Is the nature of galactic black holes similar to the nature of stellar black holes, that is, do these two objects only differ by their sizes? The classical interpretation is that stellar and galactic black holes are the same objects indeed and that galactic black holes are built by stellar black holes that swallow more and more ordinary stars.

If correct this idea leads to the conclusion that the older a galactic black hole the larger it is. This conclusion is not supported by the observation. The largest galactic black holes are, by far, observed in the very distant (young) universe We are therefore compelled to imagine that the galactic black holes are objects closely connected to the Big-Bang itself.

Let us consider a region, that we shall call a galactic black hole, in the pre Big-Bang state (defined in Section 2.3) where $bJ < 1$, surrounded by a region where $4 < bJ < 5$. In this region, because $bJ < 1$, there is no space-time, no fields and no particles. After the transformation that follows the Big-Bang has been completed we have a universe similar to our universe surrounding a no-universe region, a region where no space, no time, no fields and no particles can exist.

This region is, along our point of view, a galactic black hole. For the very same reason that makes a stellar black hole deforms its surrounding space-time, a galactic black holes creates a strong gravitational field in its vicinity.

To summarize we think that stellar black holes and galactic black holes are different objects. Their main difference is in the dimensionality d_{bh} of their internal spaces, extremely high dimensionality for stellar black holes $d_{bh} >> 1$, low dimensionality $d_{bh} \cong 0$ because $bJ < 1$, for galactic black holes. Also a stellar black hole is in thermal equilibrium with its surrounding whereas the inside of a galactic black hole remains "hot", only its boundary with our universe can cool down. So would be the mechanism that gives birth to quasars.

5. QUASARS

5.1. Generalities

Quasi-stellar radio sources (or quasars) are the brightest objects of our universe. The energy they pour in space is so enormous that it is difficult to imagine a process based upon ordinary physics that could account for the phenomenon. The observation shows that these objects are closely related to galactic black holes and it has been put forward that the quasar energy could be generated by the disintegration of stars falling in the hole. The quasar energy would originate from the energy imbedded in the remains of the stars in a spherical region placed at about 10^3 r_S(Schwarzschild radius) from the centres of galactic black holes.

There are arguments that go against that scenario. For example there exist quasars not related to any galaxy. They are called "naked quasars", HE0450-2958 is an example [11]. There also exist quasars associated with galaxies housing sparse stars populations. Another model seems to be necessary.

5.2. A New Interpretation of the Physics of Quasars

An observational important feature is the fact that quasars mainly exist in the very distant universe. Their absence in the vicinity of our galaxy is generally simply explained by the lack of stars in the centres of old enough galaxies such as our Milky Way. Another interpretation, distinct from the fall of stars, is possible. Since galactic black holes seem to be the result of a process closely related to the Big-Bang we can assume that so is the case of quasars. Quasars would the manifestation of the phase transition (a transformation) of a part of the world points belonging to a galactic black hole (where $bJ < 1$) into world points belonging to our universe (where $bJ > 1$). This process would result from a local equalization process of cosmic noises b between the (hot) galactic black hole and our own (cold) space-time. The equalization process proceeds at the spherical boundary of the galactic black hole. According to this interpretation a quasar would be a region where space-time, fields and particles are created: Far from being a trap for matter a quasar would be a source of matter instead. Quasars, according to our opinion, would be the genitors of galaxies. There are indications that the quasars are older than galaxies [11], that galaxies housing large galactic black holes have more stars. Also HE0450-2958 seems to feed the galaxy it is linked to [12]. Finally the very far away quasars are hidden behind clouds of dust [13], an observation that could be explained by the emission of newly created matter by quasars.

According to this interpretation the size of a galactic quasar must decrease with time but the erosion process is not necessarily continuous. A galactic black hole, a region of space where $bJ < 1$, can possibly house regions where $bJ > 4$. When the boundary of a galactic black hole reaches such a region the erosion process stops and the galactic black hole becomes silent. So could be the case of Sgr A*. Some faint activity could remain, however, with the possible production of electron-positron pairs; a phenomenon that has been observed.

The tremendous energy released by quasars also receives a new interpretation. We have seen in Ref. [3] that the intensity of gauge interactions, electroweak and strong, is determined by the parameter $J^{(2)}$, the energy that binds two cosmic bits. The intensity of gravitation interactions is determined by the fluctuations of world point states ϕ_i (see [3] and Section 6.2) and their intensity is then determined by the parameter $J'^{(2)}/n$. The formation of a black hole implies a rearrangement of the cosmic bits of a world point a process that requires an energy $J'^{(2)} \sum_{a=1,\cdots,n} \sigma_a = nJ'^{(2)}$ (with $\sigma_a = 1$). This means that the very large gap between the gauge and gravitation forces which is of the order of $n \cong 10^{46}$ reappears between the magnitude of quasar and gauge forces which readily explains the observed huge energy emission phenomenon.

6. DARK ENERGY

6.1. A Non-Vanishing Cosmological Constant

The measurements of the luminosity of distant supernovae show that their host galaxies are farer than the position predicted by the usual universe expansion. Some repulsive interaction between the galaxies seems to be at work [14].

The introduction of a very weak cosmological constant Λ_E in the equations of General Relativity could account for this effect. From observation the astrophysicists find that $\Lambda_E \cong 1.4 \times 10^{-52}\ m^{-2}$. We note that a modification of the Newton gravitation law has already been put forward by Milgrom to explain the anomalous rotation speeds of the outer stars of galaxies (as already seen in Section 3.1). We show that both anomalies, the stars speed anomaly and the expansion anomaly, can be explained in the single framework provided by the model of discrete spaces that we have developed so far.

6.2. A Model of Gravitation

Some arguments that we develop in this section have already been discussed in Ref. [3].

In molecules the centres of gravity of positive and negative charges are generally different so giving rise to fluctuating electric dipoles. The dipoles interact and some dipole orientations lower the energy, the closer the systems the larger the energy lowering. The result is an attractive force, the so called Van der Waals interactions. The nuclear forces are also attractive and also arise from fluctuations, the fluctuations of quarks colours.

Since the classical gravitation interactions is attractive we put forward that gravitation could arise from the coupling between fluctuations, the fluctuations of the world point polarization ϕ_i, ($\phi_i = \{\phi_1, \phi_2, \phi_3, \phi_4\}$), or more precisely the fluctuations of the Lagrangian $\Lambda(\phi_i) = k_P\phi^T{}_i\phi_i$ of world point *i*. A particular cosmic bit is not necessarily assigned to a specific dimension μ of world point *i* and may be distributed between other dimensions according to the laws of a Gaussian distribution. The amplitude fluctuations of $\phi_i^T\phi_i$ are therefore of the order of $\sqrt{n}$. Since the average norm of state ϕ_i is fixed to 1, the relative fluctuations are of the order of $\sqrt{n}/n = 1/\sqrt{n}$.

If $\phi^T\phi$ is a random Gaussian variable with standard deviation ξ, $k_p\phi^T\phi$ (where k_p is a constant) is also a random Gaussian variable with standard deviation $\sqrt{\kappa_P\zeta}$ where, according to Ref. [1], $kp = (Mc)^2$. Therefore the relevant standard deviation ζ_P of the world points Lagrangian is given by the formula

$$\zeta_P = \sqrt{\kappa_P/n} = cM/\sqrt{n}$$

To Q, another (probe) particle, with mass m we associate a fluctuation $\zeta_Q = \sqrt{\kappa_Q/n} = cm/\sqrt{n}$. The two fluctuations interact through a propagation equation, the Klein-Gordon equation for zero-mass particles. The Fourier transform of the propagator of this equation is given by

$$\pi(q,\omega) = 1/\left[\omega^2 - q^2\right]$$

Then, owing to the weakness of vertices, the interaction may be calculated by using a low order perturbation expansion. The lowest order writes

$$V(q,\omega) = \frac{cM}{\sqrt{n}}\pi(q,\omega)\frac{cm}{\sqrt{n}} = \frac{c^2}{n}Mm\pi(q,\omega)$$

The static (ω = 0) gravitation interaction writes

$$V(q,\omega=0) = -\frac{c^2}{n} Mm \frac{1}{q^2}$$

The Fourier transform of $1/q^2$ in a d-dimensional space is given by $|r|^{2-ds}$ where $d_s = d - 1$ is the spatial dimensionality. This expression yields for d_s = 3

$$V(r) \propto -\frac{c^2}{n} Mm \frac{1}{|r|}.$$

The Newton expression of attractive gravitational forces is recovered:

$$f_{\mathrm{Newton}}(r) = -\frac{\mathrm{d}V(r)}{\mathrm{d}r} = -GMm \frac{1}{|r|^2}.$$

We observe that G behaves as $1/n$. We also observe that the internal interaction matrices G_p and G_Q are of the order of $J^{(2)}$ and since $G \cong \sqrt{\kappa_p \kappa_Q}$ the gravitation interaction intensity is of the order of $J^{(2)}/n$ as stated in Section 5.2.

6.3. Milgrom Interactions and the Cosmological Constant

The dimensionality of the internal interaction matrix G^{vacuum} that is the dimensionality of space-time is given by d = Int(bJ). For the internal space of a world point to be considered as a d-dimensional vector space it is necessary to limit the fluctuations of vectors components φ_μ. Since $|\varphi_\mu|^2 = |\phi|^2/d^2$ the standard deviation ζ_d of φ_μ is given by

$$\zeta_d = d \times \zeta$$

If the fluctuations of one component φ_μ of ϕ exceed ζ_d the associated dimension μ is lost, and the internal space of world point, instead of being a (3 + 1)-dimensional, becomes a (2 + 1) dimensional space. The proportion of such world points is given by

$$\xi = \int_{-\infty}^{-d\zeta} g(x)\mathrm{d}x + \int_{d\zeta}^{+\infty} g(x)\mathrm{d}x = 2\int_{d\zeta}^{\infty} g(x)\mathrm{d}x$$

where $g(x)$ is the Gauss distribution

$$g(x)=\frac{1}{\zeta\sqrt{2\pi}}\exp\left(-\frac{x^2}{2\zeta^2}\right)$$

Then by letting $2\zeta_2 = 1$ one has

$$\xi=\frac{2}{\sqrt{\pi}}\int_d^{\infty}\exp\left(-x^2\right)\mathrm{d}x=\frac{2}{\sqrt{\pi}}\int_0^{\infty}\exp\left(-x^2\right)\mathrm{d}x-\frac{2}{\sqrt{\pi}}\int_0^{d}\exp\left(-x^2\right)\mathrm{d}x=1-\mathrm{Erf}\left(d\right)\cong\exp\left(-d^2\right)$$

Since d = 4 we find

$$\xi\ 1.12\times 10^{-7}$$

The form of the gravitation interaction associated with (2 + 1)-dimensional world points is modified because the Fourier transform of $1/q^2$ is now 2-dimensional. The potential in a 2-dimensional space becomes

$$V(r)=-\frac{c^2}{n}Mm\mathrm{Ln}\left(1/r\right)=\frac{c^2}{n}Mm\mathrm{Ln}\left(|r|\right)$$

Finally this gravitational forces is attractive and given by

$$F_{\mathrm{Milgrom}}(r)=-\frac{\mathrm{d}V(r)}{\mathrm{d}r}\cong-GMm\frac{1}{|r|}.$$

The Milgrom gravitation interaction is recovered (Section 3.1) and the gravitation interaction becomes

$$F_{N+M}=-GMm\left(\frac{1}{|r|^2}+\frac{1}{r_M|r|}\right)=-GMm\frac{1}{|r|^2}\left(1+\frac{|r|}{r_M}\right)=-GMm\frac{1}{|r|^2}\mu(x)$$

with $\mu(x)=1+|r|/r_M$. Then $\mu(r) = 1$ for $r \ll r_M$ and $\mu(r) \propto r$ for $r \gg r_M$ as assumed by Milgrom.

The disappearance of a dimension may be interpreted as the shrinking of the lengths associated to that dimension by a factor ξ. This reminds the shrinking of non observable dimensions in string theories (where 10 or 11 dimensional spaces are reduced to the classical 4-dimensional spaces) with the difference that we have here a mechanism that really puts the process at work. For example a standard length r_M is now seen as a length $\xi\, r_M$.

A world point *i* eventually loses another dimension if the fluctuations perturb simultaneously two field components $\varphi_{i,\mu}$ and $\varphi_{i,\nu}$ of the polarization ϕ_i of *i*. The probability for such a situation to occur is $\xi^2 \cong 1.2\times10^{-14}$. Then the internal space of world point becomes (1 + 1)-dimensional and the interaction potential becomes

$$V(r) \propto -c^2 mM|r|$$

The associated gravitation force is

$$F_\Lambda = -\frac{\mathrm{d}V(r)}{\mathrm{d}r} = +\frac{G}{r_E^2}mM$$

a repulsive constant force that acts as a negative pressure exactly as does the cosmological constant.

The formula gathering all various contributions to the gravitation forces is

$$F_{N+M+\Lambda}(|r|) = -GmM\left(\frac{1}{|r|^2}+\frac{1}{|r|r_M}-\frac{1}{(r_E)^2}\right)$$

In this equation $\Lambda_E = 1/r_E^2$ is the cosmological constant.

Since the motion of stars in the disks of galaxies is determined by the Mond dynamics the Milgrom parameter r_M must be of the order of, or less than, the galactic bulbs radii. The galaxy bulbs diameters φ_{Bulb} are of the order of 2×10^3 ly (light year) to 10^4 ly. We chose the value r_M but the value of r_M could be smaller.

Since the distance travelled by light in one year is 9.5×10^{15}, $r_{Bulb} \cong 10^{19}$ and, with ξ 1.1×10^{-7}, $r_E = \xi^{-1} r_M \cong 0.9\times10^{26}$. Finally, $\Lambda_E \cong 1.2\times10^{-52}$ m^{-2} which is close to the experimental observed value $\Lambda_E \cong 1.4\times10^{-52}$ m^{-2}. The agreement is striking but it must not be taken too strictly because it depends on a poorly known parameter, the Milgrom range r_M. The main interest of the derivation is that it seems to give the right orders of magnitude to the cosmological constant.

7. DISCUSSION AND CONCLUSIONS

The present paper is the last of a series of five articles devoted to the physics of discrete spaces. The purpose of preceding contributions was to show that a simple spin glass model could account for the most fundamental constructions of theoretical physics, namely the structure of space-time [1], the postulates of quantum theory [2], the gauge interactions [3], and the Standard Model of Particles [4]. Here the purpose is quite different: one uses the model to put forward an explanation to not completely understand phenomena in the field of cosmology. So is the case for the scenario of the Big-Bang, the nature of dark matter, the physics of black holes and a more detailed description of quasars.

Could the model of discrete spaces be falsified, a necessary condition, according to Popper, for a theory to deserve the adjective "scientific"?

The model is based upon three very simple and general statements that we cannot reject without jeopardizing physics itself: The universe does exist, it is not completely chaotic and it is not completely frozen. We could negate the model but this would lead to a rebuttal of at least one of well established theories. Therefore, it is difficult to say that the theory of discrete spaces that we put forward can be falsified on the account of the first four articles. This is not the case for the present contribution. For example our proposal leads to the conclusion that the search of dark matter particles is hopeless because dark matter is not made of matter. If such particles are found some aspects of the theory would have, obviously, to be revised. We think that the touchstone of the present model is in the nature of quasars because quasars are, in our approach, to be considered as the still active embers of the Big-Bang.

ACKNOWLEDGEMENTS

I would like to thank Pr Aurélien Barreau for the useful discussions we have had as regards the Pioneer effect.

REFERENCES

1. Peretto, P. (2014) Journal of Modern Physics, 8, 563-575. http://dx.doi.org/10.4236/jmp.2014.58067
2. Peretto, P. (2014) Journal of Modern Physics, 14, 1370-1386. http://dx.doi.org/10.4236/jmp.2014.514138

3. Peretto, P. (2014) Journal of Modern Physics, 18, 2049-2062. http://dx.doi.org/10.4236/jmp.2014.518201
4. Peretto, P. (2015) Journal of Modern Physics, 6, 811-828. http://dx.doi.org/10.4236/jmp.2015.66086
5. Hartle, J.B. and Hawking, S. (1983) Physical Review D, 28, 2860. http://dx.doi.org/10.1103/PhysRevD.28.2960
6. Wick (1956) Physical Review, 101, 1830.
7. Steinhardt, P.J. and Turok, N. (2004) New Astronomy Reviews, 49, 2-6.
8. Guth, A. (1982) Physical Review Letters, 49, 1110-1113. http://dx.doi.org/10.1103/PhysRevLett.49.1110
9. Tayler, J. (1991) Hidden Matter. Elis Horwood, Chichester.
10. Bekenstein, J.D. (1975) Physical Review D, 12, 3077. http://dx.doi.org/10.1103/PhysRevD.12.3077
11. Tista, M. (1975) Physics Letters B, 561, 53-56. http://dx.doi.org/10.1016/0370-2693(75)90497-9
12. Elbaz, D. (2009) Astronomy and Astrophysics, 507, 1359-1374. http://dx.doi.org/10.1051/0004-6361/200912848
13. Margain, P. (2005) Nature, 437, 381. http://dx.doi.org/10.1038/nature04013
14. Peebles, P.J. and Ratra, B. (2003) Reviews of Modern Physics, 45, 559-606. http://dx.doi.org/10.1103/RevModPhys.75.559

CHAPTER 15

The Mathematical Foundations of General Relativity Revisited

Jean-Francois Pommaret

CERMICS, Ecole des Ponts ParisTech, Marne-la-Vall_ee Cedex 02, France

ABSTRACT

The purpose of this paper is to present for the first time an elementary summary of a few recent results obtained through the application of the formal theory of partial differential equations and Lie pseudogroups in order to revisit the mathematical foundations of general relativity. Other engineering examples (control theory, elasticity theory, electromagnetism) will also be considered in order to illustrate the three fundamental results that we shall provide successively. **1) VESSIOT VERSUS CARTAN:** The quadratic terms appearing in the "Riemann tensor" according to the "Vessiot structure equations" must not be identified with the quadratic terms appearing in the well known "Cartan structure equations" for Lie groups. In particular, "curvature + torsion" (Cartan) must not be considered as a generalization of "curvature alone" (Vessiot). **2) JANET VERSUS SPENCER:** The "Ricci tensor" only depends on the nonlinear transformations (called "elations" by Cartan in 1922) that describe the "difference" existing between the Weyl group (10 parameters of the Poincaré subgroup + 1 dilatation) and the conformal group of space-time (15 parameters). It can be defined without using the indices leading to the standard contraction or trace of the Riemann tensor. Meanwhile, we shall obtain the number of components of the Riemann and Weyl tensors without any combinatoric argument on the exchange of indices. Accordingly and contrary to the "Janet sequence", the "Spencer sequence" for the conformal Killing system and its formal adjoint fully describe the Cosserat equations, Maxwell equations and Weyl equations but General Relativity is not coherent with this result. **3) ALGEBRA VERSUS GEOMETRY:** Using the powerful methods of "Algebraic Analysis", that is a mixture of homological agebra and differential geometry, we shall prove that, contrary to other equations of physics (Cauchy equations, Cosserat equations, Maxwell equations), the Einstein equations cannot be "parametrized", that is the generic solution cannot be expressed by means of the derivatives of a certain number of arbitrary potential-like functions, solving therefore negatively a 1000 $ challenge proposed by J. Wheeler in 1970. Accordingly, the mathematical foundations of electromagnetism and gravitation must be revisited within this formal framework, though striking it may look like. We insist on the fact that the arguments presented are of a purely mathematical nature and are thus unavoidable.

Keywords: General Relativity; Riemann Tensor; Weyl Tensor; Ricci Tensor; Einstein Equations; Lie Groups; Lie Pseudogroups; Differential Sequence; Spencer Operator; Janet Sequence; Spencer Sequence; Differential Module; Homological Algebra; Extension Modules; Split Exact Sequence

1. INTRODUCTION

The purpose of this paper is to present an elementary summary of a few recent results obtained through the application of the formal theory of systems of ordinary differential (OD) or partial differential (PD) equations and Lie pseudogroups in order to revisit the mathematical foundations of general relativity (GR). More elementary engineering examples (elasticity theory, electromagnetism (EM)) will also be considered in order to illustrate the quoted three fundamental results that we shall provide. The paper, based on the material of two lectures given at the department of mathematics of the university of Montpellier 2, France, in may 2013 and Firenze, Italy, in june 2013, is divided into three parts corresponding to the different formal methods used.

PART 1: In 1880 S. Lie (1842-1899) studied the groups of transformations depending on a finite number of parameters and now called Lie groups of transformations. Ten years later he discovered that these groups are only examples of groups of transformations solutions of linear or nonlinear systems of ordinary differential (OD) or partial differential (PD) equations which may even be of high order and are now called Lie pseudogroups of transformations. During the next fifty years the latter groups have only been studied by two frenchmen, namely Elie Cartan (1869-1951) who is quite famous today, and Ernest Vessiot (1865-1952) who is almost ignored today. We have proved in many books and papers that the Cartan structure equations have nothing to do with the Vessiot structure equations still not known today. Accordingly, the quadratic terms appearing in the Riemann tensor must not be identified with the quadratic terms appearing in the well known Maurer-Cartan equations for Lie groups. In particular, curvature + torsion (Cartan) must not be considered as a generalization of curvature alone (Vessiot).

PART 2: Though we consider that the first formal work on systems of PD equations is dating back to Maurice Janet (1888-1983) who introduced as early as in 1920 a differential sequence now called Janet sequence, it is only around 1970 that Donald Spencer (1912-2001) developped, in a quite independent way, the formal theory of systems of PD equations in order to study Lie pseudogroups, exactly like E. Cartan did with exterior systems. However, all the physicists who tried to understand the only book "Lie equations" that he published in 1972 with A. Kumpera, have been stopped by the fact that the examples of the Introduction (Janet sequence) have nothing to do with the core of the book (Spencer sequence). We may say that the work of Cartan is superseded by the use of the canonical Spencer sequence while the work of Vessiot is superseded by the use of the canonical Janet sequence but the link between these two sequences and thus these two works is not known today. Accordingly, the Spencer sequence for the conformal Killing system and its formal adjoint fully describe the Cosserat equations, Maxwell equations and Weyl equations but general relativity (GR) is not coherent with this result because we shall prove that the Ricci tensor only depends on the nonlinear transformations (called elations by Cartan in 1922) that describe the "difference" existing between the Weyl group (10 parameters of the Poincaré subgroup + 1 dilatation) and the conformal group of space-time (15 parameters).

PART 3: At the same time, mixing differential geometry and homological algebra but always supposing that the reader knows a lot about the work of Spencer, V.P. Palamodov (constant coefficients) and M. Kashiwara (variable coefficients) developped "algebraic analysis" in order to study the formal properties of finitely generated differential modules that do not depend on their presentation or even on a corresponding differential resolution, namely the algebraic analogue of a differential sequence. Using double duality theory, we prove that, contrary to other equations of physics (Cauchy equations, Cosserat equations, Maxwell equations), the Einstein equations cannot be "parametrized", that is the generic solution cannot be expressed by means of the derivatives of a certain number of arbitrary potential-like functions, solving therefore negatively a 1000 $ challenge proposed by J. Wheeler in 1970.

The new methods involve tools from differential geometry (jet theory, Spencer operator, δ-cohomology) and homological algebra (diagram chasing, snake theorem, extension modules, double duality). The reader may just have a look to the book [1] (review in Zbl 1079. 93001) in order to understand the amount of mathematics needed from many domains.

The following diagram summarizes at the same time the historical background and the difficulties presented in the introduction:

$$\begin{array}{ccccc} & & CARTAN & \rightarrow & SPENCER \\ & \nearrow & & & \\ LIE & & \updownarrow & ? & \updownarrow \\ & \searrow & & & \\ & & VESSIOT & \rightarrow & JANET \end{array}$$

Roughly, Cartan and followers have not been able to "quotient down to the base manifold" [2,3], a result only obtained by Spencer in 1970 through the nonlinear Spencer sequence [4-7] but in a way quite different from the one followed by Vessiot in 1903 for the same purpose [8,9]. Accordingly, the mathematical foundations of mathematical physics must be revisited within this formal framework, though striking it may look like for certain apparently well established theories such as EM (J. C. Maxwell, 1864) and GR (A. Einstein, 1915).

2. FIRST PART: FROM LIE GROUPS TO LIE PSEUDOGROUPS

If X is a manifold with local coordinates (x^i) for $i = 1,..., n = \dim(X)$, let $\mathcal{E}$ be a fibered manifold over X, that is a manifold with local coordinates (x^i, y^k) for $i = 1,..., n$ and $k = 1,...,m$ simply denoted by (x, y), projection

$$\pi : \mathcal{E} \rightarrow X : (x,y) \rightarrow (x)$$

and changes of local coordinates

$$\bar{x} = \varphi(x), \bar{y} = \psi(x,y).$$

If $\mathcal{E}$ and $\mathcal{F}$ are two fibered manifolds over X with respective local coordinates (x, y) and (x, z), we denote by $\mathcal{E} \times_X \mathcal{F}$ the fibered product of $\mathcal{E}$

and $\mathcal{F}$ over X as the new fibered manifold over X with local coordinates (x, y, z). We denote by

$$f : X \to \mathcal{E} : (x) \to (x, y = f(x))$$

a global section of $\mathcal{E}$, that is a map such that $\pi \circ f = id_X$ but local sections over an open set $U \subset X$ may also be considered when needed. We shall use for simplicity the same notation for a fibered manifold and its set of sections while setting $\dim_X(\mathcal{E}) = m$. Under a change of coordinates, a section transforms like

$$\bar{f}(\varphi(x)) = \psi(x, f(x))$$

and the derivatives transform like:

$$\frac{\partial \bar{f}^l}{\partial \bar{x}^r}(\varphi(x))\partial_i \varphi^r(x)$$
$$= \frac{\partial \psi^l}{\partial x^i}(x, f(x)) + \frac{\partial \psi^l}{\partial y^k}(x, f(x))\partial_i f^k(x)$$

We may introduce new coordinates (x^i, y^k, y_i^k) transforming like:

$$\bar{y}_r^l \partial_i \varphi^r(x) = \frac{\partial \psi^l}{\partial x^i}(x, y) + \frac{\partial \psi^l}{\partial y^k}(x, y) y_i^k$$

We shall denote by $J_q(\mathcal{E})$ the q-jet bundle of $\mathcal{E}$

with local coordinates

$$(x^i, y^k, y_i^k, y_{ij}^k, \cdots) = (x, y_q)$$

called jet coordinates and sections

$$f_q : (x) \to (x, f^k(x), f_i^k(x), f_{ij}^k(x), \cdots)$$
$$= (x, f_q(x))$$

transforming like the sections

$$j_q(f):(x)\rightarrow\left(x,f^k(x),\partial_i f^k(x),\partial_{ij} f^k(x),\cdots\right)$$
$$=\left(x,j_q(f)(x)\right)$$

where both f_q and $j_q(f)$ are over the section f of $\mathcal{E}$. Of course $J_q(\mathcal{E})$ is a fibered manifold over X with projection π_q while $J_{q+r}(\mathcal{E})$ is a fibered manifold over $J_q(\mathcal{E})$ with projection $\pi_q^{q+r}, \forall r\geq 0$.

DEFINITION 1.1: A (nonlinear) system of order q on $\mathcal{E}$ is a fibered submanifold $\mathcal{R}_q\subset J_q(\mathcal{E})$ and a solution of $\mathcal{R}_q$ is a section f of $\mathcal{E}$ such that $j_q(f)$ is a section of $\mathcal{R}_q$.

DEFINITION 1.2: When the changes of coordinates have the linear form $\bar{x}=\varphi(x),\bar{y}=A(x)y$, we say that $\mathcal{E}$ is a vector bundle over *X*. Vector bundles will be denoted by capital letters *C,E,F* and will have sections denoted by ξ,η,ζ. In particular, we shall denote as usual by $T = T(X)$ the tangent bundle of *X*, by $T^* = T^*(X)$ the cotangent bundle, by $\wedge^r T^*$ the bundle of r-forms and by SqT^* the bundle of qsymmetric covariant tensors. When the changes of coordinates have the form

$$\bar{x}=\varphi(x),\ \ \bar{y}=A(x)y+B(x)$$

we say that $\mathcal{E}$ is an affine bundle over *X* and we define the associated vector bundle *E* over *X* by the local coordinates (*x*, *v*) changing like $\bar{x}=\varphi(x)$ $\bar{v}=A(x)v$. Finally, If $\mathcal{E} = X\times X$, we shall denote by $\Pi_q = \Pi_q(X, X)$the open subfibered manifold of $J_q(X\times X)$ defined independently of the coordinate system by $\det\left(y_i^k\right)\neq 0$ with source projection

$$\alpha_q:\Pi_q\rightarrow X:\left(x,y_q\right)\rightarrow(x)$$

and target projection

$$\beta_q:\Pi_q\rightarrow X:\left(x,y_q\right)\rightarrow(y).$$

DEFINITION 1.3: If the tangent bundle $T(\mathcal{E})$ has local coordinates (*x,y,u,v*) changing like

$$\bar{u}^j=\partial_i\varphi^j(x)u^i,$$
$$\bar{v}^l=\frac{\partial\psi^l}{\partial x^i}(x,y)u^i+\frac{\partial\psi^l}{\partial y^k}(x,y)v^k,$$

we may introduce the vertical bundle $V(\mathcal{E}) \subset T(\mathcal{E})$ as a vector bundle over $\mathcal{E}$ with local coordinates (x,y,v) obtained by setting $u = 0$ and changes

$$\bar{v}^l = \frac{\partial \psi^l}{\partial y^k}(x,y)v^k .$$

Of course, when $\mathcal{E}$ is an affine bundle over X with associated vector bundle E over X, we have $V(\mathcal{E}) = \mathcal{E} \times_X E$ ..

For a later use, if $\mathcal{E}$ is a fibered manifold over X and f is a section of $\mathcal{E}$, we denote by $f^{-1}(V(\mathcal{E}))$ the reciprocal image of $V(\mathcal{E})$ by f as the vector bundle over X obtained when replacing (x,y,v) by $(x,f(x),v)$ in each chart. A similar construction may also be done for any affine bundle over $\mathcal{E}$.

We now recall a few basic geometric concepts that will be constantly used through this paper. First of all, if $\xi, \eta \in T$, we define their bracket $[\xi, \eta] \in T$ by the local formula

$$([\xi,\eta])^i(x) = \xi^r(x)\partial_r\eta^i(x) - \eta^s(x)\partial_s\xi^i(x)$$

leading to the Jacobi identity

$$[\xi,[\eta,\zeta]] + [\eta,[\zeta,\xi]] + [\zeta,[\xi,\eta]] = 0,$$
$$\forall \xi,\eta,\zeta \in T$$

allowing to define a Lie algebra and to the useful formula

$$[T(f)(\xi), T(f)(\eta)] = T(f)([\xi,\eta])$$

where $T(f): T(X) \to T(Y)$ is the tangent mapping of a map $f: X \to Y$.

When $I = \{i_1 < \cdots < i_r\}$ is a multi-index, we may set

$$dx^I = dx^{i_1} \wedge \cdots \wedge dx$$

for describing $\wedge^r T^{*i_r}$ by means of a basis and introduce the exterior derivative

$$d : \wedge^r T^* \to \wedge^{r+1} T^* : \omega = \omega_I dx^I \to d\omega$$
$$= \partial_i \omega_I dx^i \wedge dx^I$$

with $d^2 = d^\circ d \equiv 0$ in the Poincaré sequence:

$$\wedge^0 T^* \xrightarrow{d} \wedge^1 T^* \xrightarrow{d} \wedge^2 T^* \xrightarrow{d} \cdots \xrightarrow{d} \wedge^n T^* \to 0 .$$

The Lie derivative of an r-form with respect to a vector field $\xi \in T$ is the linear first order operator $\mathcal{L}(\xi)$ linearly depending on $j_1(\xi)$ and uniquely defined by the following three properties:

1) $\mathcal{L}(\xi) f = \xi . f = \xi^i \partial_i f, \forall f \in \wedge^0 T^* = C^\infty(X)$.

2) $(\xi) d = d\mathcal{L}(\xi)$.

3) $\mathcal{L}(\xi)(\alpha \wedge \beta) = (\mathcal{L}(\xi)\alpha) \wedge \beta + \alpha \wedge (\mathcal{L}(\xi)\beta), \forall \alpha, \beta \in \wedge T^*$.

It can be proved that

$$\mathcal{L}(\xi) = i(\xi)d + di(\xi)$$

where $i(\xi)$ is the interior multiplication

$$\left(i(\xi)\omega\right)_{i_1 \cdots i_r} = \xi^i \omega_{i i_1 \cdots i_r}$$

and that

$$[\mathcal{L}(\xi), \mathcal{L}(\eta)] = \mathcal{L}(\xi) \circ \mathcal{L}(\eta) - \mathcal{L}(\eta) \circ \mathcal{L}(\xi)$$
$$= \mathcal{L}([\xi, \eta]), \forall \xi, \eta \in T.$$

We now turn to group theory and start with two basic definitions:

Let G be a Lie group, that is a manifold with local coordinates (a^τ) for $\tau = 1, \ldots, p = \dim(G)$ called parameters, a composition

$$G \times G \to G : (a, b) \to ab ,$$

an inverse $G \to G : a \to a^{-1}$ and an identity $e \in G$ satisfying:

$$(ab)c = a(bc) = abc,\ aa^{-1} = a^{-1}a = e,$$
$$ae = ea = a \quad \forall a,b,c \in G.$$

DEFINITION 1.4: G is said to act on X if there is a map $X \times G \to X : (x,a) \to y = ax = f(x,a)$ such that $(ab)x = a(bx) = abx,\ \forall a,b \in G,\ \forall x \in X$ and we shall say that we have a Lie group of transformations of X. In order to simplify the notations, we shall use global notations even if only local actions are existing. It is well known that the action of G onto itself allows to introduce a purely algebraic bracket on its Lie algebra $\mathcal{G} = T_e(G)$.

DEFINITION 1.5: A Lie pseudogroup of transformations $\Gamma \subset \mathrm{aut}(X)$ is a group of transformations solutions of a system of OD or PD equations such that, if $y = f(x)$ and $z = g(y)$ are two solutions, called finite transformations, that can be composed, then

$$z = g \circ f(x) = h(x)$$

and

$$x = f^{-1}(y) = g(y)$$

are also solutions while $y = x$ is the identity solution denoted by $id = id_X$ and we shall set $id_q = j_q(id)$. In all the sequel we shall suppose that Γ is transitive that is

$$\forall x, y \in X,\ \exists f \in \Gamma,\ y = f(x).$$

It becomes clear that Lie groups of transformations are particular cases of Lie pseudogroups of transformations as the system defining the finite transformations can be obtained by eliminating the parameters among the equations

$$y_q = j_q(f)(x,a)$$

when q is large enough. The underlying system may be nonlinear and of high order. Looking for transformations "close" to the identity, that is setting $y = x + t\xi(x) + \ldots$ when $t \ll 1$ is a small constant parameter and passing to the limit $t \to 0$, we may linearize the above nonlinear system of finite Lie equations in order to obtain a linear system of infinitesimal Lie equations of the same order for vector fields. Such a system has the property that, if ξ, η are two solutions, then $[\xi, \eta]$ is also a solution. Accordingly, the set $\Theta \subset T$ of solutions of this new system satisfies $[\Theta, \Theta] \subset \Theta$ and can therefore be considered as the Lie algebra of Γ.

EXAMPLE 1.6: While the affine transformations $y = ax + b$ are solutions of the second order linear system $y_{xx} = 0$, the projective transformations

$$y = (ax + b)/(cx + d)$$

are solutions of the third order nonlinear system

$$\Psi \equiv (y_{xxx}/y_x) - \frac{3}{2}(y_{xx}/y_x)^2 = 0.$$

The sections of the corresponding linearized systems are respectively satisfying $\xi_{xx} = 0$ and $\xi_{xxx} = 0$. The generating differential invariant $\Phi \alpha\, y_{xx} / y_x$ of the affine case transforms like

$$u = \bar{u}\partial_x f + (\partial_{xx} f / \partial_x f)$$

when $\bar{x} = f(x)$ while Ψ transforms like

$$v = \bar{v}(\partial_x f)^2 + (\partial_{xxx} f / \partial_x f) - \frac{3}{2}(\partial_{xx} f / \partial_x f)^2.$$

We now sketch the discovery of Vessiot [8,9] still not known today after more than a century for reasons which are not scientific at all. Roughly, a Lie pseudogroup $\Gamma \subset \mathrm{aut}(X)$ is made by finite transformations $y = f(x)$ solutions of a (possibly nonlinear) system $\mathcal{R}_q \subset \Pi_q$ while the infinitesimal transformations $\xi \in \Theta$ are solutions of the linearized system

$$R_q = id_q^{-1}(V(\mathcal{R}_q)) \subset J_q(T)$$

as we have

$$T = id^{-1}(V(X \times X))$$

When Γ is transitive, there is a canonical epimorphism $\pi_0^q : R_q \to T$. Also, as changes of source x commute with changes of target y, they exchange between themselves any generating set of differential invariants $\{ (y_q)\}$ as in the previous example.Then one can introduce a natural bundle $\mathcal{F}$ over X, also called bundle of geomeric objects, by patching changes of coordinates of the form

$$\bar{x} = f(x), \bar{u} = \lambda\left(u, j_q\left(f(x)\right)\right)$$

thus obtained. A section ω of $\mathcal{F}$ is called a geometric object or structure on X and transforms like

$$\bar{\omega}\left(f(x)\right) = \lambda\left(\omega(x), j_q(f)(x)\right)$$

or simply $\bar{\omega} = j_q(f)^{-1}(\omega)$. This is a way to generalize vectors and tensors (q = 1) or even connections (q = 2). As a byproduct we have

$$\Gamma = \left\{ f \in \mathrm{aut}(X) \middle| j_q(f)^{-1}(\omega) = \omega \right\}$$

and we may say that Γ preserves ω. Replacing $j_q(f)$ by f_q, we also obtain

$$\mathcal{R}_q = \left\{ f_q \in \Pi_q \middle| f_q^{-1}(\omega) = \omega \right\}$$

Coming back to the infinitesimal point of view and setting

$$f_t = \exp(t\xi) \in \mathrm{aut}(X), \ \forall \xi \in T$$

we may define the ordinary Lie derivative with value in the vector bundle $F_0 = \omega^{-1}\left(V(\mathcal{F})\right)$ by the formula:

$$\begin{aligned} \mathcal{D}\xi = \mathcal{L}(\xi)\omega = \frac{\mathrm{d}}{\mathrm{d}t} j_q(f)_t^{-1}(\omega)\Big|_{t=0} \Rightarrow \Theta \\ = \left\{ \xi \in T \middle| \mathcal{L}(\xi)\omega = 0 \right\} \end{aligned}$$

and we say that $\mathcal{D}$ is a Lie operator because $\mathcal{D}\xi = 0, \mathcal{D}\eta = 0 \Rightarrow \mathcal{D}[\xi, \eta] = 0$ as we already saw.

Differentiating r times the equations of R_q that only depend on $j_1(\text{?})$, we may obtain the r- prolongation

$$R_{q+r} = J_r(R_q) \cap J_{q+r}(T) \subset J_r(J_q(T)).$$

The problem is then to know under what conditions on ω all the equations of order $q+r$ are obtained by r prolongations only, $\forall r \geq 0$ or, equivalently, $R_q$is formally integrable (FI). The solution, found by Vessiot, has been to exhibit another natural vector bundle $\mathcal{F}_1$ with local coordinates (x,u,v) over $\mathcal{F}$ with local coordinates (x,u) and to prove that an equivariant section $c:\mathcal{F} \to \mathcal{F}_1 : (x,u) \to (x,u,v=c(u))$ only depends on a finite number of constants called structure constants. The integrability conditions (IC) of R_q, called Vessiot structure equations, are of the form $I(j_1(\omega)) = c\ (\omega)$ and are invariant under any change of source.

We provide in a self-contained way and parallel manners the following five striking examples which are among the best nontrivial ones we know and invite the reader to imagine at this stage any possible link that could exist between them (A few specific definitions will be given later on).

EXAMPLE 1.7: Coming back to the last example, we show that Vessiot structure equations may even exist when $n = 1$. For this, if γ is the geometric object of the affine group $y = ax + b$ and $0 \neq \alpha = \alpha(x)dx \in T^*$ is a 1-form, we consider the object $\omega = (\alpha,\gamma)$ and get at once the two second order Medolaghi equations:

$$\mathcal{L}(\xi)\alpha \equiv \alpha\partial_x\xi + \xi\partial_x\alpha = 0,$$
$$\mathcal{L}(\xi)\gamma \equiv \partial_{xx}\xi + \gamma\partial_x\xi + \xi\partial_x\gamma = 0$$

Differentiating the first equation and substituting the second, we get the zero order equation:

$$\xi\left(\alpha\partial_{xx}\alpha - 2(\partial_x\alpha)^2 + \alpha\gamma\partial_x\alpha - \alpha^2\partial_x\gamma\right)$$
$$= 0 \Leftrightarrow \xi\partial_x\left(\frac{\partial_x\alpha}{\alpha^2} - \frac{\gamma}{\alpha}\right) = 0$$

and the Vessiot structure equation $\partial_x\alpha - \gamma\alpha = c\alpha^2$. Alternatively, setting

$$\beta = -1/\alpha \in T,$$

we get

$$\partial_x \beta + \gamma\beta = c\,.$$

With

$$\alpha = 1, \beta = -1, \gamma = 0 \Rightarrow c = 0$$

we get the translation subgroup $y = x + b$ while, with

$$\alpha = 1/x, \beta = -x, \gamma = 0 \Rightarrow c = -1$$

we get the dilatation subgroup $y = ax$. Similarly, if v is the geometric object of the projective group and we consider the new geometric object $\omega = (\gamma, v)$, we get the only Vessiot structure equation

$$\partial_x \gamma - \frac{1}{2}\gamma^2 - v = 0$$

without any structure constant.

EXAMPLE 1.8: (Principal homogeneous structure) When Γ is the Lie group of transformations made by the constant translations $y^i = x^i + a^i$ for $i = 1,..., n$ of a manifold X with $\dim(X) = n$, the characteristic object invariant by Γ is a family

$$\omega = \left(\omega^\tau = \omega_i^\tau dx^i\right) \in T^* \times_X \cdots \times_X T^*$$

of n 1-forms with $\det(\omega) \neq 0$ in such a way that

$$\Gamma = \left\{ f \in \mathrm{aut}(X) \middle| j_1(f)^{-1}(\omega) = \omega \right\}$$

where aut(X) denotes the pseudogroup of local diffeomorphisms of X, $j_q(f)$denotes the derivatives of f up to order q and $j_1(f)$ $j_1(f)$ acts in the usual way on covariant tensors. For any vector field $\xi \in T = T(X)$ the tangent bundle to X, introducing the standard Lie derivative $\mathcal{L}(\xi)$ of forms with respect to ?, we may consider the n^2 first order Medolaghi equations:

$$\Omega_i^\tau \equiv \left(\mathcal{L}(\xi)\omega\right)_i^\tau \equiv \omega_r^\tau(x)\partial_i\xi^r + \xi^r\partial_r\omega_i^\tau(x) = 0 .$$

The particular situation is found with the special choice $\omega = (dx^i)$ that leads to the involutive system $\partial_i\xi^k = 0$. Introducing the inverse matrix $\alpha = (\alpha_\tau^i) = \omega^{-1}$, the above equations amount to the bracket relations $[\xi,\alpha_\tau] = 0$ and, using crossed derivatives on the solved form

$$\partial_i\xi^k + \xi^r\alpha_\tau^k(x)\partial_r\omega_i^\tau(x) = 0 ,$$

we obtain the $n^2(n-1)/2$ zero order equations:

$$\xi^r\partial_r\left(\alpha_\rho^i(x)\alpha_\sigma^j(x)\left(\partial_i\omega_j^\tau(x) - \partial_j\omega_i^\tau(x)\right)\right) = 0 .$$

The integrability conditions (IC), that is the conditions under which these equations do not bring new equations, are thus the $n^2(n-1)/2$ Vessiot structure equations:

$$\partial_i\omega_j^\tau(x) - \partial_j\omega_i^\tau(x) = c_{\rho\sigma}^\tau\omega_i^\rho(x)\omega_j^\sigma(x)$$

with $n^2(n-1)/2$ structure constants $c = \left(c_{\rho\sigma}^\tau = -c_{\sigma\rho}^\tau\right)$.

When $X = G$, these equations can be identified with the Maurer-Cartan equations (MC) existing in the theory of Lie groups, on the condition to change the sign of the structure constants involved because we have

$$\left[\alpha_\rho, \alpha_\sigma\right] = -c_{\rho\sigma}^\tau\alpha_\tau .$$

Writing these equations in the form of the exterior system $d\omega^\tau = c_{\rho\sigma}^\tau\omega^\rho \wedge \omega$ and closing this system by applying once more the exterior derivative d, we obtain the quadratic IC:

$$c_{\mu\rho}^\lambda c_{\sigma\tau}^\mu + c_{\mu\sigma}^\lambda c_{\tau\rho}^\mu + c_{\mu\tau}^\lambda c_{\rho\sigma}^\mu = 0$$

also called Jacobi relations $J(c) = 0$.

EXAMPLE 1.9: (Riemann structure) If

$$\omega = \left(\omega_{ij} = \omega_{ji}\right) \in S_2 T^*$$

is a metric on a manifold X with dim $(X) = n$ such that det$(\omega) \neq 0$, the Lie pseudogroup of transformations preserving ω is

$$\Gamma = \left\{ f \in \mathrm{aut}(X) \middle| j_1(f)^{-1}(\omega) = \omega \right\}$$

and is a Lie group with a maximum number of $n(n+1)/2$ parameters. A special metric could be the Euclidean metric when n = 1,2,3 as in elasticity theory or the Minkowski metric when n = 4 as in special relativity [10]. The first order Medolaghi equations:

$$\begin{aligned}\Omega_{ij} &\equiv \left(\mathcal{L}(\xi)\omega\right)_{ij}\\ &\equiv \omega_{rj}(x)\partial_i \xi^r + \omega_{ir}(x)\partial_j \xi^r + \xi^r \partial_r \omega_{ij}(x) = 0\end{aligned}$$

are also called classical Killing equations for historical reasons. The main problem is that this system is not involutive unless we prolong it to order two by differentiating once the equations. For such a purpose, introducing $\omega^{-1} = (\omega^{ij})$ as usual, we may define the Christoffel symbols:

$$\begin{aligned}\gamma_{ij}^k(x) &= \frac{1}{2}\omega^{kr}(x)\left(\partial_i \omega_{rj}(x) + \partial_j \omega_{ri}(x) - \partial_r \omega_{ij}(x)\right)\\ &= \gamma_{ji}^k(x).\end{aligned}$$

This is a new geometric object of order 2 providing the Levi-Civita isomorphism $j_1(\omega) = (\omega, \partial\omega) \simeq (\omega, \gamma)$ of affine bundles and allowing to obtain the second order Medolaghi equations:

$$\begin{aligned}\Gamma_{ij}^k \equiv \left(\mathcal{L}(\xi)\gamma\right)_{ij}^k &\equiv \partial_{ij}\xi^k + \gamma_{rj}^k(x)\partial_i \xi^r + \gamma_{ir}^k(x)\partial_j \xi^r\\ &- \gamma_{ij}^r(x)\partial_r \xi^k + \xi^r \partial_r \gamma_{ij}^k(x) = 0\end{aligned}$$

Surprisingly, the following expression called Riemann tensor:

$$\rho^k_{lij}(x) \equiv \partial_i \gamma^k_{lj}(x) - \partial_j \gamma^k_{li}(x)$$
$$+\gamma^r_{lj}(x)\gamma^k_{ri}(x) - \gamma^r_{li}(x)\gamma^k_{rj}(x)$$

is still a first order geometric object and even 4 -tensor with $n^2(n^2-1)/12$ independent components satisfying the purely algebraic relations:

$$\rho^k_{lij} + \rho^k_{ijl} + \rho^k_{jli} = 0, \quad \omega_{rl}\rho^l_{kij} + \omega_{kr}\rho^r_{lij} = 0 .$$

Accordingly, the IC must express that the new first order equations $R^k_{lij} \equiv (\mathcal{L}(\xi)\rho)^k_{lij} = 0$ are only linear combinations of the previous ones and we get the Vessiot structure equations:

$$\rho^k_{lij}(x) = c\left(\delta^k_i \omega_{lj}(x) - \delta^k_j \omega_{li}(x)\right)$$

with the only structure constant C describing the constant Riemannian curvature condition of Eisenhart [11,12]. One can proceed similarly for the conformal Killing system $\mathcal{L}(\xi)\omega = A(x)\omega$ and obtain that the Weyl tensor must vanish, without any structure constant [12].

EXAMPLE 1.10: (Contact structure) We only treat the case dim(X) = 3 as the case dim(X) = $2p$+1 needs much more work [6]. Let us consider the so-called contact 1-form $\alpha = dx^1 - x^3\, dx^2$ and consider the Lie pseudogroup $\Gamma \subset \mathrm{aut}(X)$ of (local) transformations preserving ? up to a function factor, that is

$$\Gamma = \left\{ f \in \mathrm{aut}(X) \middle| j_1(f)^{-1}(\alpha) = \rho\alpha \right\}$$

where again $j_q(f)$ is a symbolic way for writing out the derivatives of f up to order q and α transforms like a 1-covariant tensor. It may be tempting to look for a kind of "object" the invariance of which should characterize ?. Introducing the exterior derivative $d\alpha = dx^2 \wedge dx^3$ as a 2-form, we obtain the volume 3-form $a \wedge da = dx^1 \wedge dx^2 \wedge dx^3$. As it is well known that the exterior derivative commutes with any diffeomorphism, we obtain sucessively:

$$j_1(f)^{-1}(d\alpha) = d\left(j_1(f)^{-1}(\alpha)\right) = d(\rho\alpha)$$
$$= \rho d\alpha + d\rho \wedge \alpha \Rightarrow j_1(f)^{-1}(\alpha \wedge d\alpha)$$
$$= \rho^2(\alpha \wedge d\alpha).$$

As the volume 3-form $\alpha \perp \delta\alpha$ transforms through a division by the Jacobian determinant

$$\Delta = \partial\left(f^1, f^2, f^3\right) / \partial\left(x^1, x^2, x^3\right) \neq 0$$

of the transformation $y = f(x)$ with inverse

$$x = f^{-1}(y) = g(y),$$

the desired object is thus no longer a 1-form but a 1- form density $\omega = (\omega_1, \omega_2, \omega_3)$ transforming like a 1- form but up to a division by the square root of the Jacobian determinant. It follows that the infinitesimal contact transformations are vector fields $\xi \in T = T(X)$ the tangent bundle of X, satisfying the 3 so-called first order Medolaghi equations:

$$\Omega_i \equiv \left(\mathcal{L}(\xi)\omega\right)_i$$
$$\equiv \omega_r(x)\partial_i\xi^r - (1/2)\omega_i(x)\partial_r\xi^r + \xi^r\partial_r\omega_i(x) = 0$$

When $\omega = (1, -x^3, 0)$, we obtain the special involutive system:

$$\partial_3\xi^3 + \partial_2\xi^2 + 2x^3\partial_1\xi^2 - \partial_1\xi^1 = 0,$$
$$\partial_3\xi^1 - x^3\partial_3\xi^2 = 0,$$
$$\partial_2\xi^1 - x^3\partial_2\xi^2 + x^3\partial_1\xi^1 - \left(x^3\right)^2\partial_1\xi^2 - \xi^3 = 0$$

with 2 equations of class 3 and 1 equation of class 2 (see later on for a precise definition) and thus only 1 compatibility conditions (CC) for the second members.

For an arbitrary ω, we may ask about the differential conditions on ⍰ such that all the equations of order r+1 are only obtained by differentiating r times the first order equations, exactly like in the special situation just considered where the system is involutive. We notice that, in a symbolic way, $\omega \perp \delta\omega$ is now a scalar $c(x)$ providing the zero order equation $\xi^r \partial_r c(x) = 0$ and the condition is $c(x) = c = cst$. The integrability condition (IC) is the Vessiot structure equation:

$$\omega_1(\partial_2\omega_3 - \partial_3\omega_2) + \omega_2(\partial_3\omega_1 - \partial_1\omega_3)$$
$$+\omega_3(\partial_1\omega_2 - \partial_2\omega_1) = c$$

involving the only structure constant c.

For $\omega = (1, -\xi^3, 0)$, we get $c = 1$. If we choose $\bar{\omega} = (1,0,0)$ leading to $\bar{c} = 0$, we may define

$$\bar{\Gamma} = \left\{ f \in \mathrm{aut}(X) \middle| j_1(f)^{-1}(\bar{\omega}) = \bar{\omega} \right\}$$

with infinitesimal transformations satisfying the involutive system:

$$\partial_3\xi^3 + \partial_2\xi^2 - \partial_1\xi^1 = 0,\ \partial_3\xi^1 = 0,\ \partial_2\xi^1 = 0$$

with again 2 equations of class 3 and 1 equation of class 2.

EXAMPLE 1.11: (Unimodular contact structure) With similar notations, let us again set

$$\alpha = dx^1 - x^3 dx^2 \Rightarrow d\alpha = dx^2 \wedge dx^3$$

but let us now consider the new Lie pseudogroup of transformations preserving α and thus $d\alpha$ too, that is preserving the mixed object

$$\omega = (\alpha, \beta) \in T^* \times_X \wedge^2 T^*$$

made up by a 1-form α and a 2-form β with

$$\gamma = \alpha \wedge \beta \neq 0 \quad \text{and} \quad d\alpha = \beta \Rightarrow d\beta = 0\,.$$

Then Γ is a Lie subpseudogroup of the one just considered in the previous example and the corresponding infinitesimal transformations now satisfy the involutive system:

$$\partial_1\xi^1 = 0,\ \partial_1\xi^2 = 0,\ \partial_1\xi^3 = 0,\ \partial_2\xi^1 + x^3\partial_3\xi^3 - \xi^3 = 0,$$
$$\partial_2\xi^2 + \partial_3\xi^3 = 0, \partial_3\xi^1 - x^3\partial_3\xi^2 = 0$$

with 3 equations of class 3, 2 equations of class 2 and 1 equation of class 1 if we exchange x^1 with x^3, a result leading now to 4 CC.

More generally, when $\omega = (\alpha, \beta)$ where α is a 1- form and ? is a 2-form satifying $\alpha \perp \beta \neq 0$, we may study the same problem as before for the general system

$$\mathcal{L}(\xi)\alpha = 0,\ \mathcal{L}(\xi)\beta = 0\,.$$

We let the reader provide the details of the tedious computation involved as it is at this point that computer algebra may be used [13]. The result, not evident at first sight, is that the 2-form $d\alpha$ must be proportional to the 2-form β, that is $d\alpha = c'(x)\beta$ and thus

$$\alpha \wedge d\alpha = c'(x)\alpha \wedge \beta$$

As $\alpha \perp \beta \neq 0$, we must have $c'(x) = c' = cst$ and thus $d\alpha = c'\beta$ .. Similarly, we get $d\beta = c''\alpha \wedge \beta$ and obtain finally the 4 Vessiot structure equations

$$d\alpha = c'\beta,\ d\beta = c''\alpha \wedge \beta$$

involving 2 structure constants $c = (c', c'')$. Contrary to the previous situation (but like in the Riemann case!) we notice that we have now 2 structure equations not containing any constant (called first kind by Vessiot) and 2 structure equations with the same number of different constants (called second kind by Vessiot), namely

$$\alpha \wedge d\alpha = c'\alpha \wedge \beta,\ d\beta = c''\alpha \wedge \beta\,.$$

Finally, closing this system by taking once more the exterior derivative, we get

$$0 - d^2\alpha = c'd\beta = c'c''\alpha \wedge \beta$$

and thus the unexpected purely algebraic Jacobi condition $c'c'' = 0$. For the special choice

$$\omega = \left(dx^1 - x^3 dx^2, dx^2 \wedge dx^3\right)$$

we get $c = (1,0)$, for the second special choice

$$\bar{\omega} = \left(dx^1, dx^2 \wedge dx^3\right)$$

we get $\bar{c} = (0,0)$ and for the third special choice

$$\bar{\bar{\omega}} = \left(\left(1/x^1\right)dx^1, x^1 dx^2 \wedge dx^3\right)$$

we get $\bar{\bar{c}} = (0,1)$.

FIRST FUNDAMENTAL RESULT: Comparing the various Vessiot structure equations containing structure constants that we have just presented and that we recall below in a symbolic way, we notice that these structure constants are absolutely on equal footing though they have in general nothing to do with any Lie algebra.

$$\begin{cases} \partial\alpha - \gamma\alpha = c\alpha^2 \\ \partial\beta + \gamma\beta = c \end{cases}$$

$$\partial\omega - \partial\omega = c\omega\omega$$

$$\partial\gamma - \partial\gamma + \gamma\gamma - \gamma\gamma = c\left(\delta\omega - \delta\omega\right)$$

$$\omega \wedge (\partial\omega - \partial\omega) = c$$

$$\begin{cases} d\alpha = c'\beta \\ d\beta = c''\alpha \wedge \beta \end{cases}.$$

Accordingly, the fact that the ones appearing in the MC equations are related to a Lie algebra is a coincidence and the Cartan structure equations have nothing to do with the Vessiot structure equations. Also, as their factors are either constant, linear or quadratic, any identification of the quadratic terms appearing in the Riemann tensor with the quadratic terms appearing in the MC equations is definitively not correct [7]. We also understand why the torsion is automatically combined with curvature in the Cartan structure equations but totally absent from the Vessiot structure equations, even though the underlying group (translations + rotations) is the same.

HISTORICAL REMARK 1.12: Despite the prophetic comments of the italian mathematician Ugo Amaldi in 1909 [12], it has been a pity that Cartan deliberately ignored the work of Vessiot at the beginning of the last century and that the things did not improve afterwards in the eighties with Spencer and coworkers (Compare MR 720863 (85 m: 12004) and MR 954613 (90e: 58166)).

3. SECOND PART: THE JANET AND SPENCER SEQUENCES

Let $\mu=(\mu_1,\cdots,\mu_n)$ be a multi-index with length

$$|\mu|=\mu_1+\cdots+\mu_n$$

class *i* if

$$\mu_1=\cdots=\mu_{i-1}=0,\mu_i\neq 0$$

and

$$\mu+1_i=(\mu_1,\cdots,\mu_{i-1},\mu_i+1,\mu_{i+1},\cdots,\mu_n).$$

We set

$$y_q=\left\{y_\mu^k \middle| 1\leq k\leq m,0\leq|\mu|\leq q\right\}$$

with $y_\mu^k=y^k$ when $|\mu|=0$ If E is a vector bundle over X with local coordinates (x,y) and $J_q(E)$ is the q-jet bundle of E with local coordinates $(x, y_q)$the Spencer operator just allows to distinguish a section ξ_q from a section $j_q(\xi)$ by introducing a kind of "difference" through the operator

$$D:J_{q+1}(E)\rightarrow T^*\otimes J_q(E):\xi_{q+1}\rightarrow j_1(\xi_q)-\xi_{q+1}$$

with local components

$$\left(\partial_i\xi^k(x)-\xi_i^k(x),\partial_i\xi_j^k(x)-\xi_{ij}^k(x),\cdots\right)$$

and more generally

$$\left(D\xi_{q+1}\right)_{\mu,i}^{k}(x)=\partial_i\xi_\mu^k(x)-\xi_{\mu+1_i}^k(x).$$

Minus the restriction of D to the kernel $S_{q+1}T^*\otimes E$ of the canonical projection

$$\pi_q^{q+1}:J_{q+1}(E)\rightarrow J_q(E)$$

can be extended to the Spencer map

$$\delta:\wedge^s T^*\otimes S_{q+1}T^*\otimes E\rightarrow\wedge^{s+1}T^*\otimes S_qT^*\otimes E$$

defined by

$$(\delta\omega)_\mu^k=dx^i\wedge\omega_{\mu+1_i}^k.$$

The kernel of D is made by sections such that

$$\xi_{q+1}=j_1\left(\xi_q\right)=j_2\left(\xi_{q-1}\right)=\cdots=j_{q+1}(\xi).$$

Finally, if $R_q\subset J_q(E)$ is a system of order q on E locally defined by linear equations

$$\Phi^\tau\left(x,y_q\right)\equiv a_k^{\tau\mu}(x)y_\mu^k=0,$$

the r-prolongation

$$\begin{aligned}R_{q+r}&=\rho_r\left(R_q\right)\\&=J_r\left(R_q\right)\cap J_{q+r}(E)\subset J_r\left(J_q(E)\right)\end{aligned}$$

is locally defined when $r = 1$ by the linear equations

$$\begin{aligned}&\Phi^\tau\left(x,y_q\right)=0,\ d_i\Phi^\tau\left(x,y_{q+1}\right)\\&\equiv a_k^{\tau\mu}(x)y_{\mu+1_i}^k+\partial_i a_k^{\tau\mu}(x)y_\mu^k=0\end{aligned}$$

and has symbol

$$g_{q+r} = R_{q+r} \cap S_{q+r}T^* \otimes E \subset J_{q+r}(E)$$

locally defined by

$$a_k^{\tau\mu}(x)\xi_{\mu+\nu}^k = 0, \ |\mu| = q, \ |\nu| = r$$

if one looks at the top order terms. If $\xi_{q+1} \in R_{q+1}$ is over $\xi_q \in R_q$, differentiating the identity

$$a_k^{\tau\mu}(x)\xi_\mu^k(x) \equiv 0$$

with respect to x^i and substracting the identity

$$a_k^{\tau\mu}(x)\xi_{\mu+1_i}^k(x) + \partial_i a_k^{\tau\mu}(x)\xi_\mu^k(x) \equiv 0,$$

we obtain the identity

$$a_k^{\tau\mu}(x)\left(\partial_i\xi_\mu^k(x) - \xi_{\mu+1_i}^k(x)\right) \equiv 0$$

and thus the restriction $D : R_{q+1} \to T^* \otimes R_q$. This first order operator induces, up to sign, the purely algebraic monomorphism $0 \to g_{q+1} \xrightarrow{\delta} T^* \otimes g_q$ on the symbol level [8,14]. The Spencer operator has never been used in GR though an accelerometer in a rocket merely measures one of the components of the Spencer operator involving second order jets.

DEFINITION 2.1: $R_q$is said to be formally integrable (FI) when the restriction

$$\pi_{q+r}^{q+r+1} : R_{q+r+1} \to R_{q+r}$$

is an epimorphism $\forall r \geq 0$. In that case, the Spencer form $R_{q+1} \subset J_1(R_q)$ is a canonical equivalent formally integrable first order system on R_q with no zero order equations.

DEFINITION 2.2: R_q is said to be involutive when it is formally integrable and the symbol g_q is involutive, that is all the sequences

$$\cdots \xrightarrow{\delta} \wedge^{s} T^{*} \otimes g_{q+r} \xrightarrow{\delta} \cdots$$

are exact $\forall 0 \leq s \leq n, \forall r \geq 0$. Equivalently, using a linear change of local coordinates if necessary, we may successively solve the maximum number $\beta_q^n, \beta_q^{n-1}, \cdots, \beta_q^1$ of equations with respect to the leading or principal jet coordinates of strict order q and class n, n-1,...,1. Then R_q is involutive if R_{q+1} is obtained by only prolonging the β_q^i equations of class i with respect to $d_1,\ldots,d_i$ for $i = 1, \ldots, n$. In that case, such a prolongation procedure allows to compute in a unique way the principal jets from the parametric other ones and may also be applied to nonlinear systems as well [8,15].

When R_q is involutive, the linear differential operator

$$D: E \xrightarrow{j_q} J_q(E) \xrightarrow{\Phi} J_q(E)/R_q = F_0$$

of order q with space of solutions $\Theta \subset E$ is said to be involutive and one has the canonical linear Janet sequence [8]:

$$0 \rightarrow \Theta \rightarrow E \xrightarrow{\mathcal{D}} F_0 \xrightarrow{\mathcal{D}_1} F_1 \xrightarrow{\mathcal{D}_2} \cdots \xrightarrow{\mathcal{D}_n} F_n \rightarrow 0$$

with Janet bundles

$$F_r = \wedge^r T^* \otimes J_q(E) / \left(\wedge^r T^* \otimes R_q + \delta\left(\wedge^{r-1} T^* \otimes S_{q+1} T^* \otimes E \right) \right).$$

Each operator $\mathcal{D}_{r+1} : F_r \rightarrow F_{r+1}$ is first order involutive as it is induced by

$$D: \wedge^r T^* \otimes J_{q+1}(E) \rightarrow \wedge^{r+1} T^* \otimes J_q(E) : \alpha \otimes \xi_{q+1}$$
$$\rightarrow d\alpha \otimes \xi_q + (-1)^r \alpha \wedge D\xi_{q+1}$$

and generates the compatibility conditions (CC) of the preceding one. As the Janet sequence can be cut at any place, the numbering of the Janet bundles has nothing to do with that of the Poincaré sequence, contrary to what many people believe in GR.

Similarly, we have the involutive first Spencer operator

$$D:\wedge^r T^* \otimes J_{q+1}(E) \to \wedge^{r+1} T^* \otimes J_q(E) : \alpha \otimes \xi_{q+1}$$
$$\to d\alpha \otimes \xi_q + (-1)^r \alpha \wedge D\xi_{q+1}$$

of order one induced by $D: R_{q+1} \to T^* \otimes R_q$. Introducing the Spencer bundles

$$C_r = \wedge^r T^* \otimes R_q / \delta\left(\wedge^{r-1} T^* \otimes g_{q+1}\right),$$

the first order involutive $(r+1)$ Spencer operator $D_{r+1} : C_r \to C_{r+1}$ is induced by

$$D:\wedge^{r+1} T^* \otimes R_{q+1} \to \wedge^{r+1} T^* \otimes R_q$$

and we obtain the canonical linear Spencer sequence [8,14]:

$$0 \to \Theta \overset{j_q}{\to} C_0 \overset{D_1}{\to} C_1 \overset{D_2}{\to} C_2 \overset{D_3}{\to} \cdots \overset{D_n}{\to} C_n \to 0$$

as the Janet sequence for the first order involutive system $R_{q+1} \subset J_1(R_q)$. Introducing the other Spencer bundles

$$C_r(E) = \wedge^r T^* \otimes J_q(E) / \delta\left(\wedge^{r-1} T^* \otimes S_{q+1} T^* \otimes E\right)$$

With $C_r \subset C_r(E)$, the linear Spencer sequence is induced by the linear hybrid sequence:

$$0 \to E \overset{j_q}{\to} C_0(E) \overset{D_1}{\to} C_1(E) \overset{D_2}{\to} C_2(E) \overset{D_3}{\to} \cdots \overset{D_n}{\to} C_n(E) \to 0$$

which is at the same time the Janet sequence for j_q and the Spencer sequence for

$$J_{q+1}(E) \subset J_1(J_q(E))$$

[8,14]. Such a sequence projects onto the Janet sequence and we have the following commutative diagram with exact columns:

$$
\begin{array}{ccccccccccccccc}
 & & & & & & 0 & & 0 & & 0 & & 0 & & \\
 & & & & & & \downarrow & & \downarrow & & \downarrow & & \downarrow & & \\
 & & 0 & \to & \Theta & \xrightarrow{j_q} & C_0 & \xrightarrow{D_1} & C_1 & \xrightarrow{D_2} & C_2 & \xrightarrow{D_3}\dots\xrightarrow{D_n} & C_n & \to & 0 \\
 & & & & & & \downarrow & & \downarrow & & \downarrow & & \downarrow & & \\
 & & 0 & \to & E & \xrightarrow{j_q} & C_0(E) & \xrightarrow{D_1} & C_1(E) & \xrightarrow{D_2} & C_2(E) & \xrightarrow{D_3}\dots\xrightarrow{D_n} & C_n(E) & \to & 0 \\
 & & & & \| & & \downarrow \Phi_0 & & \downarrow \Phi_1 & & \downarrow \Phi_2 & & \downarrow \Phi_n & & \\
0 & \to & \Theta & \to & E & \xrightarrow{\mathcal{D}} & F_0 & \xrightarrow{\mathcal{D}_1} & F_1 & \xrightarrow{\mathcal{D}_2} & F_2 & \xrightarrow{\mathcal{D}_3}\dots\xrightarrow{\mathcal{D}_n} & F_n & \to & 0 \\
 & & & & & & \downarrow & & \downarrow & & \downarrow & & \downarrow & & \\
 & & & & & & 0 & & 0 & & 0 & & 0 & &
\end{array}
$$

In this diagram, only depending on the linear differential operator $\mathcal{D} = \Phi \circ j_q$, the epimorhisms

$$\Phi_r : C_r(E) \to F_r \quad \text{for} \quad 0 \leq r \leq n$$

are induced by the canonical projection

$$\Phi = \Phi_0 : C_0(E) = J_q(E) \to J_q(E)/R_q = F_0$$

if we start with the knowledge of $R_q \subset J_q(E)$ or from the knowledge of an epimorphism

$$\Phi : J_q(E) \to F_0$$

if we set $R_q = \ker(\Phi)$. In the theory of Lie equations considered, $E = T$, $R_q \subset J_q(T)$ is a transitive involutive system of infinitesimal Lie equations of order q and the corresponding operator $\mathcal{D}$ is a Lie operator. As an exercise, we invite the reader to draw this diagram in the affine and projective 1-dimensional cases.

EXAMPLE 2.3: If we restrict our study to the group of isometries of the euclidean metric ω in dimension $n \geq 2$, exhibiting the Janet and the Spencer sequences is not easy at all, even when $n = 2$, because the corresponding Killing operator $\mathcal{D}\xi = \mathcal{L}(\xi)\omega = \Omega \in S_2T^*$, involving the Lie derivative $\mathcal{L}$ and providing twice the so-called infinitesimal deformation tensor ε of continuum mechanics, is not involutive. In order to overcome this problem, one must differentiate once by considering also the Christoffel symbols γ and add the operator

$$\mathcal{L}(\xi)\gamma = \Gamma \in S_2T^* \otimes T .$$

Now, one can prove that the Spencer sequence for Lie groups of transformations is locally isomorphic to the tensor product of the Poincaré sequence by the Lie algebra of the underlying Lie group [7,8]. Hence, if two Lie groups $G \subset \hat{G}$ act on X, it follows from the definition of the Janet and Spencer bundles that the Spencer sequence for G is embedded into the Spencer sequence for $\hat{G}$ while the Janet sequence for G projects onto the Janet sequence for $\hat{G}$ but the common differences are isomorphic to $\wedge^r T^* \otimes (\hat{\mathcal{G}}/\mathcal{G})$. This rather philosophical comment, namely to replace the Janet sequence by the Spencer sequence, must be considered as the crucial key for understanding the work of the brothers E. and F. Cosserat in 1909 [7,17-19] or H. Weyl in 1918 [7,16], the best picture being that of Janet and Spencer playing at see-saw. Indeed, when n = 2, one has 3 parameters (2 translations + 1 rotation) and the following commutative diagram which only depends on the left commutative square:

$$\begin{array}{ccccccccccccc}
 & & & & & & 0 & & 0 & & 0 & & \\
 & & & & & & \downarrow & & \downarrow & & \downarrow & & \\
 & & 0 & \longrightarrow & \Theta & \xrightarrow{j_2} & 3 & \xrightarrow{D_1} & 6 & \xrightarrow{D_2} & 3 & \longrightarrow 0 & \text{Spencer} \\
 & & & & & & \downarrow & & \downarrow & & \downarrow & & \\
 & & 0 & \longrightarrow & 2 & \xrightarrow{j_2} & 12 & \xrightarrow{D_1} & 16 & \xrightarrow{D_2} & 6 & \longrightarrow 0 & \\
 & & & & \| & & \downarrow \Phi_0 & & \downarrow \Phi_1 & & \downarrow \Phi_2 & & \\
0 & \longrightarrow & \Theta & \longrightarrow & 2 & \xrightarrow{\mathcal{D}} & 9 & \xrightarrow{\mathcal{D}_1} & 10 & \xrightarrow{\mathcal{D}_2} & 3 & \longrightarrow 0 & \text{Janet} \\
 & & & & & & \downarrow & & \downarrow & & \downarrow & & \\
 & & & & & & 0 & & 0 & & 0 & &
\end{array}$$

In this diagram, there is no way to compare $\mathcal{D}_1$ (curvature alone as in Vessiot) with D_2 (curvature + torsion as in Cartan).

For proving that the adjoint of $\mathcal{D}_1$ provides the Cosserat equations which can be parametrized by the adjoint of D_2, we may lower the upper indices by means of the constant euclidean metric and look for the factors of ⍰$_1$, ⍰$_2$ and $\xi_{1,2} = -\ \xi_{2,1}$ in the integration by parts of the sum:

$$\sigma^{11}\left(\partial_1\xi_1 - \xi_{1,1}\right) + \sigma^{12}\left(\partial_2\xi_1 - \xi_{1,2}\right) + \sigma^{21}\left(\partial_1\xi_2 - \xi_{2,1}\right)$$
$$+\sigma^{22}\left(\partial_2\xi_2 - \xi_{2,2}\right) + \mu^r\left(\partial_r\xi_{1,2} - \xi_{1,2r}\right)$$

in order to obtain:

$$\partial_1\sigma^{11} + \partial_2\sigma^{12} = f^1,\ \partial_1\sigma^{21} + \partial_2\sigma^{22} = f^2,$$
$$\partial_1\mu^1 + \partial_2\mu^2 + \sigma^{12} - \sigma^{21} = m$$

Finally, we get the nontrivial first order parametrization

$$\sigma^{11} = \partial_2\phi^1,\ \sigma^{12} = -\partial_1\phi^1,$$
$$\sigma^{21} = -\partial_2\phi^2,\ \sigma^{22} = \partial_1\phi^2,$$
$$\mu^1 = \partial_2\phi^3 + \phi^1,\ \mu^2 = -\partial_1\phi^3 - \phi^2$$

by means of the three arbitrary functions ϕ^1, ϕ^2, ϕ^3, in a coherent way with the Airy second order parametrization obtained if we set

$$\phi^1 = \partial_2\phi,\ \phi^2 = \partial_1\phi,\ \phi^3 = -\phi$$

when $\mu^1 = 0$, $\mu^2 = 0$ as we shall see in the third part.

The link between the FI of R_q and the CC of $\mathcal{D}$ is expressed by the following diagram that may be used inductively:

$$\begin{array}{ccccccccc}
 & & 0 & & 0 & & 0 & & CC \\
 & & \downarrow & & \downarrow & & \downarrow & & \downarrow \\
0 & \rightarrow & g_{q+r} & \rightarrow & S_{q+r}T^*\otimes E & \overset{\sigma_r(\Phi)}{\rightarrow} & S_rT^*\otimes F_0 & \rightarrow & coker(\sigma_r(\Phi)) & \rightarrow 0 \\
 & & \downarrow & & \downarrow & & \downarrow & & \downarrow \\
0 & \rightarrow & R_{q+r} & \rightarrow & J_{q+r}(E) & \overset{\rho_r(\Phi)}{\rightarrow} & J_r(F_0) & \rightarrow & coker(\rho_r(\Phi)) & \rightarrow 0 \\
 & & \downarrow & & \downarrow \pi^{q+r}_{q+r-1} & & \downarrow \pi^r_{r-1} & & \downarrow \\
0 & \rightarrow & R_{q+r-1} & \rightarrow & J_{q+r-1}(E) & \overset{\rho_{r-1}(\Phi)}{\rightarrow} & J_{r-1}(F_0) & \rightarrow & coker(\rho_{r-1}(\Phi) & \rightarrow 0 \\
 & & & & \downarrow & & \downarrow & & \downarrow \\
 & & FI & & 0 & & 0 & & 0
\end{array}$$

The "snake theorem" [8,20] then provides the long exact connecting sequence:

$$0 \to g_{q+r} \to R_{q+r} \to R_{q+r-1} \to \mathrm{coker}(\sigma_r(\Phi)) \to \mathrm{coker}(\rho_r(\Phi)) \to \mathrm{coker}(\rho_{r-1}(\Phi)) \to 0 .$$

If we apply such a diagram to first order Lie equations with no zero or first order CC, we have $q = 1$, $E = T$ and we may apply the Spencer δ-map to the top row obtained with $r = 2$ in order to get the commutative diagram:

$$\begin{array}{ccccccccc}
 & & 0 & & 0 & & 0 & & \\
 & & \downarrow & & \downarrow & & \downarrow & & \\
0 & \to & g_3 & \to & S_3T^* \otimes T & \to & S_2T^* \otimes F_0 & \to & F_1 \to 0 \\
 & & \downarrow \delta & & \downarrow \delta & & \downarrow \delta & & \\
0 & \to & T^* \otimes g_2 & \to & T^* \otimes S_2T^* \otimes T & \to & T^* \otimes T^* \otimes F_0 & \to & 0 \\
 & & \downarrow \delta & & \downarrow \delta & & \downarrow \delta & & \\
0 & \to & \wedge^2 T^* \otimes g_1 & \to & \wedge^2 T^* \otimes T^* \otimes T & \to & \wedge^2 T^* \otimes F_0 & \to & 0 \\
 & & \downarrow \delta & & \downarrow \delta & & \downarrow & & \\
0 & \to & \wedge^3 T^* \otimes T & = & \wedge^3 T^* \otimes T & \to & 0 & & \\
 & & \downarrow & & \downarrow & & & & \\
 & & 0 & & 0 & & & &
\end{array}$$

with exact rows and exact columns but the first that may not be exact at $\wedge^2 T^* \otimes g_1$. We shall denote by $B^2(g_1)$ the coboundary as the image of the central δ, by $Z^2(g_1)$ the cocycle as the kernel of the lower δ and by $H^2(g_1)$ the Spencer δ-cohomology at $\wedge^2 T^* \otimes g_1$ as the quotient.

In the classical Killing system, $g_1 \subset T^* \otimes T$ is defined by

$$\omega_{rj}(x)\xi_i^r + \omega_{ir}(x)\xi_j^r = 0$$
$$\Rightarrow \xi_r^r = 0, g_2 = 0, g_3 = 0$$

Applying the previous diagram, we discover that the Riemann tensor is a section of the bundle

$$\mathrm{Riemann} = F_1 = H^2(g_1) = Z^2(g_1)$$

with

$$\begin{aligned}&\dim(\text{Riemann})\\&=\left(n^2(n+1)^2/4\right)-\left(n^2(n+1)(n+2)/6\right)\\&=\left(n^2(n-1)^2/4\right)-\left(n^2(n-1)(n-2)/6\right)\\&=n^2\left(n^2-1\right)/12\end{aligned}$$

by using the top row or the left column. Though we discover the two properties of the Riemann tensor through the chase involved, we have no indices and cannot therefore exhibit the Ricci tensor of GR by means of the usual contraction or trace.

Let us proceed the same way with the conformal Killing system

$$\begin{aligned}\hat{\Omega}_{ij}&\equiv\left(\mathcal{L}(\xi)\hat{\omega}\right)_{ij}\\&\equiv\hat{\omega}_{rj}\partial_i\xi^r+\hat{\omega}_{ir}\partial_j\xi^r-\frac{2}{n}\omega_{ij}\partial_r\xi^r+\xi^r\partial_r\hat{\omega}_{ij}=0\end{aligned}$$

obtained by introducing

$$\hat{\omega}_{ij}=\omega_{ij}\Big/\left|\det(\omega)\right|^{\frac{1}{n}}$$

or, equivalently, by eliminating $A(x)$ in

$$\mathcal{L}(\xi)\omega=A(x)\omega\,.$$

Now $\hat{g}_1$ is defined by

$$\omega_{rj}\xi^r_i+\omega_{ir}\xi^r_j-\frac{2}{n}\omega_{ij}\xi^r_r=0$$

but we have $\hat{g}_3=0,\forall n\geq 3$ with $H^2(\hat{g}_2)=0,\forall n\geq 4$ and the Weyl tensor is a section of the bundle

$$\text{Weyl}=\hat{F}_1=H^2(\hat{g}_1)=Z^2(\hat{g}_1)\Big/\delta\left(T^*\otimes\hat{g}_2\right)$$

with

$$\dim(\text{Weyl}) = n(n+1)(n+2)(n-3)/12 .$$

Similarly, we have no indices and cannot therefore exhibit the Ricci tensor. However, when $n = 4$, among the components of the Spencer operator we have

$$\partial_i \xi^r_{rj} - \xi^r_{rij} = \partial_i \xi^r_{rj}$$

and thus

$$\partial_i \xi^r_{rj} - \partial_j \xi^r_{ri} = F_{ij} .$$

Such a result allows to recover the electromagnetic field in the image of the Spencer operator D_1 and Maxwell equations by duality along the way proposed by Weyl in [16] but the use of the Spencer operator provides the only possibility to exhibit a link with Cosserat equations.

Comparing the classical and conformal Killing systems by using the inclusions

$$R_1 \subset \hat{R}_1 \Rightarrow g_1 \subset \hat{g}_1 ,$$

we finally obtain the following commutative and exact diagram where a diagonal chase allows to identify *Ricci* with

$$S_2T^* \subset T^* \otimes T^* \simeq T^* \otimes \hat{g}_2$$

and to split the right column [7,12,20]:

$$\begin{array}{ccccccccccc}
 & & & & & & & & 0 & & \\
 & & & & & & & & \downarrow & & \\
 & & & & & & 0 & & Ricci & & \\
 & & & & & & \downarrow & & \downarrow & & \\
 & & & & 0 & \rightarrow & Z^2(g_1) & \rightarrow & Riemann & \rightarrow 0 & \\
 & & & & \downarrow & & \downarrow & & \downarrow & & JANET \\
 & & 0 & \rightarrow & T^* \otimes \hat{g}_2 & \xrightarrow{\delta} & Z^2(\hat{g}_1) & \rightarrow & Weyl & \rightarrow 0 & \\
 & & & & \downarrow & & \downarrow & & \downarrow & & \\
0 \rightarrow & S_2T^* & & \xrightarrow{\delta} & T^* \otimes T^* & \xrightarrow{\delta} & \wedge^2 T^* & \rightarrow & 0 & & \\
 & & & & \downarrow & & \downarrow & & & & \\
 & & & & 0 & & 0 & & & &
\end{array}$$

SECOND FUNDAMENTAL RESULT: The Ricci tensor only depends on the "difference" existing between the clasical Killing system and the conformal Killing system, namely the n second order jets (elations once more). The Ricci tensor, thus obtained without contracting the indices as usual, may be embedded in the image of the Spencer operator made by 1-forms with value in 1-forms that we have already exhibited for describing EM. It follows that the foundations of both EM and GR are not coherent with jet theory and must therefore be revisited within this new framework.

4. THIRD PART: ALGEBRAIC ANALYSIS

EXAMPLE 3.1: Let a rigid bar be able to slide along an horizontal axis with reference position x and attach two pendula, one at each end, with lengths l_1 and l_2, having small angles θ_1 and θ_2 with respect to the vertical. If we project Newton law with gravity g on the perpendicular to each pendulum in order to eliminate the tension of the threads and denote the time derivative with a dot, we get the two equations:

$$\ddot{x} + l_1\ddot{\theta}_1 + g\theta_1 = 0, \ \ddot{x} + l_2\ddot{\theta}_2 + g\theta_2 = 0$$

As an experimental fact, starting from an arbitrary movement of the pendula, we can stop them by moving the bar if and only if $l_1 \neq l_2$ and we say that the system is controllable.

More generally, we can bring the OD equations describing the behaviour of a mechanical or electrical system to the Kalman form $\dot{y} = Ay + bu$ with input $u = (u^1,\dots, u^p)$ and output $y = (y^1,\dots, y^m)$. We say that the system is controllable if, for any given

$$y(0), \ y(T), \ T < \infty,$$

one can find $u(t)$ such that a coherent trajectory $y(t)$ may be found. In 1963 [21], R. E. Kalman discovered that the system is controllable if and only if

$$rk\left(B, AB, \cdots, A^{m-1}B\right) = m \ .$$

Surprisingly, such a functional definition admits a formal test which is only valid for Kalman type systems with constant coefficients and is thus far from being intrinsic. In the PD case, the Spencer form will replace the Kalman form.

EXAMPLE 3.2:

$$\dot{y} - \dot{u} = 0 \Rightarrow y(t) - u(t) = c = cst$$
$$\Rightarrow u(T) - u(0) = y(T) - y(0)$$

can always be achieved and the system is thus controllable in the sense of the definition but $z = y - u \Rightarrow \dot{z} = 0$ is not controllable in the sense of the test.

EXAMPLE 3.3:

$$\dot{y}^1 - a(t)y^2 - \dot{y}^3 = 0,\ y^1 - \dot{y}^2 + \dot{y}^3 = 0 .$$

Any way to bring this system to Kalman form provides the controllability condition $a(a\text{-}1) \neq 0$ if $a = cst$ but nothing can be said if $a = a(t)$. Also, getting y^1 from the second equation and substituting in the first, we get the second order OD equation

$$\ddot{y}^2 - \ddot{y}^3 - \dot{y}^3 - a(t)y^2 = 0$$

for which nothing can be said at first sight.

PROBLEM 1: Is a SYSTEM of OD or PD equations "controllable" (answer must be YES or NO) and how can we define controllability?

Now, if a differential operator $\xi \xrightarrow{\mathcal{D}} \eta$ is given, a direct problem is to find (generating) compatibility conditions (CC) as an operator $\eta \xrightarrow{\mathcal{D}_1} \zeta$ such that

$$\mathcal{D}\xi = \eta \Rightarrow \mathcal{D}_1\eta = 0 .$$

Conversely, the inverse problem will be to find $\theta \xrightarrow{\mathcal{D}_{-1}} \xi$ such that $\mathcal{D}$ generates the CC of $\mathcal{D}_{-1}$ and we shall say that $\mathcal{D}$ is parametrized by $\mathcal{D}_{-1}$. Of course, solving the direct problem (Janet, Spencer) is necessary for solving the inverse problem.

EXAMPLE 3.4: When n = 2, the Cauchy equations for the stress in continuum mechanics are

$$\partial_1\sigma^{11} + \partial_2\sigma^{21} = 0,\ \partial_1\sigma^{12} + \partial_2\sigma^{22} = 0$$

with $\sigma^{12} = \sigma^{21}$.

Their parametrization

$$\sigma^{11} = \partial_{22}\phi,\ \sigma^{12} = \sigma^{21} = -\partial_{12}\phi,\ \sigma^{22} = \partial_{11}\phi$$

has been discovered by Airy in 1862 and ? is called the Airy function. When n = 3, Maxwell and Morera discovered a similar parametrization with 3 potentials (exercise).

EXAMPLE 3.5: When n = 4, the Maxwell equations dF = 0 where $F \in \wedge^2 T^*$ is the EM field are parametrized by dA = F where $A \in T^*$ is the 4-potential. The second set of Maxwell equations can also be parametrized by the so-called pseudopotential which is a pseudovector density (exercise).

EXAMPLE 3.5: If n = 4, ω is the Minkowski metric and ϕ = GM/r is the gravitational potential, then $\phi/c^2 \ll 1$ and a perturbation Ω of ω may satisfy in vacuum the 10 second order Einstein equations for the 10 Ω:

$$\omega^{rs}\left(d_{ij}\Omega_{rs} + d_{rs}\Omega_{ij} - d_{ri}\Omega_{sj} - d_{sj}\Omega_{ri}\right)$$
$$-\omega_{ij}\left(\omega^{rs}\omega^{uv}d_{rs}\Omega_{uv} - \omega^{ru}\omega^{sv}d_{rs}\Omega_{uv}\right) = 0$$

The parametrizing challenge has been proposed in 1970 by J. Wheeler for 1000 \$ and solved negatively in 1995 by the author who only received 1 \$.

PROBLEM 2: Is an OPERATOR parametrizable (answer must be YES or NO) and how can we find a parametrization?

Let A be a unitary ring, that is

$$1, a, b \in A \Rightarrow a + b, ab \in A, 1a = a$$

and even an integral domain, that is

$$ab = 0 \Rightarrow a = 0 \quad \text{or} \quad b = 0\,.$$

We say that $M = {}_A M$ is a left module over A if

$$x, y \in M \Rightarrow ax,\ x + y \in M,\ \forall a \in A$$

and we denote by $\hom_A(M, N)$ $\hom_A(M, N)$ the set of morphisms

$f: M \to N$such that $f(ax) = af(x)$.

DEFINITION 3.6: We define the torsion submodule

$$t(M)=\{x\in M \mid \exists 0\neq a\in A, ax=0\}\subseteq M \ .$$

There is a sequence

$$0\rightarrow t(M)\rightarrow M\overset{\varepsilon}{\rightarrow}\hom_A(\hom_A(M,A),A)$$

where the morphism ε is defined by

$$\varepsilon(x)(f)=f(x),\forall x\in M,\forall f\in\hom_A(M,A)$$

because we have at once

$$af(x)=f(ax)=f(0)=0\Rightarrow t(M)\subseteq\ker(\varepsilon).$$

PROBLEM 3: Is a MODULE M torsion-free, that is $t(M) = 0$ (answer must be YES or NO) and how can we test such a property?

In the remaining of this paper we shall prove that the three problems are indeed identical and that only the solution of the third will provide the solution of the two others [1,22-24].

Let $\mathbb{Q}\subset K$ be a differential field, that is a field

$$(a\in K\Rightarrow 1/a\in K)$$

with n commuting derivations $\{\partial_1,\cdots,\partial_n\}$ with

$$\partial_i\partial_j=\partial_j\partial_i=\partial_{ij},\ \forall i,j=1,\cdots,n$$

such that

$$\partial_i(a+b)=\partial_i a+\partial_i b$$

and

$$\partial_i(ab)=(\partial_i a)b+a\partial_i b,\ \forall a,b\in K\ .$$

Using an implicit summation on multiindices, we may introduce the (noncommutative) ring of differential operators

$$D = K\left[d_1,\cdots,d_n\right] = K\left[d\right]$$

with elements $P = a^{\mu}\delta_{\mu}$ such that $|\mu| < \infty$ and $d_i a = a d_i + \partial_i a$. Now, if we introduce differential indeterminates $y = (y^1,\ldots, y^m)$, we may extend $d_i y^k_{\mu} = y^k_{\mu+1_i}$ to

$$\Phi^{\tau} \equiv a^{\tau\mu}_k y^k_{\mu} \xrightarrow{d_i} d_i\Phi^{\tau} \equiv a^{\tau\mu}_k y^k_{\mu+1_i} + \partial_i a^{\tau\mu}_k y^k_{\mu}$$

For $\tau = 1,\ldots, p$. Therefore, setting

$$Dy^1 + \cdots + dy^m = Dy \simeq D^m ,$$

we obtain by residue the differential module or D- module $M = Dy / D\Phi$. Introducing the two free differential modules $F_0 \simeq D^m$, $F_1 \simeq D^p$, we obtain equivalently the free presentation $F_1 \xrightarrow{D} F_0 \rightarrow M \rightarrow 0$. More generally, introducing the successive CC as in the preceding section, we may finally obtain the free resolution of M, namely the exact sequence

$$\cdots \xrightarrow{D_2} F_2 \xrightarrow{D_1} F_1 \xrightarrow{D} F_0 \rightarrow M \rightarrow 0 .$$

The "trick" is to let $\mathcal{D}$ act on the left on column vectors in the operator case and on the right on row vectors in the module case. Homological algebra has been created for finding intrinsic properties of modules not depending on their presentation or even on their resolution.

EXAMPLE 3.7: In order to understand that different presentations may nevertheless provide isomorphic modules, let us consider the linear inhomogeneous system $Py \equiv d_{22}y = u$, $Q \equiv d_{12}y - y = v$ with $K = \mathbb{Q}$. Differentiating twice, we get $y = d_{11}u - d_{12}v - v$ and the two fourth order CC:

$$A \equiv d_{1122}u - d_{1222}v - d_{22}v - u = 0,$$
$$B \equiv d_{1112}u - d_{1122}v - d_{11}u = 0 \Rightarrow E$$
$$\equiv d_{12}B - d_{11}A + B = 0.$$

However, as $PQ = QP$, we also get the CC $C \equiv d_{12}u\text{-}d_{22}v\text{-}u = 0$ and the two resolutions:

$$0 \to D \to D^2 \to D^2 \to M \to 0,$$
$$0 \to D \to D^2 \to M \to 0$$

where we can identify the two differential modules involved on the right with D because:

$$A \equiv d_{12}C + C,$$
$$B \equiv d_{11}C \Leftrightarrow C \equiv d_{22}B - d_{12}A + B = 0.$$

We now exhibit another approach by defining the formal adjoint of an operartor P and an operator matrix $\mathcal{D}$:

DEFINITION 3.8:

$$P = a^{\mu} d_{\mu} \in D \overset{ad}{\leftrightarrow} ad(P) = (-1)^{|\mu|} d_{\mu} a^{\mu} \in D$$
$$\langle \lambda, \mathcal{D}\xi \rangle = \langle ad(\mathcal{D})\lambda, \xi \rangle + \text{div}(\cdots)$$

from integration by part, where λ is a row vector of test functions and $\langle\ \rangle$ the usual contraction.

PROPOSITION 3.9: If we have an operator $E \overset{\mathcal{D}}{\to} F$, we obtain by duality an operator

$$\wedge^n T^* \otimes E^* \overset{ad(\mathcal{D})}{\longleftarrow} \wedge^n T^* \otimes F^*$$

where E^* is obtained from E by inverting the transition matrix and EM provides a fine example of such a procedure [10].

Now, with operational notations, let us consider the two differential sequences:

$$\xi \overset{\mathcal{D}}{\to} \eta \overset{\mathcal{D}_1}{\to} \zeta$$

$$\nu \overset{ad(\mathcal{D})}{\longleftarrow} \mu \overset{ad(\mathcal{D}_1)}{\longleftarrow} \lambda$$

where $\mathcal{D}_1$ generates all the CC of $\mathcal{D}$. Then

$$\mathcal{D}_1 \circ \mathcal{D} \equiv 0 \Leftrightarrow ad(\mathcal{D}) \circ ad(\mathcal{D}_1) \equiv 0$$

but $ad(\mathcal{D})$ may not generate all the CC of $ad(\mathcal{D}_1)$.

EXAMPLE 3.10: With $\partial_{22}\xi = \eta^2$, $\partial_{12}\xi = \eta^1$ for $\mathcal{D}$, we get $\partial_1\eta^2 - \partial_2\eta^1 = \xi$ for $\mathcal{D}_1$. Then $ad(\mathcal{D}_1)$ is defined by $\mu^2 = -\partial_1\lambda$, $\mu^1 = \partial_2\lambda$ while $ad(\mathcal{D})$ is defined by $\nu = \partial_{12}\mu^1 + \partial_{22}\mu^2$ but the CC of $ad(\mathcal{D}_1)$ are generated by $\nu' = \partial_1\mu^1 + \partial_2\mu^2$. Passing to the module framework, we obtain the sequences:

$$\begin{array}{ccccccc} D & \overset{\mathcal{D}_1}{\rightarrow} & D^2 & \overset{\mathcal{D}}{\rightarrow} & D & \rightarrow M \rightarrow 0 \\ D & \overset{ad(\mathcal{D}_1)}{\leftarrow} & D^2 & \overset{ad(\mathcal{D})}{\leftarrow} & D & \end{array}$$

THEOREM 3.11: The cohomology $ext^1(M)$ at D^2 of the lower sequence does not depend on the resolution of M and is a torsion module called the first extension module of M.

Exactly like we defined the differential module M from $\mathcal{D}$, let us define the differential module N from $ad(\mathcal{D})$. The proof of the next theorem is quite tricky and out of the scope of this paper [1,22-24]:

MAIN THEOREM 3.12: $ext^1(N) = t(M) = \ker(\varepsilon)$

FORMAL TEST 3.13: The double duality test needed in order to check whether $t(M) = 0$ or not and to find out a parametrization if $t(M) = 0$ has 5 steps which are drawn in the following diagram where $ad(\mathcal{D}_{-1})$ generates the CC of $ad(\mathcal{D})$ and $\mathcal{D}'$ generates the CC of $\mathcal{D}_{-1}$:

$$\begin{array}{ccccccccc} & & & & & & \eta' & & 5 \\ & & & & & \overset{\mathcal{D}'}{\nearrow} & & & \\ 4 & \theta & \overset{\mathcal{D}_{-1}}{\rightarrow} & \xi & \overset{\mathcal{D}}{\rightarrow} & & \eta & & 1 \\ & & & & & & & & \\ 3 & \nu & \overset{ad(\mathcal{D}_{-1})}{\leftarrow} & \mu & \overset{ad(\mathcal{D})}{\leftarrow} & & \lambda & & 2 \end{array}$$

THEOREM 3.14: $\mathcal{D}$ parametrized by

$$\mathcal{D}_{-1} \Leftrightarrow \mathcal{D} = \mathcal{D}' \Leftrightarrow t(M) = 0 \Leftrightarrow \text{ext}^1(N) = 0 .$$

COROLLARY 3.15: If $n = 1$ and $\mathcal{D}$ is surjective, then $t(M) = 0$ if and only if $ad(\mathcal{D})$ is injective [24,25].

EXAMPLE 3.16: (Kalman test revisited) If we multiply the Kalman system $-y + Ay + Bu = 0$ on the left by a test row vector λ, we obtain:

$$ad(\mathcal{D}) \begin{cases} y & \rightarrow & \dot{\lambda} + \lambda A = 0 \\ u & \rightarrow & \lambda B = 0 \end{cases}$$

Differentiating the zero order equations and using the first order ones, we get $\lambda AB = 0$ and so on. Using the Cayley-Hamilton theorem, we stop at $\lambda A^{\mu-1}B = 0$ and find back exactly the Kalman test but in a completely different intrinsic framework.

EXAMPLE 3.17: (Double pendulum revisited) Using two test functions λ^1 and λ^2, we get:

$$ad(\mathcal{D}) \begin{cases} x & \rightarrow & \ddot{\lambda}^1 + \ddot{\lambda}^2 = 0 \\ \theta_1 & \rightarrow & l_1\ddot{\lambda}^1 + g\lambda^1 = 0 \\ \theta_2 & \rightarrow & l_2\ddot{\lambda}^2 + g\lambda^2 = 0 \end{cases}$$

and obtain at once the zero order equation

$$l_1\lambda^2 + \lambda_2\lambda^1 = 0.$$

Differentiating twice and substituting, we also get $(l_1/ l_2)\, \lambda^2 + (l_2/ l_1)\, \lambda^1 = 0$and $ad(\mathcal{D})$ is injective if and only if $l_2 - l_1 \neq 0$.

EXAMPLE 3.18: (Airy parametrization revisited) When $n = 2$, we may study the infinitesimal deformation $\varepsilon \in S_2T^*$ by means of the Killing operator $\mathcal{D}\xi \equiv \mathcal{L}(\xi)\omega = 2\varepsilon$ when w is the euclidean metric. Then $ad(\mathcal{D})$ provides (up to sign and factor 2) the Cauchy equations $\partial_r\sigma^{rj} + \gamma^j_{rs}\sigma^{rs} = 0$ for the stress tensor density [12,16,26]. The following diagram describes the Poincaré scheme:

$$\begin{array}{lccccc} GEOMETRY & 2 & \overset{Killing}{\longrightarrow} & 3 & \overset{Riemann}{\longrightarrow} & 1 \\ \updownarrow & & & & & \\ PHYSICS & 2 & \overset{Cauchy}{\longleftarrow} & 3 & \overset{Airy}{\longleftarrow} & 1 \end{array}$$

Accordingly, the second order Airy parametrization is nothing else than the adjoint of the only Riemann CC involved, namely $\partial_{11}e_{22} + \partial_{22}e_{11} - 2\partial_{12}e_{12} = 0$ which is the linearization of the Riemann tensor of Example 1.9.

EXAMPLE 3.19: (Einstein equations revisited) Contrary to the Ricci operator, the Einstein operator is selfadjoint because it comes from a variational procedure, the sixth terms being exchanged between themselves under *ad*. For example, we have:

$$\lambda^{ij}\left(\omega^{rs}d_{ij}\Omega_{rs}\right)\overset{ad}{\longleftrightarrow}\left(\omega^{rs}d_{ij}\lambda^{ij}\right)\Omega_{rs}=\left(\omega_{ij}d_{rs}\lambda^{rs}\right)\Omega^{ij}$$

and the adjoint of the first operator is the sixth. Accordingly, one has the following diagram where $\mathcal{D} \neq \mathcal{D}'$:

$$\begin{array}{ccccc} & & & & 20 \\ & & & \overset{Riemann}{\nearrow} & \\ 4 & \overset{Killing}{\longrightarrow} & 10 & \overset{Einstein}{\longrightarrow} & 10 \\ & & & & \\ 4 & \overset{Cauchy}{\longleftarrow} & 10 & \overset{Einstein}{\longleftarrow} & 10 \end{array}$$

THIRD FUNDAMENTAL RESULT: Comparing this diagram to the previous one proves that Einstein equations are not coherent with Janet and Spencer sequences as conformal geometry has not been introduced in this last part.

EXERCISE 3.20: Prove that

$$\ddot{y}^2 - \dddot{y}^3 - \dot{y}^3 - a(t)y^2 = 0$$

is controllable if and only if $a + a^2 - a \neq 0$ (Riccati) and find a parametrization.

EXERCISE 3.21: Prove that the infinitesimal contact transformations of Example 1.10 admit the injective parametrization

$$-x^3\partial_3\theta+\theta=\xi^1,\ -\partial_3\theta=\xi^2,$$
$$\partial_2\theta-x^3\partial_1\theta=\xi^3\Rightarrow\xi^1-x^3\xi^2=\theta.$$

5. CONCLUSION

Jean-Francois Pommaret The mathematical foundations of General Relativity leading to Einstein equations are always presented in textbooks or papers without any reference to conformal geometry. However, comparing the classical Killing equations to the conformal Killing equations while constructing corresponding differential sequences, the Ricci tensor appears as the kernel of the canonical projection of the Riemann tensor onto the Weyl tensor. After obtaining such a result in a purely intrinsic way, that is without using indices, we have been able to introduce "diagram chasing" in order to relate for the first time electromagnetism and gravitation to the Spencer δ-cohomology of the classical and conformal Killing symbols. Accordingly, the mathematical foundations of general relativity are not coherent with jet theory and must therefore be revisited within this new framework along the lines we have sketched. Finally, the fact that Einstein equations cannot be parametrized, contrary to most other equations of physics or engineering, also brings a deep structural question on these equations that will have to be solved in the future by means of algebraic analysis.

REFERENCES

1. J.-F. Pommaret, "Partial Differential Control Theory," Kluwer, Dordrecht, 2001. doi:10.1007/978-94-010-0854-9
2. E. Cartan, Annales Scientifiques de l'École Normale Supérieure, Vol. 21, 1904, pp. 153-206.
3. E. Cartan, Annales Scientifiques de l'École Normale Supérieure, Vol. 40, 1923, pp. 325-412.
4. H. Goldschmidt, Journal of Differential Geometry, Vol. 6, 1972, pp. 357-373, Kumpera and D. C. Spencer, "Lie Equations," Annals of Mathematics Studies 73, Princeton University Press, Princeton, 1972.
5. J.-F. Pommaret, "Differential Galois Theory," Gordon and Breach, New York, 1983.
6. J.-F. Pommaret, "Spencer Operator and Applications: From Continuum Mechanics to Mathematical Physics," In: Y. Gan, Ed., Continuum Mechanics-Progress in Fundamentals and Engineering Applications, InTech, 2012. http://www.intechopen.com/books/

continuum-mechanics-progress-in-fundamentals-and-engineering-applications/spencer-operator-and-applications-from-continuum-mechanics-to-mathematical-physics

7. J.-F. Pommaret, "Partial Differential Equations and Group Theory," Kluwer, 1994.doi:10.1007/978-94-017-2539-2
8. E. Vessiot, Annales scientifiques de l'École Normale Supérieure, Vol. 20, 1903, pp. 411-451.
9. V. Ougarov, "Théorie de la Relativité Restreinte," MIR, Moscow, 1969.
10. L. P. Eisenhart, "Riemannian Geometry," Princeton University Press, Princeton, 1926.
11. J.-F. Pommaret, "Lie Pseudogroups and Mechanics," Gordon and Breach, New York, 1988.
12. Lorenz, "Jet Groupoids, Natural Bundles and the Vessiot Equivalence Method," Acta Applicandae Mathematicae, Vol. 1, 2008, pp. 205-213. doi:10.1007/s10440-008-9193-7
13. D. C. Spencer, Bulletin of the American Mathematical Society, Vol. 75, 1965, pp. 1-114.
14. M. Janet, Journal de Mathématiques Pures et Appliquées, Vol. 8, 1920, pp. 65-151.
15. H. Weyl, "Space, Time, Matter," Berlin, 1918.
16. E. Cosserat and F. Cosserat, "Théorie des Corps Déformables," Hermann, Paris, 1909.
17. J.-F. Pommaret, Acta Mechanica, Vol. 149, 2001, pp. 23-39. doi:10.1007/BF01261661
18. J.-F. Pommaret, Acta Mechanica, Vol. 215, 2010, pp. 43- 55. doi:10.1007/s00707-010-0292-y
19. J. J. Rotman, "An Introduction to Homological Algebra," Academic Press, Cambridge, 1979.
20. R. E. Kalman, Y. C. Yo and K. S. Narenda, Journal of Differential Equations, Vol. 1, 1963, pp. 189-213.
21. M. Kashiwara, Mémoires de la Société Mathématique de France, Vol. 63, 1995, pp. 1-72.
22. V. P. Palamodov, "Linear Differential Operators with Constant Coefficients," Grundlehren der Mathematischen Wissenschaften 168, Springer, 1970. doi:10.1007/978-3-642-46219-1
23. J.-F. Pommaret, "Algebraic Analysis of Control Systems Defined by Partial Differential Equations," Advanced Topics in Control Systems Theory, Springer, Lecture Notes in Control and Information Sciences 311, 2005, pp. 155-223.
24. E. Kunz, "Introduction to Commutative Algebra and Algebraic Geometry," Birkhäuser, 1985.
25. W. Pauli, "Theory of Relativity," Pergamon Press, London, 1958.

INDEX